Automatic Target Recognition

Third Edition

Tutorial Texts Series

- *Applications of Lock-in Amplifiers in Optics*, Gerhard Kloos, Vol. TT117
- *Systems Engineering for Astronomical Telescopes*, Paul A. Lightsey and Jonathan W. Arenberg, Vol. TT116
- *Low-Level Light Therapy: Photobiomodulation*, Michael R. Hamblin, Ying-Ying Huang, Cleber Ferraresi, James D. Carroll, and Lucas Freitas de Freitas, Vol. TT115
- *Fiber Bragg Gratings: Theory, Fabrication, and Applications*, Marcelo M. Werneck, Regina C. Allil, and Fábio V. B. de Nazaré, Vol. TT114
- *Powering Laser Diode Systems*, Grigoriy A. Trestman, Vol. TT112
- *Optics Using MATLAB®*, Scott W. Teare, Vol. TT111
- *Plasmonic Optics: Theory and Applications*, Yongqian Li, Vol. TT110
- *Design and Fabrication of Diffractive Optical Elements with MATLAB®*, A. Vijayakumar and Shanti Bhattacharya, Vol. TT109
- *Energy Harvesting for Low-Power Autonomous Devices and Systems*, Jahangir Rastegar and Harbans S. Dhadwal, Vol. TT108
- *Practical Electronics for Optical Design and Engineering*, Scott W. Teare, Vol. TT107
- *Engineered Materials and Metamaterials: Design and Fabrication*, Richard A. Dudley and Michael A. Fiddy, Vol. TT106
- *Design Technology Co-optimization in the Era of Sub-resolution IC Scaling*, Lars W. Liebmann, Kaushik Vaidyanathan, and Lawrence Pileggi, Vol. TT104
- *Special Functions for Optical Science and Engineering*, Vasudevan Lakshminarayanan and L. Srinivasa Varadharajan, Vol. TT103
- *Discrimination of Subsurface Unexploded Ordnance*, Kevin A. O'Neill, Vol. TT102
- *Introduction to Metrology Applications in IC Manufacturing*, Bo Su, Eric Solecky, and Alok Vaid, Vol. TT101
- *Introduction to Liquid Crystals for Optical Design and Engineering*, Sergio Restaino and Scott Teare, Vol. TT100
- *Design and Implementation of Autostereoscopic Displays*, Byoungho Lee, Soon-gi Park, Keehoon Hong, and Jisoo Hong, Vol. TT99
- *Ocean Sensing and Monitoring: Optics and Other Methods*, Weilin Hou, Vol. TT98
- *Digital Converters for Image Sensors*, Kenton T. Veeder, Vol. TT97
- *Laser Beam Quality Metrics*, T. Sean Ross, Vol. TT96
- *Military Displays: Technology and Applications*, Daniel D. Desjardins, Vol. TT95
- *Interferometry for Precision Measurement*, Peter Langenbeck, Vol. TT94
- *Aberration Theory Made Simple, Second Edition*, Virendra N. Mahajan, Vol. TT93
- *Modeling the Imaging Chain of Digital Cameras*, Robert D. Fiete, Vol. TT92
- *Bioluminescence and Fluorescence for* In Vivo *Imaging*, Lubov Brovko, Vol. TT91
- *Polarization of Light with Applications in Optical Fibers*, Arun Kumar and Ajoy Ghatak, Vol. TT90
- *Digital Fourier Optics: A MATLAB Tutorial*, David G. Voeltz, Vol. TT89
- *Optical Design of Microscopes*, George Seward, Vol. TT88
- *Analysis and Evaluation of Sampled Imaging Systems*, Richard H. Vollmerhausen, Donald A. Reago, and Ronald Driggers, Vol. TT87
- *Nanotechnology: A Crash Course*, Raúl J. Martin-Palma and Akhlesh Lakhtakia, Vol. TT86
- *Direct Detection LADAR Systems*, Richard Richmond and Stephen Cain, Vol. TT85
- *Optical Design: Applying the Fundamentals*, Max J. Riedl, Vol. TT84
- *Infrared Optics and Zoom Lenses, Second Edition*, Allen Mann, Vol. TT83
- *Optical Engineering Fundamentals, Second Edition*, Bruce H. Walker, Vol. TT82
- *Fundamentals of Polarimetric Remote Sensing*, John Schott, Vol. TT81
- *The Design of Plastic Optical Systems*, Michael P. Schaub, Vol. TT80
- *Radiation Thermometry: Fundamentals and Applications in the Petrochemical Industry*, Peter Saunders, Vol. TT78
- *Matrix Methods for Optical Layout*, Gerhard Kloos, Vol. TT77
- *Fundamentals of Infrared Detector Materials*, Michael A. Kinch, Vol. TT76
- *Practical Applications of Infrared Thermal Sensing and Imaging Equipment, Third Edition*, Herbert Kaplan, Vol. TT75
- *Bioluminescence for Food and Environmental Microbiological Safety*, Lubov Brovko, Vol. TT74

(For a complete list of Tutorial Texts, see http://spie.org/publications/books/tutorial-texts.)

Automatic Target Recognition

Third Edition

Bruce J. Schachter

Tutorial Texts in Optical Engineering
Volume TT118

SPIE PRESS
Bellingham, Washington USA

Library of Congress Cataloging-in-Publication Data

Names: Schachter, Bruce J. (Bruce Jay), 1946- author.
Title: Automatic target recognition / Bruce J. Schachter.
Other titles: Tutorial texts in optical engineering ; v. TT 118.
Description: Third edition. | Bellingham, Washington : SPIE Press, [2018] | Series: Tutorial texts in optical engineering ; volume TT 118 | Includes bibliographical references and index.
Identifiers: LCCN 2018005261 | ISBN 9781510618565 (softcover ; alk. paper) | ISBN 1510618562 (softcover) | ISBN 9781510618572 (pdf) | ISBN 1510618570 (pdf) | ISBN 9781510618589 (epub) | ISBN 1510618589 (epub) | ISBN 9781510618596 (Kindle/Mobi) | ISBN 1510618597 (Kindle/Mobi)
Subjects: LCSH: Radar targets. | Optical pattern recognition. | Algorithms. | Image processing.
Classification: LCC TK6580 .S33 2018 | DDC 623.7/348–dc23 LC record available at https://lccn.loc.gov/2018005261

Published by

SPIE
P.O. Box 10
Bellingham, Washington 98227-0010 USA
Phone: +1 360.676.3290
Fax: +1 360.647.1445
Email: books@spie.org
Web: http://spie.org

Printed in the United States of America.
First Printing.
For updates to this book, visit http://spie.org and type "TT118" in the search field.

SPIE.

Introduction to the Series

Since its inception in 1989, the Tutorial Texts (TT) series has grown to cover many diverse fields of science and engineering. The initial idea for the series was to make material presented in SPIE short courses available to those who could not attend and to provide a reference text for those who could. Thus, many of the texts in this series are generated by augmenting course notes with descriptive text that further illuminates the subject. In this way, the TT becomes an excellent stand-alone reference that finds a much wider audience than only short course attendees.

Tutorial Texts have grown in popularity and in the scope of material covered since 1989. They no longer necessarily stem from short courses; rather, they are often generated independently by experts in the field. They are popular because they provide a ready reference to those wishing to learn about emerging technologies or the latest information within their field. The topics within the series have grown from the initial areas of geometrical optics, optical detectors, and image processing to include the emerging fields of nanotechnology, biomedical optics, fiber optics, and laser technologies. Authors contributing to the TT series are instructed to provide introductory material so that those new to the field may use the book as a starting point to get a basic grasp of the material. It is hoped that some readers may develop sufficient interest to take a short course by the author or pursue further research in more advanced books to delve deeper into the subject.

The books in this series are distinguished from other technical monographs and textbooks in the way in which the material is presented. In keeping with the tutorial nature of the series, there is an emphasis on the use of graphical and illustrative material to better elucidate basic and advanced concepts. There is also heavy use of tabular reference data and numerous examples to further explain the concepts presented. The publishing time for the books is kept to a minimum so that the books will be as timely and up-to-date as possible. Furthermore, these introductory books are competitively priced compared to more traditional books on the same subject.

When a proposal for a text is received, each proposal is evaluated to determine the relevance of the proposed topic. This initial reviewing process has been very helpful to authors in identifying, early in the writing process, the need for additional material or other changes in approach that would serve to strengthen the text. Once a manuscript is completed, it is peer reviewed to ensure that chapters communicate accurately the essential ingredients of the science and technologies under discussion.

It is my goal to maintain the style and quality of books in the series and to further expand the topic areas to include new emerging fields as they become of interest to our reading audience.

James A. Harrington
Rutgers University

Contents

Preface *xiii*

1 Definitions and Performance Measures **1**

1.1 What is Automatic Target Recognition (ATR)? 1

1.1.1 Buyers and sellers 4

1.2 Basic Definitions 4

1.3 Detection Criteria 10

1.4 Performance Measures for Target Detection 13

1.4.1 Truth-normalized measures 13

1.4.1.1 Assigned targets and confusers (AFRL COMPASE Center terminology) 15

1.4.2 Report-normalized measure 15

1.4.3 Receiver operating characteristic curve 15

1.4.4 P_d versus FAR curve 18

1.4.5 P_d versus list length 19

1.4.6 Other factors that can enter the detection equation 19

1.4.7 Missile terminology 19

1.4.8 Clutter level 20

1.5 Classification Criteria 21

1.5.1 Object taxonomy 21

1.5.2 Confusion matrix 25

1.5.2.1 Compound confusion matrix 26

1.5.3 Some commonly used terms from probability and statistics 26

1.6 Experimental Design 29

1.6.1 Test plan 31

1.6.2 ATR and human subject testing 32

1.7 Characterizations of ATR Hardware/Software 33

References 34

2 Target Detection Strategies **35**

2.1 Introduction 35

2.1.1 What is target detection? 36

2.1.2 Detection schemes 36

2.1.3 Scale 38

2.1.4 Polarity, shadows, and image form 38
2.1.5 Methodology for algorithm evaluation 40
2.1.5.1 Evaluation criteria for production systems 40
2.1.5.2 Target detection: machine versus human 41
2.2 Simple Detection Algorithms 41
2.2.1 Triple-window filter 41
2.2.2 Hypothesis testing as applied to an image 42
2.2.3 Comparison of two empirically determined means: variations on the T-test 43
2.2.4 Tests involving variance, variation, and dispersion 46
2.2.5 Tests for significance of hot spot 48
2.2.6 Nonparametric tests 49
2.2.6.1 Percent-bright tests 50
2.2.7 Tests involving textures and fractals 51
2.2.8 Tests involving blob edge strength 51
2.2.9 Hybrid tests 52
2.2.10 Triple-window filters using several inner-window geometries 53
2.3 More-Complex Detectors 53
2.3.1 Neural network detectors 53
2.3.2 Discriminant functions 54
2.3.3 Deformable templates 55
2.4 Grand Paradigms 55
2.4.1 Geometrical and cultural intelligence 56
2.4.2 Neuromorphic paradigm 57
2.4.3 Learning-on-the-fly 57
2.4.4 Integrated sensing and processing 58
2.4.5 Bayesian surprise 59
2.4.6 Modeling and simulation 60
2.4.7 SIFT and SURF 61
2.4.8 Detector designed to operational scenario 61
2.5 Traditional SAR and Hyperspectral Target Detectors 62
2.5.1 Target detection in SAR imagery 63
2.5.2 Target detection in hyperspectral imagery 64
2.6 Conclusions and Future Direction 66
References 67
Appendices 69

3 Target Classifier Strategies 77

3.1 Introduction 77
3.1.1 Parables and paradoxes 77
3.2 Main Issues to Consider in Target Classification 80
3.2.1 Issue 1: Concept of operations 81
3.2.2 Issue 2: Inputs and outputs 81
3.2.3 Issue 3: Target classes 82

3.2.4 Issue 4: Target variations 83
3.2.5 Issue 5: Platform issues 84
3.2.6 Issue 6: Under what conditions does a sensor supply useful data? 85
3.2.7 Issue 7: Sensor issues 86
3.2.8 Issue 8: Processor 87
3.2.9 Issue 9: Conveying classification results to the human-in-the-loop 87
3.2.10 Issue 10: Feasibility 88
3.3 Feature Extraction 92
3.4 Feature Selection 96
3.5 Examples of Feature Types 100
3.5.1 Histogram of oriented gradients 101
3.5.2 Histogram of optical flow feature vector 103
3.6 Examples of Classifiers 103
3.6.1 Simple classifiers 104
3.6.1.1 One-class classifiers 104
3.6.1.2 Two-class linear classifiers 104
3.6.1.3 Support vector machine 105
3.6.2 Basic classifiers 108
3.6.2.1 Single-nearest-neighbor classifier 108
3.6.2.2 Naïve Bayes classifier 110
3.6.2.3 Perceptron 111
3.6.2.4 Learning vector quantization family of algorithms 113
3.6.2.5 Feedforward multilayer perceptron trained with backpropagation of error 114
3.6.2.6 Mean-field theory networks 114
3.6.2.7 Model-based classifiers 116
3.6.2.8 Map-seeking circuits 116
3.6.2.9 Ensemble classifiers 118
3.6.3 Contest-winning and newly popular classifiers 118
3.6.3.1 Hierarchical temporal memory 120
3.6.3.2 Long short-term memory recurrent neural network 120
3.6.3.3 Convolutional neural network 121
3.6.3.4 Sentient ATR 124
3.7 Discussion 125
References 131

4 Unification of Automatic Target Tracking and Automatic Target Recognition 135

4.1 Introduction 135
4.2 Categories of Tracking Problems 138
4.2.1 Number of targets 138
4.2.2 Size of targets 139

4.2.3 Sensor type 140
4.2.4 Target type 141
4.3 Tracking Problems 141
4.3.1 Point target tracking 141
4.3.2 Video tracking 146
4.3.2.1 Correlation tracking (video data) 147
4.3.2.2 Feature-vector-aided tracking (video data) 148
4.3.2.3 Mean-shift-based moving object tracker (video tracking) 149
4.4 Extensions of Target Tracking 149
4.4.1 Activity recognition (AR) 150
4.4.2 Patterns-of-life and forensics 152
4.5 Collaborative ATT and ATR (ATT↔ATR) 153
4.5.1 ATT data useful to ATR 153
4.5.2 ATR data useful to ATT 154
4.6 Unification of ATT and ATR (ATT∪ATR) 155
4.6.1 Visual pursuit 155
4.6.2 A bat's echolocation of flying insects 157
4.6.3 Fused ATT∪ATR 158
4.6.3.1 Spatiotemporal target detection 158
4.6.3.2 Forecast of features and classes 161
4.6.3.3 Detection-to-track association 163
4.6.3.4 Track maintenance 163
4.6.3.5 Incorporation of higher-level knowledge 164
4.6.3.6 Implementation 165
4.7 Discussion 165
References 166

5 Multisensor Fusion **169**

5.1 Introduction 169
5.2 Critical Fusion Issues Related to ATR 172
5.3 Levels of Fusion 176
5.3.1 Data-level fusion 176
5.3.2 Feature-level fusion 184
5.3.2.1 Sensor selection for feature-level sensor fusion 184
5.4 Multiclassifier Fusion 187
5.4.1 Fusion of classifiers making hard decisions 189
5.4.1.1 Majority voting 189
5.4.1.2 Combined class rankings 190
5.4.1.3 Borda count method 190
5.4.1.4 Condorcet criterion 190
5.4.2 Fusion of classifiers making soft decisions 191
5.4.2.1 Simple Bayes average 191
5.4.2.2 Bayes belief integration 192

5.4.2.3 Trainable classifier as a combiner 193
5.4.2.4 Dempster–Shafer theory 193
5.5 Multisensor Fusion Based on Multiclassifier Fusion 194
5.6 Test and Evaluation 195
5.7 Beyond Basic ATR Fusion 196
5.7.1 Track fusion 196
5.7.2 Multifunction RF systems 199
5.7.2.1 Multifunction RF 199
5.7.2.2 Cognitive radar 200
5.7.3 Autonomous land vehicles 201
5.7.4 Intelligent preparation of the battlespace 202
5.7.4.1 Dissemination and integration 203
5.7.5 Zero-shot learning 204
5.7.5.1 Manned–unmanned teaming 205
5.7.5.2 Expert systems 207
5.8 Discussion 209
References 210

6 Next-Generation ATR 215

6.1 Introduction 215
6.2 Hardware Design 216
6.2.1 Hardware recommendations for next-generation neuromorphic ATR 223
6.2.1.1 What *shouldn't* be copied from biology? 224
6.2.1.2 What *should* be copied from biology? 225
6.3 Algorithm/Software Design 228
6.3.1 Classifier architecture 229
6.3.1.1 Decision tree 229
6.3.1.2 Embodied and situated (ES) 232
6.3.1.3 Adaptivity and plasticity (PI) 233
6.3.2 Embodied, situated, plastic RNN [**M** = ES-PI-RNN($\mathbb{Q}$)] coupled with a controller **C** 233
6.3.2.1 Training the controller **C** 236
6.3.3 Software infrastructure 236
6.3.4 Test results 237
6.4 Potential Impact 237
References 238

7 How Smart is Your Automatic Target Recognizer? 241

7.1 Introduction 241
7.2 Test for Determining the Intelligence of an ATR 243
7.2.1 Does the ATR understand human culture? 244
7.2.2 Can the ATR deduce the gist of a scene? 244
7.2.3 Does the ATR understand physics? 245

7.2.4 Can the ATR participate in a pre-mission briefing? 247
7.2.5 Does the ATR possess deep conceptual understanding? 247
7.2.6 Can the ATR adapt to the situation, learn on-the-fly, and make analogies? 248
7.2.7 Does the ATR understand the rules of engagement? 249
7.2.8 Does the ATR understand order of battle and force structure? 251
7.2.9 Can the ATR control platform motion? 252
7.2.10 Can the ATR fuse information from a wide variety of sources? 253
7.2.11 Does the ATR possess metacognition? 254
7.3 Sentient Versus Sapient ATR 255
7.4 Discussion: Where is ATR Headed? 256
References 260

Appendix 1: Resources **261**

Appendix 2: Questions to Pose to the ATR Customer **295**

Appendix 3: Acronyms and Abbreviations **299**

Index *305*

Preface

An automatic target recognizer (ATR) is a real-time or near-real-time image/signal-understanding system. An ATR is presented with a stream of data. It outputs a list of the targets that it has detected and recognized in the data provided to it. A complete ATR system can also perform other functions such as image stabilization, preprocessing, mosaicking, target tracking, activity recognition, multi-sensor fusion, sensor/platform control, and data packaging for transmission or display.

In the early days of ATR, there were fierce debates between proponents of signal processing and those in the emerging field of computer vision. Signal processing fans were focused on more advanced correlation filters, stochastic analysis, estimation and optimization, transform theory, and time-frequency analysis of nonstationary signals. Advocates of computer vision said that signal processing provides some nice tools for our toolbox, but what we really want is an ATR that works as well as biological vision. ATR designers were less interested in processing signals than understanding scenes. They proposed attacking the ATR problem through artificial intelligence (AI), computational neuroscience, evolutionary algorithms, case-based reasoning, expert systems, and the like. Signal processing experts are interested in tracking point-like targets. ATR engineers want to track a target with some substance to it, identify what it is, and determine what activity it is engaged in. Signal processing experts keep coming up with better ways to compress video. ATR engineers want more intelligent compression. They want the ATR to tell the compression algorithm which parts of the scene are more important and hence deserving of more bits in the allocation. ATR, in and of itself, can be thought of as a data reduction technique. The ATR takes in a lot of data and outputs relatively little data. Data reduction is necessary due to bandwidth limitations of the data link and workload limits of the time-strapped human operator. People are very good at analyzing video until fatigue sets in or they get distracted. They don't want to be like the triage doctor at the emergency ward, assessing everything that comes in the door, continually assigning priorities to items deserving further attention. Pilots and ground station operators want a machine to relieve their burden as long as it rarely makes a mistake. Trying to do this keeps ATR engineers employed. As often told to the author, pilots and image analysts are not looking for machines to replace them entirely. However, such decisions will be made higher up in the chain of command as ATR technology progresses.

The human vision system is not "designed" to analyze certain kinds of data such as rapid step-stare imagery, complex-valued signals that arise in radars, hyperspectral imagery, 3D LADAR data, or fusion of signal data with various forms of precise metadata. ATR shines when the sustained data rate is too high or too prolonged for the human brain, or the data is not well suited for presentation to humans. Nevertheless, most current ATRs operate with humans-in-the-loop. Humans, at present, are much better than ATRs at tasks requiring consultation, comprehension, and judgement. Humans still make the final decision and determine the action to be taken. This means that ATR output, which is statistical and multi-faceted by nature, has to be presented to the human decision makers in an easily understood form. This is a difficult man–machine interface problem. Marching toward the future, more autonomous robotic systems will necessarily rely more on ATRs to substitute for human operators, possibly serving as the "brains" of entire robotic platforms. We leave this provocative topic to the end of the book.

Systems engineers took notice once ATRs became deployable. Systems engineers are grounded in harsh reality. They care little about the debate between signal processing and computer vision. They don't want to hear about an ATR being brain-like. They are not interested in which classification paradigm performs 1% better than the next. They care about the concept of operations (ConOps) and how it directs performance and functionality. They care about mission objectives and mission requirements. They want to identify all possible stakeholders, form an integrated product team, determine key performance parameters (KPPs), and develop test and evaluation (T&E) procedures to determine if performance requirements are met. Self-test is the norm for published papers and conference talks. Independent test and evaluation, laboratory blind tests, field tests, and software regression tests are the norm for determining if a system is deployable. The systems engineer's focus is broader than ATR performance. Systems engineers want the entire system, or system of systems, to work well, including platform, sensors, ATR, and data links. They want to know what data can be provided to the ATR and what data the ATR can provide to the rest of the system. They want to know how one part of the system affects all other parts of the system. Systems designers care a lot about size, weight, power, latency, current and future costs, logistics, timelines, mean time between failure, and product repair and upgrade. They want to know the implications of system capture by the enemy.

At one time, ATR was the sole charge of the large defense electronics companies, working closely with the government labs. Only the defense companies and government have fleets of data collection aircraft, high-end sensors, and access to foreign military targets. Although air-to-ground has been the focus of much ATR work, ATR actually covers a wide range of sensors, operating within or between the layers of space, air, ocean/land surface, and undersea/underground. Although the name ATR implies recognition of targets, ATR engineers have broader interests. ATR groups tackle any type of military problem involving the smart processing of imagery or signals. The government (or government-funded prime contractor) is virtually the only customer. So, some of the ATR engineer's time is spent reporting to the government, participating in joint data collections, taking part in government-sponsored tests, and proposing new programs to the government.

Since the 1960s, the field of ATR has advanced in parallel with similar work in the commercial sector and academia, involving industrial automation, medical imaging, surveillance and security, video analytics, and space-based imaging. Technologies of interest to both the commercial and defense sector include low-power processors, novel sensors, increased system autonomy, people detection, robotics, rapid search of vast amounts of data (big data), undersea inspection, and remote medical diagnosis. The bulk of funding in some of these areas has recently shifted from the defense to the commercial sector. More money is spent on computer animation for Hollywood movies than for the synthesis of forward-looking infrared (FLIR) and synthetic aperture radar (SAR) imagery. The search engine companies are investing much more in neural networks compared to the defense companies. Well-funded brain research programs are investigating the very basis of human vision and cognitive processing. The days of specialized military processors (e.g., VHSIC) are largely over. Reliance is now on chips in high-volume production: multi-core processors (e.g., Intel and ARM), FPGAs (e.g., Xilinx and Intel/Altera), and GPUs (e.g., Nvidia and AMD). Highly packaged sensors (visible, FLIR, LADAR, and radar) combined with massively parallel processors are advancing rapidly for the automotive industry to meet new safety standards (e.g., Intel/MobilEye). Millions of systems will soon be produced per year. Current advanced driver assistance systems (ADAS) can detect pedestrians, animals, bicyclists, road signs, traffic lights, cars, trucks, and road markers. These are a lot like ATR tasks. The rapid advancement of ADAS will lead to driverless cars.

Some important differences between ATRs and commercial systems are worth noting. ATRs generally have to detect and recognize objects at much longer ranges than commercial systems. Enemy detection and recognition are non-cooperative processes. Although a future car might have a LADAR, radar, or FLIR sensor, it won't have one that can produce high-quality data from a 20,000-ft range. An ADAS will detect a pedestrian but won't report if he is carrying a rifle. Search engine companies need to search large volumes of data with an image-based search, but they don't have the metadata to help the search, such as is available on military platforms. That being said, the cost and innovation rate of commercial electronics can't be matched by military systems. The distinction between commercial and military systems is starting to blur in some instances. Cell phones now include cameras, inertial measurement units, GPS, computers, algorithms, and transmitters/receivers. Slightly rugged versions of commercial cell phones and tablet computers are starting to be used by the military, even with ATR apps. "Toy" drones are approaching the sophistication of the smallest military unmanned air vehicles. They are now produced in volumes of a million per year. ATR engineers are in tune with advances in the commercial sector and their applicability to ATR. Even their hobbies tend to focus on technology, e.g., hobbies such as quadcopters, novel cameras, 3D printers, computers, phone apps, robots, etc.

ATR is not limited to a device; it is also a field of research and development. ATR technology can be incorporated into systems in the form of self-contained hardware, FPGA code, or higher-level language code. ATR groups can help add autonomy to many types of systems. ATR can be viewed very narrowly or very broadly, borrowing concepts from a wide variety of fields. Papers on ATR are often of the form: "Automatic Target Recognition using XXX," where the XXX can be any technology such as super-resolution, principal component analysis, sparse coding, singular value decomposition, Eigen templates, correlation filters, kinematic priors, adaptive boosting, hyperdimensional manifolds, Hough transforms, foveation, etc. In the more ambitious papers, the XXX is a mélange of technologies, such as fuzzy-rule-based expert systems, wavelet neural genetic networks, fuzzy morphological associative memory, optical holography, deformable wavelet templates, hierarchical support vector machines, Bayesian recognition by parts, etc. Get the picture? Nearly any type of technology, everything but the kitchen sink, can be thrown at the ATR problem, with scant large-scale independent competitive test results to indicate which approach really works best, supposing that "best" can be defined and measured. This book is not a comprehensive survey of every technology that has ever been applied to ATR. This book covers some of the basics of ATR. While some of the topics in this book can be found in textbooks on pattern recognition and computer vision, this book focuses on their application to military problems as well as the unique requirements of military systems.

The topics covered in the book are organized in the way one would design an ATR. The first step is to understand the military problem and make a list of potential solutions to the problem. A key issue is the availability of sufficiently comprehensive sets of data to train and test the potential solutions. This involves developing a sound test plan, specifying procedures and equations, and determining who is going to do the testing. Testing isn't open ended. Exit criteria are needed to determine when a given test activity has been successfully completed. The next steps in ATR design are choosing the detector and classifier. The detector focuses attention on the regions-of-interest in the imagery requiring additional scrutiny. The classifier further processes these regions-of-interest and is the decision engine for class assignment. It can operate at any or all levels of a decision tree, from clutter rejection to identifying a specific vehicle or activity. Detected targets are often tracked. Target tracking has historically been treated as a separate subject from ATR, mainly because point-like targets contain too little information to apply an ATR. However, as sensor resolution improves, the engineering disciplines of target tracking and ATR are starting to merge. The ATR and tracker can be united for efficiency and performance. The fifth chapter covers the basics of multisensor fusion. Then it broadens the topic to a variety of other forms of fusion. A strawman design is provided for a more advanced ATR, but with no claim that this is the only way to construct a next-generation ATR. The strawman design should be thought of as a brainstormed simple draft proposal intended to generate discussion of its advantages and disadvantages, and to trigger the generation of new and better proposals. Future ATRs will have to combine data from multiple sources. The last chapter points out how primitive current ATRs really are, as compared to biological systems. It suggests ways for measuring the intelligence of an ATR. This goes far beyond the basic performance measurement techniques covered in Chapter 1. The first appendix lists the many resources available to the ATR engineer. Many of the listed agencies supply training and testing data, perform blind tests, and sponsor research into compelling new sensor and ATR designs. The second appendix advances the notion that a problem that is well described is half solved. The third appendix explains the acronyms and abbreviations used in the book.

CHAPTER 1: ATR technology has benefited from a significant investment over the last 50 years. However, the once-accepted definitions and evaluation criteria have been displaced by the march of technology. The first chapter updates the language for describing ATR systems and provides well-defined criteria for evaluating such systems. This will move forward collaboration between ATR developers, evaluators, and end-users.

ATR is used as an umbrella term for a broad range of military technology beyond just the recognition of targets. In a more general sense, ATR means *sensor data exploitation*. Two types of definitions are included in the first chapter. One type defines fundamental concepts. The other type defines basic performance measures. In some cases, definitions consist of a list of alternatives. This approach enables choices to be made to meet the needs of particular programs. The important point to keep in mind is that within the context of a particular experimental design, a set of protocols should be adopted to best fit the situation, applied, and then kept constant throughout the evaluation. This is especially important for competitive testing.

The definitions given in Chapter 1 are intended for evaluation of end-to-end ATR systems as well as the prescreening and classifier stages of the systems.

Sensor performance and platform characteristics are excluded from the evaluation. It is recognized that sensor characteristics and other operational factors affect the imagery and associated metadata. A thorough understanding of data quality, integrity, synchrony, availability, and timeline are important for ATR development, test, and evaluation. Data quality should be quantified and assessed. However, methods for doing so are not covered in this book. The results and validity of ATR evaluation depend on the representativeness and comprehensiveness of the development and test data. The adequacy of development and test data is primarily a budgetary issue. The ATR engineer should understand and be able to convey the implications of limited, surrogate, or synthetic data. The ATR engineer should be able to damp down naïve proposals centered around the use of an off-the-shelf deep-learning neural network as a miraculous cure to the alleged ATR affliction.

Chapter 1 formalizes definitions and performance measures associated with ATR evaluation. All performance measures must be accepted as ballpark predictions of actual performance in combat. More carefully formulated experiments will provide more meaningful conclusions. The final measure of effectiveness takes place in the battlefield.

CHAPTER 2: Hundreds of simple target detection algorithms were tested on mid- and longwave FLIR images, as well as X-band and Ku-band SAR images. Each algorithm is briefly described. Indications are given as to which performed well. Some of these simple algorithms are loosely derived from standard tests of the difference of two populations. For target detection, these are typically populations of pixel grayscale values or features derived from them. The statistical tests are often implemented in the form of sliding triple-window filters. Several more-elaborate algorithms are also described with their relative performances noted. These algorithms utilize neural networks, deformable templates, and adaptive filtering. Algorithm design issues are broadened to cover system design issues and concepts of operation.

Since target detection is such a fundamental problem, it is often used as a test case for developing technology. New technology leads to innovative approaches for attacking the problem. Eight inventive paradigms, each with deep philosophical underpinnings, are described in relation to their effect on target detector design.

CHAPTER 3: Target classification algorithms have generally kept pace with developments in the academic and commercial sectors since the 1970s. However, most recently, investment into object classification by Internet companies and various large-scale projects for understanding the human brain has far outpaced that of the defense sector. The implications are noteworthy.

There are some unique characteristics of the military classification problem. Target classification is not solely an algorithm design problem, but is part of a larger system design task. The design flows down from a

ConOps and KPPs. Required classification level is specified by contract. Inputs are image and/or signal data and time-synchronized metadata. The operation is often real-time. The implementation minimizes size, weight, and power (SWaP). The output must be conveyed to a time-strapped operator who understands the rules of engagement. It is assumed that the adversary is actively trying to defeat recognition. The target list is often mission dependent, not necessarily a closed set, and can change on a daily basis. It is highly desirable to obtain sufficiently comprehensive training and testing data sets, but costs of doing so are very high, and data on certain target types are scarce or nonexistent. The training data might not be representative of battlefield conditions, suggesting the avoidance of designs tuned to a narrow set of circumstances. A number of traditional and emerging feature extraction and target classification strategies are reviewed in the context of the military target classification problem.

CHAPTER 4: The subject being addressed is how an automatic target tracker (ATT) and an ATR can be fused so tightly and so well that their distinctiveness becomes lost in the merger. This has historically not been the case outside of biology and a few academic papers. The biological model of ATT∪ATR arises from dynamic patterns of activity distributed across many neural circuits and structures (including those in the retinae). The information that the brain receives from the eyes is "old news" at the time that it receives it. The eyes and brain forecast a tracked object's future position, rather than relying on the perceived retinal position. Anticipation of the next moment—building up a consistent perception—is accomplished under difficult conditions: motion (eyes, head, body, scene background, target) and processing limitations (neural noise, delays, eye jitter, distractions). Not only does the human vision system surmount these problems, but it has innate mechanisms to exploit motion in support of target detection and classification. Biological vision doesn't normally operate on snapshots. Feature extraction, detection, and recognition are spatiotemporal. When scene understanding is viewed as a spatiotemporal process, target detection, target recognition, target tracking, event detection, and activity recognition (AR) do not seem as distinct as they are in current ATT and ATR designs. They appear as similar mechanisms taking place at varying time scales. A framework is provided for unifying ATT, ATR, and AR.

CHAPTER 5: Predatory animals detect, stalk, recognize, track, chase, home in on, and if lucky, catch their prey. Stereo vision is generally their most important sensor asset. Most predators also have a good sense of hearing. Some predators can smell their prey from a mile away. Most creatures combine data from multiple sensors to eat or avoid being eaten. Different creatures use different combinations of sensors, including sensors that detect vibration, infrared radiation, various spectral bands, polarization, Doppler, and magnetism.

Biomimicry suggests that a combination of diverse sensors works better than use of a single sensor type. Sensor fusion intelligently combines sensor data from disparate sources such that the resulting information is in some ways superior to the data from a single source. Chapter 5 provides techniques for low-level, mid-level, and high-level information fusion. Other forms of fusion are also of interest to the ATR engineer. Multifunction fusion combines functions normally implemented by separate systems into a single system. Zero-shot learning (ZSL) is a way of recognizing a target without having trained on examples of the target. ZSL provides a vivid description of a detected target as a fusion of its semantic attributes. The commercial world is embracing multisensor fusion for driverless cars. New sensor and processor designs are emerging with applicability to autonomous military vehicles.

CHAPTER 6: Traditional feedforward neural networks, including multilayer perceptrons (MLPs) and the newly popular convolutional neural networks (CNNs), are trained to compute a function that maps an input vector to an output vector. The N-element output vector can convey estimates of the probabilities of N target classes. Nearly all current ATRs perform target classification using feedforward neural networks. These can be shallow or deep. The ATR detects a candidate target, transforms it to a feature vector, and then processes the vector unidirectionally, step by step; the number of steps is proportional to the number of layers in the neural network. Signals travel one way from input to output. A recurrent neural network (RNN) is an appealing alternative. Its neurons send feedback signals to each other. These feedback loops allow RNNs to exhibit dynamic temporal behavior. The feedback loops also establish a type of internal memory. While feedforward neural networks are generally trained in a supervised fashion by backpropagation of output error, RNNs are trained by backpropagation through time.

Although feedforward neural networks are said to be inspired by the architecture of the brain, they do not model many abilities of the brain, such as natural language processing and visual processing of spatiotemporal data. Feedback is omnipresent in the brain, endowing both short-term and long-term memory. The human brain is thus an RNN—a network of neurons with feedback connections. It is a dynamical system. The brain is plastic, adapting to the current situation. The human vision system not only learns patterns in sequential data, but even processes still frame (snapshot) data quite well with its RNN, jerking the eyes in saccades to shift focus over key points on a snapshot, turning the snapshot into a movie.

An improved type of RNN, called long short-term memory (LSTM), was developed in the 1990s by Jürgen Schmidhuber and his former Ph.D. student Sepp Hochreiter. LSTM and its many variants are now the predominant RNN. LSTM is said to be in use in billions of commercial devices.

Brains don't come in a box like a desktop computer or supercomputer. All natural intelligence is embodied and situated. Many military systems, such as unmanned air vehicles and robot ground vehicles, are embodied and situated. The body (platform) maneuvers the sensor systems to view the battlespace from different situations. An ATR based on an RNN, that is embodied and situated [ES], adaptive and plastic [Pl], and of limited precision (e.g., 16-bit floating point), will be denoted by the model **M**=ES-Pl-RNN($\mathbb{Q}_{16}$). A recurrent ATR is more powerful in many ways than a standard ATR. Both computationally more powerful and biologically more plausible than other types of ATRs, an RNN-based ATR understands the notion of events that unfold over time. Its design can benefit from ongoing advances in neuroscience.

Professor Schmidhuber has made an additional improvement to his model. He tightly couples a controller **C** to a model **M**. Both can be RNNs or composite designs incorporating RNNs. Following Schmidhuber's lead, we propose a strawman ATR that couples a controller **C** to our model **M** = ES-Pl-RNN($\mathbb{Q}_{16}$) to form a complete system (**C** ∪ **M**) that is more powerful in many ways than a standard ATR. **C** ∪ **M** can learn a never-ending sequence of tasks, operate in unknown environments, realize abstract planning and reasoning, perform experiments, and retrain itself on-the-fly. This next-generation ATR is suitable for implementation on two chips: a single custom low-power chip (<1 W) for effecting **M**, hosted by a standard processor serving as the controller **C**. A heterogeneous chip design incorporating high-speed I/O, multicore ARM processors, logic gates, GPU, codec, and neural section is also appropriate. This next-generation ATR is applicable to various military systems, including those with extreme size, weight, and power constraints.

CHAPTER 7: ATRs have been under development since the 1960s. Advances in computer processing, computer memory, and sensor resolution are easy to evaluate. However, the time horizon of the truly smart ATR seems to be receding at a rate of one year per year. One issue is that there has never been a way to measure the intelligence of an ATR. This is fundamentally different from measuring detection and classification performance. The description of what constitutes an ATR, and in particular a smart ATR, keeps changing. Early ATRs did little more than detect fuzzy bright spots in first-generation FLIR video or ten-foot-resolution SAR data. Sensors are getting better, computers are getting faster, and the ATR is expected to take over more of the workload. With unmanned systems there is no human onboard to digest information. The ATR is compelled to transmit only the most important information over a limited-bandwidth data link. The ATR or robotic system can be viewed as a substitute for a human. What constitutes intelligence in artificial humans has long been debated, starting with stories of golems, continuing to the Turing test, and including current dire predictions of super-intelligent robots superseding humans. Chapter 7 provides a Turing-like test for judging the intelligence of an ATR.

APPENDIX 1: The first appendix lists the many resources available to the ATR engineer and includes a brief historical overview of the technologies involved in ATR development.

APPENDIX 2: A successful project starts with a clear description of the problem to be solved. However, a well-defined ATR problem is surprisingly hard to come by. The second appendix provides some questions to pose to a customer to help get a project going.

APPENDIX 3: The third appendix defines all of the acronyms and abbreviations used in this book.

Acknowledgments

Special thanks to the United States Army Night Vision and Electronic Sensors Directorate (NVESD), Air Force, Navy, DARPA, and Northrop Grumman for supporting this work over the years. This book benefited from critique and suggestions made by the reviewers and SPIE staff.

Author's Disclaimer

The views and opinions expressed in this book are solely those of the author in his private capacity and do not represent those of any company, the United States Federal Government, any entity of the U.S. Federal Government, or any private organization. Links to organizations are provided solely as a service to our readers. Links do not constitute an endorsement by any organization or the Federal Government, and none should be inferred. While extensive efforts have been made to verify statements and facts presented in this book, any factual errors or errors of opinion are solely those of the author. No position or endorsement by the U.S. Federal Government, any entity of the Federal government, or any other organization regarding the validity of any statement of fact presented in this book should be inferred.

Author's Contact Information

Comments on this book are welcome. The author can be contacted at Bruce.Jay.Schachter@gmail.com.

Bruce J. Schachter
February 2018

Chapter 1
Definitions and Performance Measures

1.1 What is Automatic Target Recognition (ATR)?

ATR is often used as an umbrella term for the entire field of military image exploitation. ATR Working Group (ATRWG) workshops cover a wide range of topics including image quality measurement, geo-registration, target tracking, similarity measures, and progress in various military programs. In a narrower sense, ATR refers to the automatic (unaided) processing of sensor data to locate and classify targets. ATR can refer to a set of algorithms, as well as software and hardware to implement the algorithms. As a hardware-oriented description, ATR stands for automatic target recognition system or automatic target recognizer. ATR can also refer to an operating mode of a sensor or system such as a radar. Several similar terms follow:

- AiTR: Aided target recognition. This term emphasizes that a human is in the decision-making loop. The function of the machine is to reduce the workload of the human operator. Most ATR systems can be viewed as AiTR systems in the broader context.
- ATC/R: Aided target cueing and recognition.
- ATD/C: Automatic target detection and classification.
- ATT: Automatic target tracking.
- ISR: Intelligence, surveillance, and reconnaissance.
- NCTR: Non-cooperative target recognition.
- PED: Processing, exploitation, and dissemination.
- SDE: Sensor data exploitation.
- STA: Surveillance and target acquisition.

This chapter sets the stage for the rest of the book. It defines the terms and evaluation criteria critical to ATR design and test. However, every ATR project is different. The terms and criteria presented here will need to be modified to meet the unique circumstances of individual programs. Consider a competition to choose an ATR for a particular military platform. Multiple

ATRs can only be evaluated fairly within a consistent framework—consistent in definition of terms, evaluation criteria, and developmental and test data. All parties being tested must have equal knowledge of test conditions and an equal ability to negotiate changes to the conditions. Regrettably, perfect fairness is impossible to achieve. Bias occurs because important factors cannot be controlled. One ATR developer might be the manufacturer of the sensor and know all about it; this developer is able to collect large amounts of data and tune up the ATR and sensor so that they work well together. Another developer might have influence over the test site, target set, test plan, and performance requirements. Another developer might manufacture the host platform (e.g., aircraft), the processor box, and have a long history of working with the end-user on ConOps. Another developer might simply have more time and money to prepare for a competitive test. When ATR components are tested, bias often arises when a developer has an investment in a favorite approach and gives short shrift to tuning up competing approaches. The definitions and performance measures provided here give all stakeholders a common language for discussion and can help to make competitive tests somewhat fairer, but not absolutely fair.

ATR can be used as a generic term to cover a broad range of military data exploitation technologies and tasks. These include image fusion, target tracking, minefield detection, as well as technologies for specific missions such as persistent surveillance and suppression of enemy air defenses. The term can be broadened to cover homeland security tasks such as border monitoring, building protection, and airport security. It can include environmental efforts such as detection of fires, whales, radioactive material, and gas plumes. Commercial applications similar to the military ATR problem are grouped under the name *video analytics*. These include parking lot security, speed cameras, and advanced signage. Internet companies are making huge investments in image-based search engines and face recognition. Industrial automation and medical applications of machine vision and pattern recognition use the same basic technology. This chapter focuses on the narrower military problem, epitomized by the basic ATR architecture depicted in Fig. 1.1. This architecture consists of two main components: a front-end anomaly detector (prescreener) and a back-end classifier. The classifier completes the detection/clutter-rejection process. The classifier can also assign a target category to a detected object. The performances of the two primary ATR components can be measured separately. Alternatively, the ATR can be treated as a single *black box*. In this latter case, the only concerns are the inputs and outputs. The inner workings of the ATR, that is, how it transforms the inputs to the outputs, might not be of interest to a team evaluating an ATR's technology readiness level (TRL).

Figure 1.1 shows the input data as a 2D image plus ancillary information. This will be the case in point used in this book. Other types of ATRs might

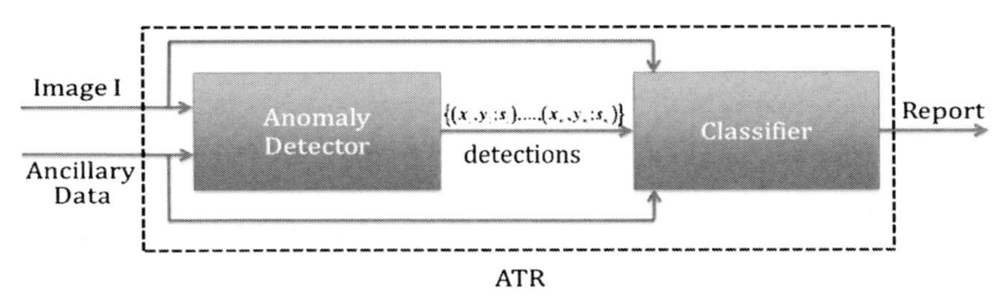

Figure 1.1 Basic traditional ATR architecture. The classifier stage stands for any or all levels of target classification that can take place, as well as supporting processing such as feature extraction and segmentation.

process different types of data, such as 1D or 3D signals, data from multiple sensors, or data in compressed form, just to give a few examples. Some ATRs do not fit the archetype shown in Fig. 1.1.

An ATR might process each input frame of data independently from the next frame, as in a synthetic aperture radar (SAR) or an infrared-step-stare system. A triggered unattended ground system can rarely generate but a single frame of data. Or, the ATR could process video data, using temporal information to help make its decisions.

An ATR often receives ancillary information. The nature of the ancillary data depends on the sensor type and system design. For an electro-optical/infrared (EO/IR) system on a helicopter, this type of metadata includes inertial data, latitude, longitude, altitude, velocity, time, date, digital terrain elevation map, laser range, bad pixel list, and focal plane array nonuniformity. The ATR can also receive target handoff information from another sensor on the same or different platform. The ATR can receive commands requesting that it look for certain targets, switch modes, or render itself useless upon capture. The ATR could as well send commands to the sensor, such as to change integration time, switch modes, or slew in a certain direction.

The borderline between the sensor and ATR is not clear cut. Either the ATR or the EO/IR sensor might perform image correction, frame averaging, stabilization, enhancement, quality measurement, tracking, or image mosaicing. The ATR might implement its own unique version of SAR autofocus and SAR image formation. Some customary ATR functions, such as image compression, could also be handled by other platform components such as a data link or storage system. In the future, the ATR could be just another function within a sensor system, analogous to face detection in handheld color cameras. Or, the future ATR might be given an expanded role to serve as the brains of a robotic platform.

The output of an ATR is a report. The report provides information about targets located and/or tracked, the ATR's health and status, an assessment of the quality of input data, etc. The report could be in the form of graphical overlays for display. The ATR might also output image data for storage or transmission over a data link, perhaps stitching together frames or compressing target areas to higher fidelity than background areas.

1.1.1 Buyers and sellers

"Academic exercise" is a somewhat pejorative term, meaning something with little or no relevance beyond academe. An ATR study or algorithm is an academic exercise if the authors (1) have no power to implement their approach, and (2) if their recommendations are divorced from such considerations as ConOps, performance requirements, specific sensors and sensor modes, metadata, cost, timelines, logistics, competing technologies, countermeasures, independent test and evaluation (T&E), and the DoD procurement process.

This book treats ATR as a product rather than as an academic exercise. There are buyers and sellers. The buyers and sellers need to use common terminology when discussing a transaction. It is up to a buyer to describe in excruciating detail the specifications and key performance requirements of the product being procured. It is natural for sellers to describe their products in best possible terms. It is up to the buyer to do the requisite independent T&E and due diligence to determine if the seller's product meets all requirements. The following discussion should help.

1.2 Basic Definitions

Image: A 2D array of pixels.

Discrete image samples (pixels) can be single valued, representing grayscale pictures. Unless otherwise noted in the text, pixels will be regarded as 8- to 20-bit integers. For certain other sensor types, pixels can be vector quantities: dual-band, third-generation IR (2 band); visual color or commercial IR color (CIR) (3 band); multispectral (4–16 band); or hyperspectral (17–1000 band). Image samples can also be complex-valued signals, as in radar or sonar data, or they can be matrix-valued from polarization cameras. A radar can have multiple modes of operation, each producing different kinds of data. Image samples can have embedded information. For example, the most significant bit might be a good/bad pixel indicator. Some ATRs grab and digitize frames or fields of analog video. Ancillary information can be embedded in the first few lines of each frame of data. Alternatively, a file of ancillary information might be associated with each frame of image data or with multiple frames. The temporal synchronization of sensor data and metadata is a critical issue. ATR systems can also operate on 1D signal data or 3D LADAR data. For example, an ATR might process the Automatic Identification System (AIS) data broadcast by commercial ships.

The ATR and a human viewer will generally receive sensor data from different paths. For example, the ATR might receive 14-bit/pixel image data at 120 frames per second, while the human might view 8-bit/pixel video with annotation overlays at 30 frames per second. The ATR might receive complex-valued SAR data, while the human views magnitude SAR data.

Operating conditions (OCs): All factors that might affect how well a given ATR performs.

OCs characterize:

- targets (articulation, damage, operating history, etc.),
- sensor (type, spectral band, operating mode, depression angle, etc.),
- environment (background, clutter level, atmosphere, etc.),
- ATR (settings, *a priori* target probability assumptions, etc.) and
- interactions (tree lines, revetments, etc.).[1]

OCs are the independent conditions of the experimental design. A bin (experimental bin or OC bin) is data, such as a set of test images that meet some pattern of OCs. For example, one bin might include only day images, and another bin only night images. Even simple terms such as day and night should be clearly defined.

Ground truth: Reference data available from a data collection.

This information is generally of two types:

(1) scenario information: climatic zone, weather, time, date, sun angle; target locations, types, conditions, etc.
(2) sensor information: sensor location, pointing angles, operating mode, characteristics, etc.

Ground truth is a term used in various fields to refer to the absolute truth of something. Thus, it can refer to truth about ships and space targets, not just ground targets. Although ground truth might provide target location, velocity, direction, and range [for example, by global positioning system (GPS) transponder on each target], it will not indicate the pixels in the scene that are on target. Determining which pixels are on target is not as easy as it might at first seem, as illustrated in Fig. 1.2.

Target: Any object of military interest.

Traditional targets are strategic and tactical military craft. This will be the case in point used in this text. However, today, the list can also include improvised explosive devices (IEDs), enemy combatants, human activities, muzzle flashes, fixed sites, commercial vehicles, land minefields, tunnels, undersea mines, and technicals (commercial vehicles modified to contain armament).

Figure 1.2 Pixels on target must be labeled according to a set of rules.

Image truth target location or region: A single reference pixel on target or set of pixels on target (target region) as estimated by an image analyst, using ground truth when available.

Bounding box: Rectangle around all of the target or the main body of the target.

For forward-looking imagery, the bounding box is generally rectilinearly oriented (Fig. 1.3). For down-looking imagery, the bounding box will be at an angle with respect to the axes of the image (Fig. 1.4).

For forward-looking imagery, ground truth target location will generally be pinned to the ground surface rather than at the center of the grayscale mass of the vehicle. This is because the range to the point that the target touches the ground is different from the range along the view-ray through the target center to the ground. This truth will have an associated target location error (TLE) in geographical and pixel coordinates. The TLE for database targets can only be specified statistically. The truthing process might indicate the set of pixels on the target, known as the *target region*. These pixels can match the shape of the target, or be more crudely specified as a rectangular (as in Fig. 1.3) or elliptical region. The target region is generally, but not always, contiguous. A target region can even be smaller than a single pixel, as for the case of low-resolution hyperspectral imagery.

Target report: Report output by ATR generally providing, as a minimum: location in the image of detection (by its reference pixel), the equivalent

Figure 1.3 Illustration of a bounding box (a) around the entire target or (b) around the main body of the target.

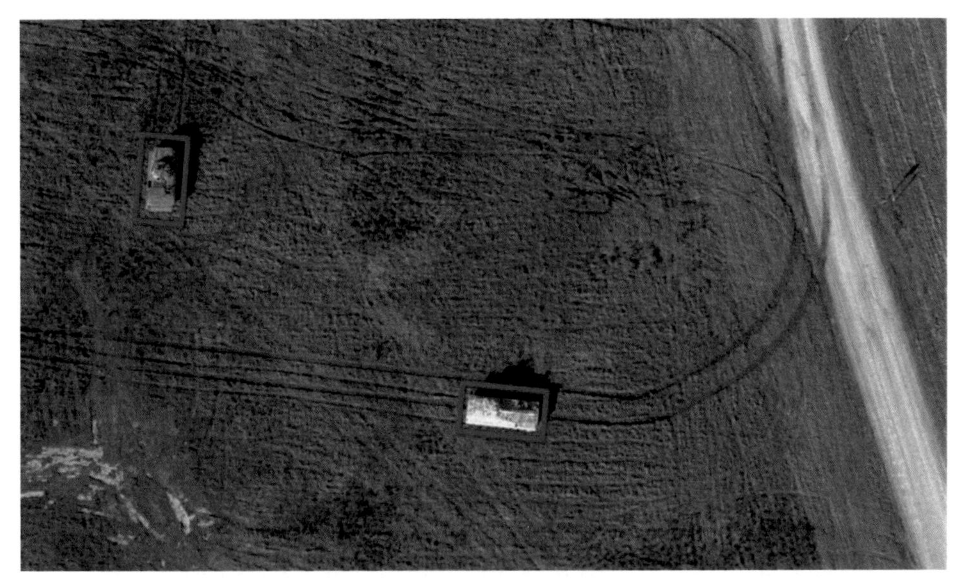

Figure 1.4 Boxes around targets in overhead imagery can be at any angle.

location as latitude and longitude on an earth map, various categories of classification assigned to the target, and associated probability estimates.

The information contained in the target report can be quite extensive, but only parts of it can be disseminated due to mission and bandwidth. A popular protocol is MITRE's Cursor-on-Target (CoT). The CoT event data model defines an XML data schema for exchanging time-sensitive position of moving objects between systems: "what, "when," and "where" information.

Target location and/or region as reported by ATR: Estimated target reference pixel p_{ATR} or region R_{ATR} as provided in an ATR's report.

The ATR will report a target location. This can be the target's geometric center, the center of grayscale mass, the center of the rectangle about the target, the brightest point on the target, or a point where the target touches the ground. The ATR might estimate the pixels on target through a segmentation process. ATR engineers should understand the scoring process and end-user requirements, so as to know how best to report a target.

Target detection: Correct association of target location p_{ATR} or target region R_{ATR}, as reported by the ATR, with the corresponding target location p_t or target region R_t in the truth database.

Detection criterion: The rule used to score whether an ATR's reported target location or region sufficiently matches the location or region given in the truth database.

Note that the truth database can contain mitigating circumstances for which the ATR is given a pass if it doesn't detect particular targets. Such circumstances can be: target out of range, not discernable by eye, mostly obscured, half-off image, covered by camouflage netting, etc. Such objects are referred to as *non-spec targets*.

Once a tracker locks onto a detected target, performance is measured by rules associated with trackers rather than detectors. Tracker evaluation criteria are well established, but are not covered in this book.

Specific kinds of detections include:

- **Multiple detection**: Detections on a target, beyond the first or strongest one reported (for a single frame).
- **Group detection**: A single detection on an assemblage of objects in close proximity, such as a huddle of combatants.
- **Event detection**: Detection of an occurrence, such as: a missile ready to launch, persons unloading a truck, or a person planting an IED.
- **Flash detection**: Detection of the location in image coordinates of a muzzle flash.
- **Muzzle blast detection**: Detection in geographic coordinates of the origin of the auditory (sound) and non-auditory (overpressure wave) components after a muzzle flash.
- **Change detection**: Detection of something in an image that wasn't perceived at that location at a previous point in time.
- **Detection of disturbed earth**: Place where an IED or landmine might have been buried.
- **Standoff detection**: Detection of a dangerous object from a safe distance.
- **Brownout detection**: Detection of the presence of a dust cloud degrading the visual environment.
- **Extended-object detection**: Detection of something very long with no obvious beginning or end, such as power lines, a pipeline, a tunnel, a string of landmines or an underwater cable.
- **Fingerprinting**: Detection not of a *type* of vehicle, but one *particular* vehicle, for example, the car with the bombers in it.

Caveats and ambiguities

Although we will use the common term *image truth*, we note that what is normally referred to as image truth is more realistically *expert opinion*, rather than absolute omnipotent truth. Image truth is often produced by one or more image analysts using ground truth information when obtainable. Image truth can include supporting information such as target aspect angle, image quality near target, and the number of pixels on target. Image truth can also include the truther's opinion of clutter level. Although the image truthing process currently involves significant manual labor, work is underway to automate parts or all of the process.

Image truth is best produced during a data collection rather than months afterwards. Image truth will contain errors. For example, in IR imagery, some parts of the target will fade into the background. Exhaust can heat up the ground. A puff of smoke or kicked up dust can obscure the target or appear to be part of the target. Before starting image truthing, it must be made clear what should get labeled as part of a target. Consider possible cases of ambiguity (some of which are illustrated in Fig. 1.2): a bush in front of the target, antenna, flag, chain, open space within target's convex hull, gun and backpack carried by dismounted combatant (dismount), target behind another target, vehicle transported on truck bed, camel carrying combatant or weapons, decoy, netting draped over and off of target, hulk of vehicle, fuel supply vehicle adjacent to target vehicle, object towed by target, target shadow, false data produced by turbulence, dust trail behind moving vehicle, aircraft's contrail, ship's wake, etc.

If a scene is synthetically generated, pixels on target are known. Even then, a decision must be made about how to label a pixel that is part target and part background.

Specifying region on target, whether by man or machine, is a nice concept in theory. However, in practice, it might not be possible in some circumstances. Parts of an object in thermal imagery might be darker (colder) than the background, and other parts much hotter, but much of the target can be of similar temperature to the background. This can happen if the object is a vehicle (Fig. 1.5) or dismount. In such cases, it is not obvious which groups of pixels combine to form the region on target. In the visible band, painted camouflage patterns on vehicles and camouflage uniforms can make segmentation of vehicles or soldiers problematic, depending on the background color and texture. In overhead visible band imagery, dark vehicles tend to blend in with

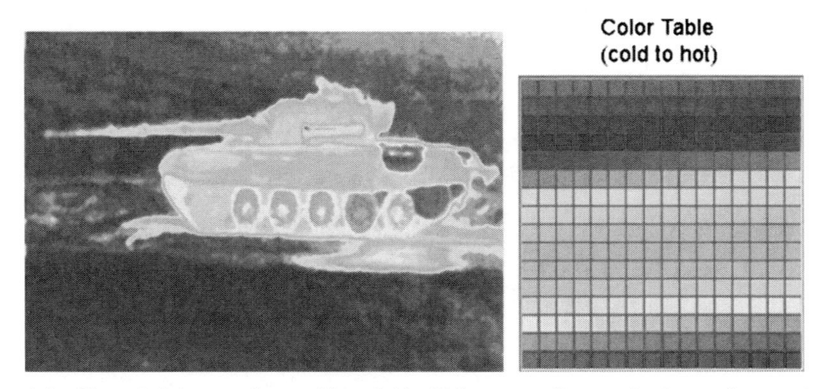

Figure 1.5 Target data may be multimodal in IR imagery. Some pixels can be much hotter than the background, while other pixels match the background temperature. (Shown in false color in electronic book formats.)

shadows. In SAR, it is difficult to determine target region when vehicles are tucked into a tree line.

1.3 Detection Criteria

It is quite challenging to precisely and unambiguously stipulate what is meant by *target detection*. Let us first consider some relevant terms:

$|R|$ = cardinality of R = the number of pixels in region R.
Let R_t = region on target as indicated by truth data.
R_{ATR} = region on target as reported by ATR.
p_t = point (or reference pixel) on target according to the truth data, i.e., the target reference pixel.
p_{ATR} = point (or reference pixel) on target as reported by ATR.
$\|a - b\|$ = distance between points a and b.

First, let us suppose that the ATR outputs a single detection point per object and the truth database contains a single detection point per target. Let

$A = \{p_{ATR}\}$ denote the set of detection points output by the ATR, and
$T = \{p_t\}$ denote the set of detection points in the truth database.

The set C of correct detections output by the ATR is such that each detection in C matches a target in the truth database T according to some match criterion. Here, we will define several common detection criteria, illustrated in Fig. 1.6.

Minimum distance criterion: If the minimum distance between an ATR reported target point and the nearest target point in the truth database is less

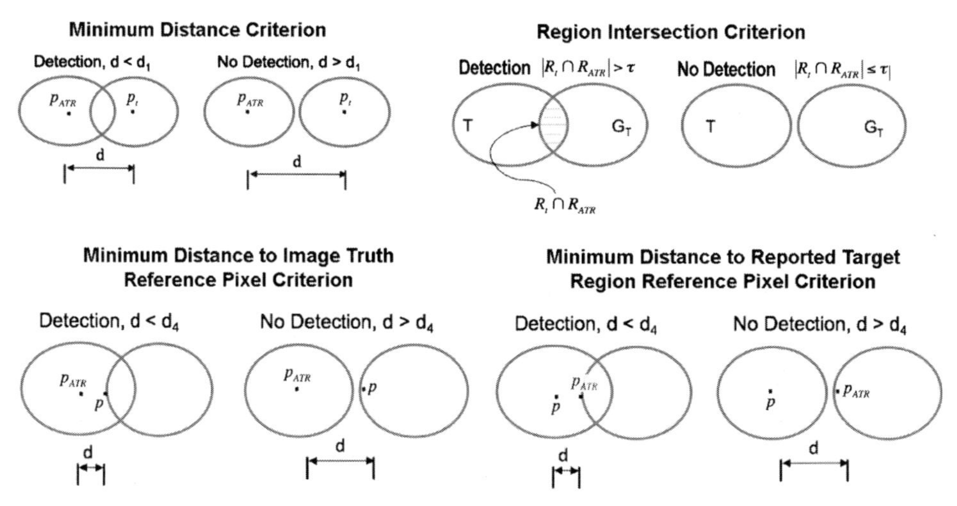

Figure 1.6 Illustration of several detection criteria.

than a preselected value d, then the ATR has detected a valid target, as defined by

$$p_t \in C \text{ iff } \min_{p_t \in T} \|p_{ATR} - p_t\| \leq d.$$

The number of correct detections is given by $|C|$.

This definition allows for, at most, one correct detection on each target in the truth database T. Note that for the purpose of scoring an ATR, if the exact same vehicle appears in more than one image, it is usually considered a different target for each image in which it appears. Detecting the exact same target in two different images results in two correct detections. It is possible, but unlikely, that one detection point reported by the ATR will result in a count of two or more correct detections, for example, if part of a target is in front of another target.

Another way of defining a correct detection is if a target region in a truth database intersects the corresponding target region output by the ATR.

Region intersection criterion: $|R_t \cap R_{ATR}| > \tau$, where τ is a threshold.

Alternatively, a target indicated by a single reference pixel can be said to be detected if the reference pixel falls within a target region. This leads to two additional detection criteria.

Minimum distance to image truth reference pixel criterion: $\min_{p \in R_t} \|p - p_{ATR}\| < \tau$.

Minimum distance to reported target region reference pixel criterion: $\min_{p \in R_{ATR}} \|p - p_t\| < \tau$.

The opposite of a true detection is a *false alarm*. Defining a false alarm involves the concept of clutter.

Clutter object: Non-target object with characteristics similar to those of a target object. A clutter object can be natural or manmade. A clutter object can be ephemeral, such as a hot patch of ground or an opening between two trees.

False alarm: A detection reported by the ATR that does not correspond to any target in the truth database, according to some agreed-on detection criterion.

With this definition, it is possible for the ATR to get penalized for multiple false alarms by reporting multiple detections on the same clutter object within a single image. The ATR could also be penalized for a false alarm by reporting a point between two adjacent targets. This penalty could be relaxed for a particular system if its purpose is to extract (substantially larger than target) regions-of-interest to display to a human operator. For this type of system, two adjacent vehicles would be displayed to the operator for further decision, even if the detection point is between them. The same would be true for a huddle of combatants. It generally won't be necessary to detect each and every person in the huddle. However, we have seen government tests where detection of each person in a close group was required.

Region-of-interest (ROI): A rectangular image chip about a detected object. It may be scaled as a function of range.

Several examples of ROIs are shown in Fig. 1.7. It is preferable that the detected object be near the center of the ROI.

False alarm rate: A measure of the frequency of occurrence of false alarms in a reference context.

False alarm rate (FAR) is measured differently for forward-looking sensors compared to downward-looking sensors. This is because with a forward-looking sensor it is not possible to determine the ground area covered by the image. Instead, a measure of the solid angle of the optics is used. For example, a sensor might have a horizontal field of view of 2 deg and a vertical field of view of 1.5 deg, or equivalently, 3 square deg. Various measures of FAR can be defined based on different reference contexts:

$$\text{Pixel FAR: } FAR = \frac{N_{FA}}{N_{MP}} = \frac{\text{number of false alarms}}{\text{number of megapixels processed}}.$$

$$\text{Frame FAR: } FAR = \frac{N_{FA}}{N_{FP}} = \frac{\text{number of false alarms}}{\text{number of image frames processed}}.$$

$$\text{Area FAR: } FAR = \frac{N_{FA}}{N_{km}} = \frac{\text{number of false alarms}}{\text{sum of ground area covered by images processed}}.$$

$$\text{Temporal FAR: } FAR = \frac{N_{FA}}{\Delta_{time}} = \frac{\text{number of false alarms}}{\text{time interval}}.$$

$$\text{Angular FAR: } FAR = \frac{N_{FA}}{N_{SD}} = \frac{\text{number of false alarms}}{\text{number of square degrees processed}}.$$

Figure 1.7 Examples of ROIs from several types of sensors: (a) IR, (b) visible, (c) SAR, and (d) sonar.

Ambiguities in measures of false alarm rate
There is some ambiguity in each of these measures. Ambiguity should be eliminated in a particular project based on more precise definitions that take into account the important issues for the project. For example, the ATR might not be able to process and detect targets toward the edges of images, meaning that the total number of pixels in the image set differs from the number of pixels fully processed. The ATR might not process the region above the skyline. Frames might overlap, such that the sum of the ground area processed for the ensemble of frames is greater than the actual ground area covered. Frames or parts of frames might be discarded due to insufficient image quality, or some areas may be outside of range limits. The left side of a non-target might appear in one frame and the right side in the next step-stare frame. With video data, the same rock could produce 30 false alarms per second. Should this be counted as a single false alarm or as multiple false alarms?

There are many other types of ambiguities that must be resolved. If, for example, a military truck or plane is considered a target, should detection of the equivalent civilian truck or plane be treated as a correct detection or as a false alarm? Should the detection of a military vehicle outside the list of targets sought be treated as a correct detection or a false alarm? That is, is the target set open or closed? Some of these issues can be resolved by defining a "don't care" class. Detection of a *don't care object* has no effect on scoring. A calibration target is always a don't care object. A calibration target could be a corner reflector in a radar test, a thermal target board in an IR test, or a color panel in a hyperspectral test.

One problem with scoring a very large geographic area is that objects outside of a military compound will not have associated ground truth. Could there be a military vehicle or similar civilian vehicle on the road to the military base or used as decoration at an armory? If the objective is to verify a rate of 0.001 false alarms per km^2, an area nearly the size of Connecticut would need to be truthed. This is not feasible.

1.4 Performance Measures for Target Detection

Performance measures include truth-normalized measures, report-normalized measures, and various graphical depictions.

1.4.1 Truth-normalized measures

Probability: A way of expressing knowledge or belief that an event will occur *or* has occurred.

There is often some confusion resulting from the dual nature of the definition of *probability* as "has occurred" or "will occur." ATR engineers use probability rather loosely to mean measured ATR performance over a particular database. Probability is not an unqualified prediction about the future. It is only a measure of performance in a controlled experiment. It only serves as a prediction of future performance to the extent that the future data has characteristics similar to the data processed.

Probability of detection: The probability that the ATR associates a non-redundant detection with a target in the truth database:

$$P_d = \frac{|C|}{|T|} = \frac{\text{number of correct target detections}}{\text{number of ground truth targets}}. \tag{1.1}$$

P_d by itself does not have much utility. It is always possible to declare a detection at each pixel in an image and achieve $P_d = 100\%$. It is the tradeoff between missed detections and false alarms that counts. In textbook terms, this is the normal tradeoff between Type I and Type II errors (Table 1.1).

Probability of a miss: $P_{miss} = 1 - P_{det}$.

Probability of false alarm: The number of false detections normalized by the number of opportunities for false alarm:

$$P_{FA} = \frac{|F|}{|O|} = \frac{\text{number of false alarms}}{\text{number of false alarm opportunities}}. \tag{1.2}$$

The number of false alarm opportunities is an imprecise concept. It is often determined as follows. A polygonal tile is chosen to match the average size of a ground truth target. The size can be a function of range. The number of tiles required to cover the image set is considered to be the number of false alarm opportunities. This doesn't make much sense for forward-looking scenes containing a vanishing point, trees, and sky.

Suppose that we are just testing the back end of the ATR over ROIs. For a fixed database of target and clutter ROIs, the number of false alarm opportunities is then the number of clutter ROIs in the test database. This use of the term P_{FA} makes more sense.

Table 1.1 Tradeoff between Type I and Type II errors.

		Decision	
		Target	**Clutter blob**
Truth	Target	Correct detection (true positive)	Missed detection (Type II error)
	Clutter blob	False alarm (Type I error)	Clutter rejection (true negative)

1.4.1.1 Assigned targets and confusers (AFRL COMPASE Center terminology)

An ATR assigns *cues* (ID labels) to objects in the test database that sufficiently match signatures in a target data library. The function of this ATR is then solely to assist or "cue the operator." The ATR is referred to as an *automatic target cuer* (ATC). An *assigned target* is a particular target type selected from the target library by the human operator. The ATR can be directed to find that specific target type (or perhaps several assigned target types). This defines the Mission of the Day. A confuser is a target-like object intentionally inserted into an experiment to determine whether or not it confuses the ATR. The ATR correctly rejects a confuser that does not meet its decision criteria for an assigned target. *Bad actors* are confusers that, for a given assigned target, inordinately contribute to the cue error rate.

Cue correct rate (CCR): The ratio of the number of *correct* assigned target cues to the total number of assigned target cues.

Confuser rejection rate (CRR): The percent of confusers that are rejected, i.e., determined not to be an assigned target.

1.4.2 Report-normalized measure

Probability of detection report reliability: The probability that a detection reported by the ATR is a true target:

$$P_{DR} = \frac{|C|}{|A|}. \tag{1.3}$$

1.4.3 Receiver operating characteristic curve

Suppose that the ATR associates a detection strength (score) with each raw detection. Figure 1.8(a) shows sample probability density curves for true targets (true positives) and non-targets (true negatives) versus computed detection strength. This ATR makes soft decisions. It is not declaring objects to be targets. It is only assigning a degree of *targetness* to detected objects. Suppose a threshold τ is set after all of the target reports are generated. If only those detections with strength above the threshold are reported target decisions, then this ATR will have a fixed probability of detection versus probability of false alarm for this test set. If the threshold is adjusted up and down, then a P_d versus P_{fa} curve results [Fig. 1.8(b)]. This is known as the ATR's receiver operating characteristic (ROC) curve. The ROC curve applies to a particular test set and as such is not a prediction of future performance on data of a different nature. Thus, a ROC curve is simply defined as in the definition that follows.

ROC curve: Plot of P_d versus P_{fa}.

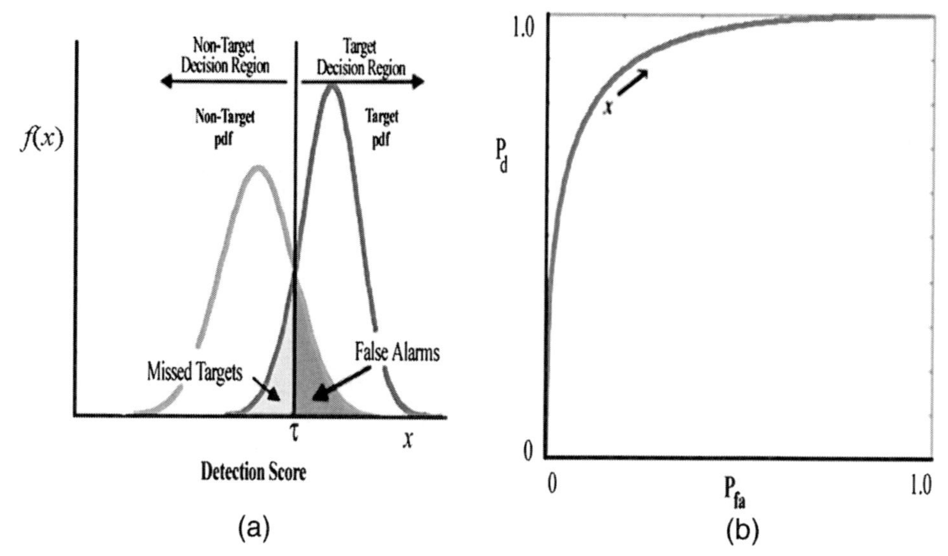

Figure 1.8 (a) Target and non-target probability distribution function (pdf). (b) Probability distribution functions transformed into an operating curve. (See Refs. 2 and 3 for in-depth discussions.)

ROC analysis was developed in the 1950s for evaluating radar systems. The term has since been applied to other types of systems, without regard to its original more specific meaning. Each point on the ROC curve represents a different tradeoff (cost ratio) between false positives and false negatives. The ROC plot thus provides a convenient gestalt of the tradeoff between detection and false alarm performance. If two ROC curves do not intersect (except at their endpoints), then the ATR corresponding to the higher curve performed better than the other. If two ROC curves intersect once, then one ATR performed better at low P_{fa}, and the other performed better at higher P_{fa}.

If a particular ATR can only make a hard decision, then this ATR has no ROC curve, even though it invariably has internal settings that can be adjusted in software. Regardless of the case, there is no assurance that setting a particular threshold on an operational system will produce a pre-specified performance level on new data.

The concept of a ROC curve obtained by adjusting a single threshold doesn't hold up to scrutiny. If the ATR were actually designed to operate at an extremely low false alarm rate, algorithmic changes would be required for best performance. If the ATR were to operate at a very high false alarm rate, changes would also be needed, such as reporting a longer list of raw detections out of the pre-screener. As illustrated in Fig. 1.9, The ROC curve is better suited for comparing different ATR back-end final detectors over a well-chosen set of target and clutter ROIs.

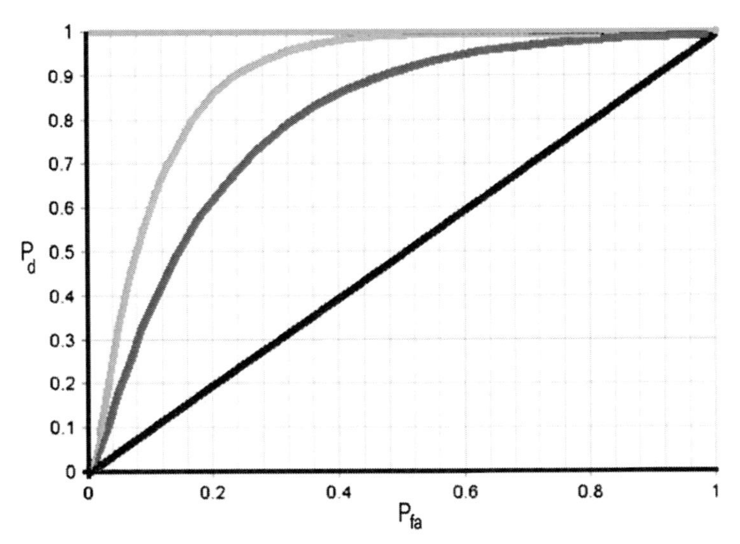

Figure 1.9 ROC curves for four ATRs: The top line represents a perfect ATR. The ATRs with performance indicated by progressively lower curves performed progressively worse. The bottom line indicates an ATR that can't tell the difference between a true target and a non-target.

A good place to find in-depth analyses of ATR performance evaluation methodologies is in Air Force Institute of Technology Ph.D. dissertations, which can be found on-line at the Defense Technical Information Center (DTIC®). ROC curves, the area under ROC curves (AUCs), and alternative ways of comparing ROC curves are provided by Alsing and Bassham.[2,3] The top curve in Fig. 1.9 has a higher AUC than the bottom curve. Two ROC curves can be compared by their AUCs. AUCs make no assumptions concerning target and non-target distributions. Alsing says that if the comparison of classifiers is to be independent of the decision threshold, AUC is a reasonable "metric."[2] However, the AUC measure is not quite a proper metric. Two ROC curves with totally different shapes can have the same AUC value. This violates the *definiteness* property of true metrics. As an alternative, Alsing suggests the use of a multinomial procedure to evaluate competing classifiers.[2] Rather than simultaneously comparing the classifiers over the entire test data set, a multinomial selection procedure compares the performance of each classifier on each data point using some scoring measure.

Bassham analyzes several variants of the ROC curve including localization, frequency, and expected utility, as well as an inverse ROC curve called a response analysis characteristic curve. He also analyzes methods for comparing ROC curves, including:

- average metric distance: the average distance between two ROC curves, using some distance metric,
- area under ROC curve that is above diagonal chance line,

- Kolmogorov method: nonparametric confidence bounds are constructed around ROC curves based on Kolmogorov theory.[3]

Bassham's thesis has a good discussion of the performance measures used during ATR development compared to measures that capture the operational effectiveness of ATRs.

Another variation on the ROC curve is obtained by making whatever internal changes are necessary for the ATR to presumably function best at different rates of false alarm. This generates a collection of $\{P_d, P_{fa}\}$ pairs for a test set. Connecting performance points will produce a P_d versus P_{fa} curve, which is not necessarily well behaved.

1.4.4 P_d versus FAR curve

Figure 1.10 gives an example of a P_d versus FAR curve. Like the ROC curve, each point on this curve corresponds to a different detection threshold. That is, the ATR is run on a set of data and reports a set of detections, each with an associated strength. In this example, FAR is plotted per square degree of sensor viewing angle, but any measure of FAR could be used. FAR is often plotted on a log scale.

Many such plots would characterize a single system test. For an EO/IR system, these may cover different OCs: fields of view, ranges to target, times of day, clutter level, target types, etc. One would compute separate curves to report performance over dismounts and vehicles. For a SAR system, conditions requiring different P_d versus FAR curves include sensor resolution, clutter level, and target categories.

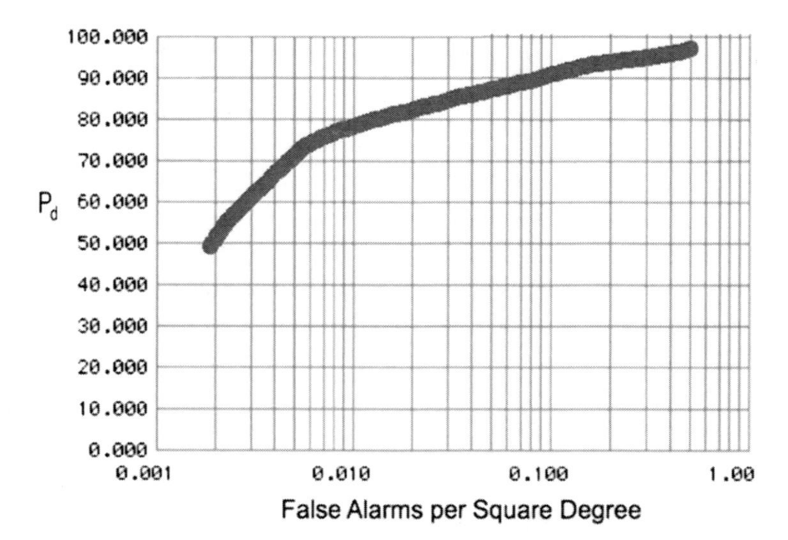

Figure 1.10 Example of a P_d vs. FAR curve.

1.4.5 P_d versus list length

Suppose that the ATR follows the simple model of Fig. 1.1. Suppose that the front-end anomaly detection stage outputs n raw detections per image frame ordered by strength. The processing requirement of the back-end of the ATR is directly proportional to n. The back-end of the ATR has limited processing capacity, which sets an upper bound on n. The longer the list of raw detections out the front-end of the ATR the harder it is for the back-end classifier to reject every non-target. P_d can be plotted against n to determine if there is a point of diminishing returns. Various front-end detectors can be compared in this manner (Fig. 1.11).

1.4.6 Other factors that can enter the detection equation

The equations given so far are for the basic case. A complete equation for a specific project can include other terms relevant to the experimental design and ConOps, such as:

- number of redundant detections on targets,
- number of detections on decoys,
- number of detections on don't care objects,
- number of detections on targets of unknown type,
- number of front-end detections for which the back-end of the ATR makes no decision (this is the opposite of requiring a forced decision),
- number of detections on objects specifically put into the database as confuser objects, and
- number of detections on *spec targets*, i.e., targets meeting specified criteria.

1.4.7 Missile terminology

Determining the effectiveness of a missile or missile defense system involves modeling and simulation (M&S). Verification and validation of the M&S are

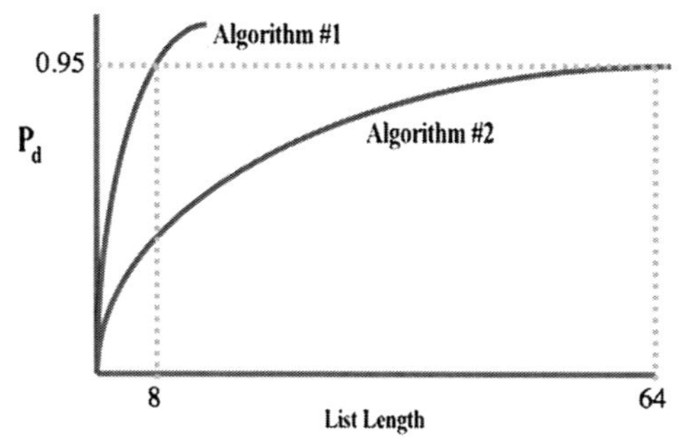

Figure 1.11 Plot of P_d versus list length for two front-end detectors. One detector requires a list of length 8 to detect 95% of targets, while the other requires a list of length 64.

extremely complex. The single missile kill chain is represented by a sequence of events. Each event has its own probability of failure. Each step in the chain decreases the final probability of kill.

Engineers developing missiles and other types of munitions tend to use terminology that differs from that used by the ATR community. A simple example follows:

$$P_{ssk} = P_h \times P_d \times R_m \times R_w,$$

where
P_{ssk} is the probability of a single shot kill,
P_h is the probability of a hit,
P_d is the probability of detecting the target,
R_m is the reliability of the missile, and
R_w is the reliability of the weapon.

1.4.8 Clutter level

Target detection performance should be reported with regard to the clutter level of the database over which performance is measured. There have been many attempts over the years to develop an equation or procedure for characterizing clutter level in images as an alternative to expert opinion. None of these attempts have been thoroughly successful. The problem is one of circular reasoning. If an ATR's P_d versus FAR curve is bad, then the clutter level must have been high as perceived by that particular ATR. A different ATR using different features and algorithms may perform better on the same data. To this ATR, the clutter level is low. Furthermore, if an algorithm can measure clutter level, then it must be able to distinguish targets from clutter, which itself defines an ATR.

Richard Sims did some of the best work in this area.[4] His signal-to-clutter-ratio (SCR) metric is given by[4]

$$SCR = \sum_{i=N_{c+1}}^{N} \frac{\lambda_i}{1-\lambda_i} + \sum_{i=1}^{N_c} \frac{1-\lambda_i}{\lambda_i}. \tag{1.4}$$

This equation derives from eigenvalues of the Karhunen–Love decomposition of a target-sized image region—referred to as the Fukunaga–Koontz transform. The first term encompasses all eigenvalues, denoted by λ_i, where the target dominates. The second term is a measure of the useful information where clutter dominates. The reader is referred to referenced papers for details.[4,5]

1.5 Classification Criteria

Classifier categorization is often represented graphically, while performance is given by a table.

1.5.1 Object taxonomy

In the context of ATR, ontology is a subject of study involving the categories of objects relevant to an experiment, mission, or battlespace. The product of the study, called an *ontology*, is a catalog (a.k.a. library) of the objects assumed to exist in a domain of interest from a military perspective, as well as more precise specification of the basic categories of the objects and their relationships to each other. Such objects can be grouped, related within a hierarchy, and subdivided according to their similarities or differences.

The first stage of a modeling is called *conceptualization*. An ontology is a formal explicit specification of the conceptualization. It provides a shared vocabulary that can be used to model the types of objects in a military domain and their relationships—with sufficient specificity to develop and test an ATR. Ontologies consist of concepts that can be structured hierarchically, thus forming a *taxonomy*.

Taxonomy: Objects arranged in a tree structure according to hyponymy (is a) relations.

For example, a T-72 *is a* tank. A taxonomy places all of the objects into a hierarchy and clarifies the possible labels for an object at various category levels. An example is given in Fig. 1.12.

A taxonomy is a structure for classification. A classifier assigns detected objects to categories based upon the taxonomy. In practice, the classification categories are predetermined, but not exhaustive. That is, the taxonomy will not cover all military vehicle types in the world but should include all vehicles of interest to a military program, mission, or experiment. The categories are

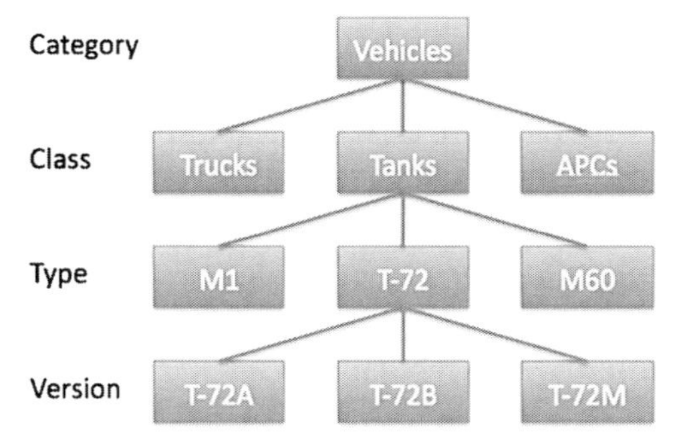

Figure 1.12 Example of a taxonomy.

exclusive. For any level of the taxonomy, an object can be assigned to category A or B, but not both.

A label of *other* can be included at any level of the taxonomy. For example, at the Type level, *other* can mean any other type of main battle tank not explicitly listed.

Decision tree: A visualization of the complex decision making taking place within an ATR, illustrating possible decisions and possible outcomes, and modeled on the hierarchy of the taxonomy.

The decision tree illustrates all possible classification outcomes and the paths by which they can be reached. While the taxonomy answers the question, "What is a T-72?" the decision tree answers the question, "What kinds of tanks are of interest?"

A decision tree is used as both a visual aid and an analytical tool. It uses its tree-like graph to model the flow of decisions. The decision tree emanates from a starting point (usually a root node at the top of the diagram) and continues through a series of branches and nodes until a final result is reached at the bottom end (leaf of upside-down tree). It illustrates how an ATR's classifier could go about making its pronouncements, step-by-step, coarse-to-fine. (However, we will give some examples later as to why the decision tree structure might not properly model the operation of a particular ATR's classifier.)

At any level of the decision tree, a classifier can make a *declaration*, which is a decision to provide a label. The label corresponds to a node name within the taxonomy. A classifier might, for example, declare a detected object to be a T-72 but may not be able to specify the version of T-72. In this case, all of the levels of the decision tree above the T-72 node would be declared, but not those below the T-72 node.

It is common practice to label the levels of the decision tree by names corresponding to the specificity of the decisions. Thus, in order of increasing specificity, names such as detection, classification, recognition, and identification are commonly used. Such terms must be clearly defined in the context of a particular program. (In this usage, "classification" refers to a specific level of the decision tree rather than the overall operation of the ATR's classifier stage.) Decision trees can be quite different for different programs. In the Dismount Identification Friend or Foe program, dismounts were said to be *identified* if it was determined that they were carrying large weapons rather than confuser objects such as 2×4 lumber or farm tools. For a classifier used to screen people entering a military base, identification could mean naming the person. Several examples of decision trees are given in Fig. 1.13.

Taxonomies and decision trees: ambiguities and exceptions

A node at each level of a taxonomy has only one parent node. It may not be clear how to arrange the taxonomy when this condition is violated, that is,

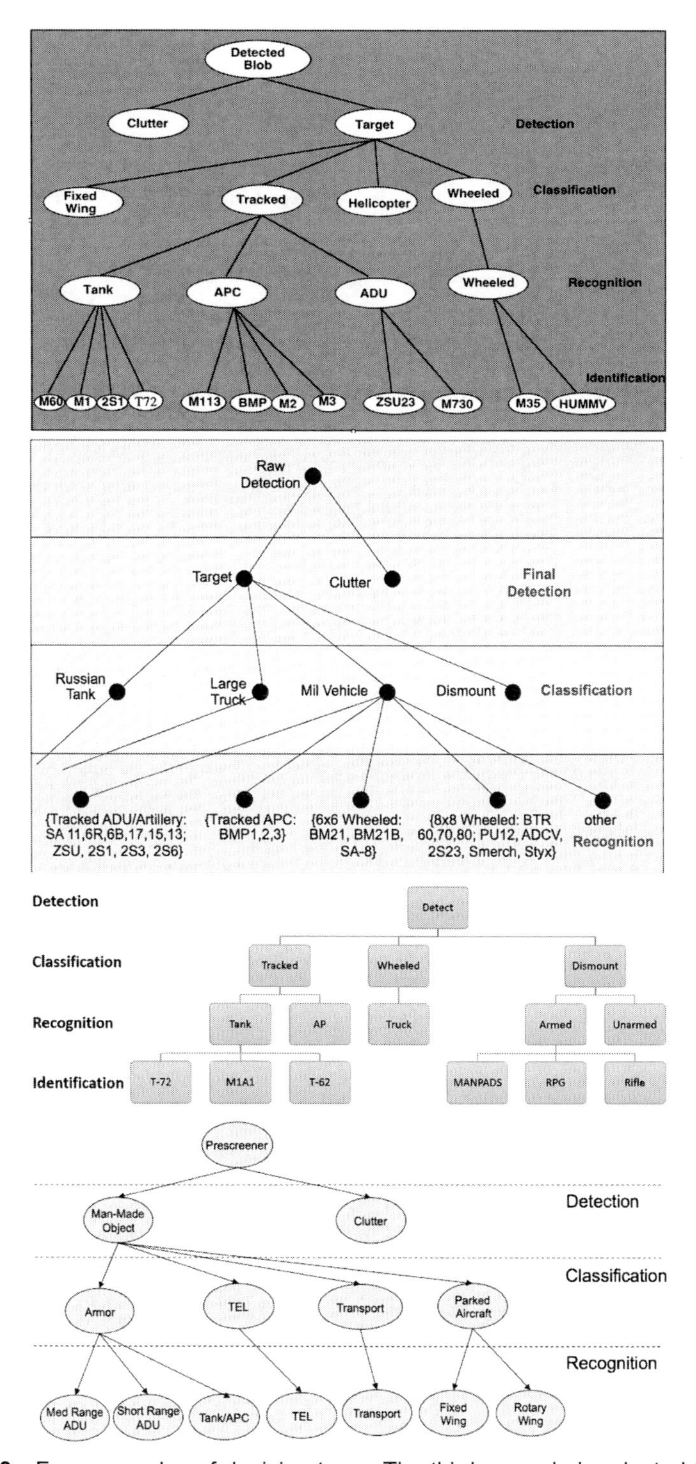

Figure 1.13 Four examples of decision trees. The third example is adapted from Ref 6.

when some vehicles fit into more than one category per Venn diagram (see Fig. 1.14). Several ambiguous cases (from an ATR perspective) follow:

- friend or foe (e.g., enemy T-72 versus NATO T-72),
- tracked or wheeled (e.g., SA-19 on tracks versus SA-19 on wheels), or
- scout car or Air Defense Unit (ADU) (e.g., BRDM scout car with SA-9 ADU weapon).

It is generally the responsibility of those funding a program to specify the targets of interest and their taxonomy. However, the funding organization might not have a good grasp of this issue. An often heard comment is "just try your ATR on the data and see what happens," not understanding that the ATR must be trained to operate with a specified taxonomy. Another common problem occurs when the funding organization supplies test data with targets not fitting the stipulated taxonomy and then complains that obvious targets are not being reported.

A taxonomy is easily transformed into a decision tree. However, a particular ATR's decision process might not fit that of the decision tree corresponding to the taxonomy. For example, a template matcher might only operate at the Type level. Class is obtained by generalizing from Type. Another ATR might utilize separate neural network classifiers for each level of the taxonomy. There is no guarantee that a decision made at the Class level will correspond to a generalization of the decision made at the Type level. An ATR might receive data from multiple sensors, some of which support decisions at one level of the tree and others that support decisions at other levels. For this case, the decision making process might be much more

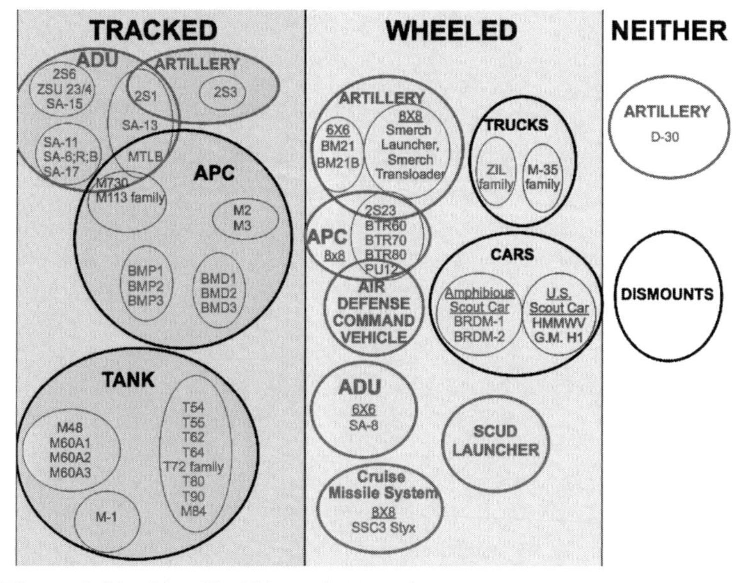

Figure 1.14 Simplified Venn diagram for some common targets types.

complex than indicated by a simple decision tree. Decisions can also change over time as a target is tracked, the sensor and ATR switch modes, or off-board information is received.

1.5.2 Confusion matrix

An *error matrix* quantifies the discrepancy between fact and the ATR's opinion. Each column of the matrix represents instances of a reported category, while each row represents the actual category. With a strict interpretation, measures of error are realizable only when truth is known absolutely.[7] In an ATR test, this would occur with synthetically generated data.

In ATR tests with field-collected data, *truth* is more fittingly called *expert opinion*. Human *truthers* provide their expert opinion of a reference point or pixels on a target, using available ground truth (instrumented data) to help determine target type and location. Target labels can be error prone for targets that are imaged but not intentionally put into the test site, such as vehicles at a military base outside of the planned test site. ATR test results are reported in a *confusion matrix* rather than an *error matrix*. The confusion matrix measures the ability of an ATR to generalize from its training data to the test data, within the accuracy of the image truth. (In this context, training includes the development of templates for a template matcher or storing of vectors for a nearest neighbors classifier.) The ATR's classification performance is evaluated from the data in the confusion matrix.

A confusion matrix characterizes the ATR classifier stage's ability to assign categories to detected objects. If the ATR has an adjustable detection threshold or other adjustable internal parameters, then the confusion matrix addresses performance at those particular settings. For example, the front-end detection stage of the ATR might be set up to operate so that only very strong targets are detected. The classifier stage of the ATR will then have an easier time assigning these strong detections to categories than if the front-end detection stage were less restrictive, i.e., operating "full throttle."

Probability of (correct) classification: Number of objects correctly classified divided by total number of objects classified.

For Table 1.2, $P_c = \dfrac{a+e+i}{a+b+c+d+e+f+g+h+i}$.

The form of the confusion matrix in Table 1.2 indicates that this test is only over objects known to be targets. Otherwise, there would be an additional row and column labeled "non-target."

Ambiguities and caveats associated with confusion matrices

An ATR can be properly designed and trained through use of assumptions of the *a priori* probabilities of target categories (as well as those of target versus clutter blobs). It also must make assumptions about operating conditions (OCs). This is not to say that the designers of the ATR, or testing

Table 1.2 Example of a confusion matrix (APC is armored personnel carrier).

		Reported by ATR		
		Tank	Truck	APC
Truth (actually, expert opinion)	Tank	*a*	*b*	*c*
	Truck	*d*	*e*	*f*
	APC	*g*	*h*	*i*

organization, actually think through these assumptions. Best performance is achieved when these assumptions match the actualities of the test data. A fair test is one in which all parties being tested have equal knowledge of the proportion of targets of each category in the test set, as well as OCs covered by the test data. In an operational setting, there will be intelligence information on enemy forces, so it is not unreasonable to provide such information for pre-production ATR testing. An operational IR ATR will, for example, know if it is day or night. A typical unfair test is one in which the provided training data is for night and the test data is for day. The equation given for P_c is implicitly making the assumption that the sum of the entries in the rows of the confusion matrix are the priors for the populations of interest. Assumptions about the priors can only be avoided by not aggregating the elements of the confusion matrix; however, this still doesn't solve the problem of assumptions about priors used in training the ATR. Prior assumptions used in training and scoring the ATR should be reported along with test results.

1.5.2.1 Compound confusion matrix

A compound confusion matrix reports results for more than one level of a decision tree. Consider the example shown in Fig. 1.15. The form of this confusion matrix indicates that the ATR's back-end classifier stage is to be tested on target and clutter ROIs. Several performance results are given based on the cells of the confusion matrix.

Decision trees and confusion matrices can be quite complex, as illustrated in the multilevel example given in Fig. 1.16.

1.5.3 Some commonly used terms from probability and statistics

Let us review some terms and introduce a few others. Suppose that the ATR declares a detected target to be a tank with a score of 0.8. This score can be considered a probability estimate with certain restrictions. The sum of all possible outcomes at a given level of specificity, called the sample space of the experiment, must be 1.0.

At the Target Class level of the decision tree, the ATR output is more specifically an *a posteriori* probability estimate vector, where each element of the vector corresponds to a target class. But, sometimes only the maximum

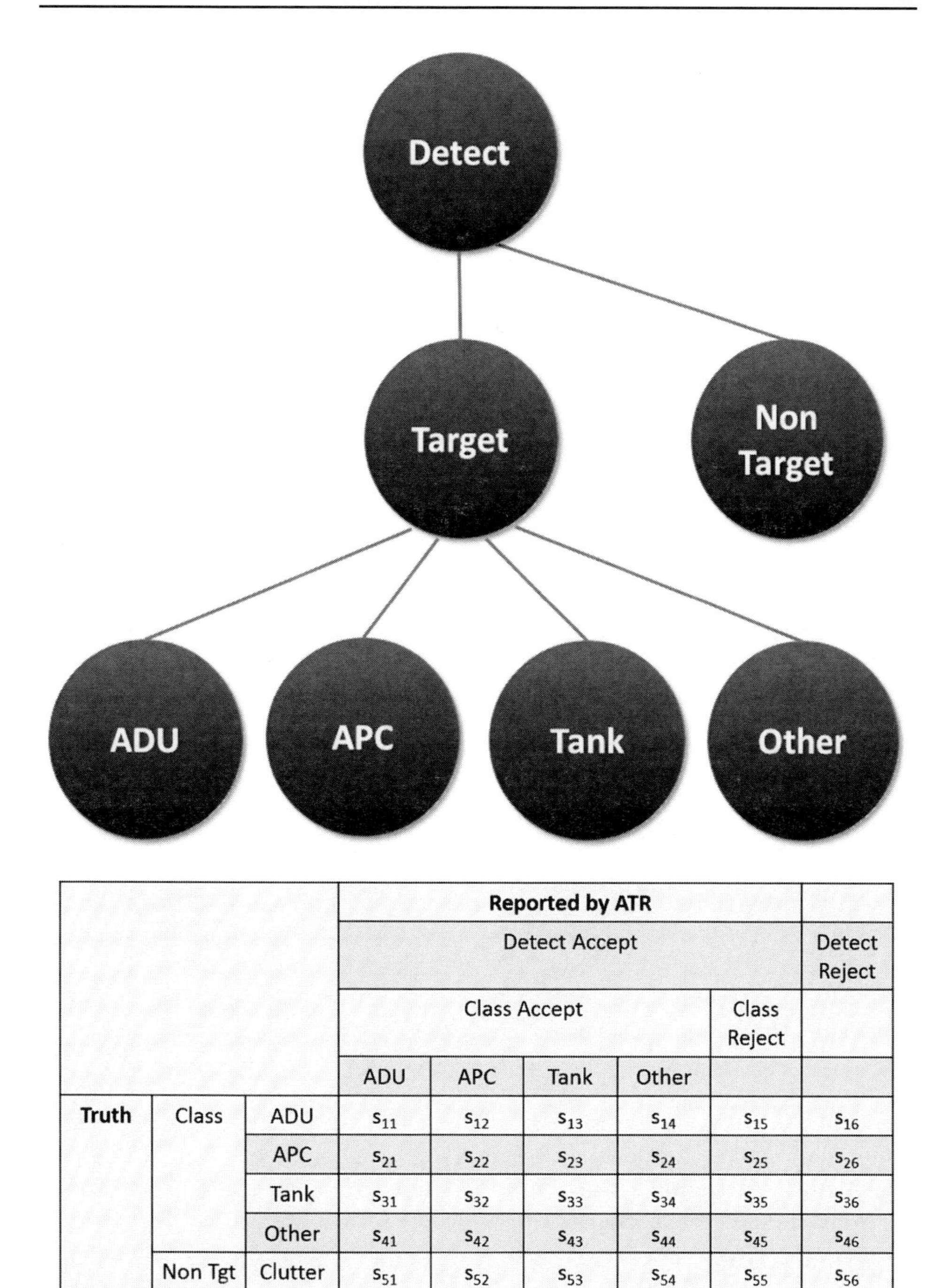

			Reported by ATR					
			Detect Accept					Detect Reject
			Class Accept				Class Reject	
			ADU	APC	Tank	Other		
Truth	Class	ADU	s_{11}	s_{12}	s_{13}	s_{14}	s_{15}	s_{16}
		APC	s_{21}	s_{22}	s_{23}	s_{24}	s_{25}	s_{26}
		Tank	s_{31}	s_{32}	s_{33}	s_{34}	s_{35}	s_{36}
		Other	s_{41}	s_{42}	s_{43}	s_{44}	s_{45}	s_{46}
	Non Tgt	Clutter	s_{51}	s_{52}	s_{53}	s_{54}	s_{55}	s_{56}

Figure 1.15 Decision tree (top) and corresponding compound confusion matrix (bottom).

element of the vector is reported along with class label. These probability estimates are often based on the assumption that the *a priori* probabilities of all allowable target classes are equal and that the training data is in some sense

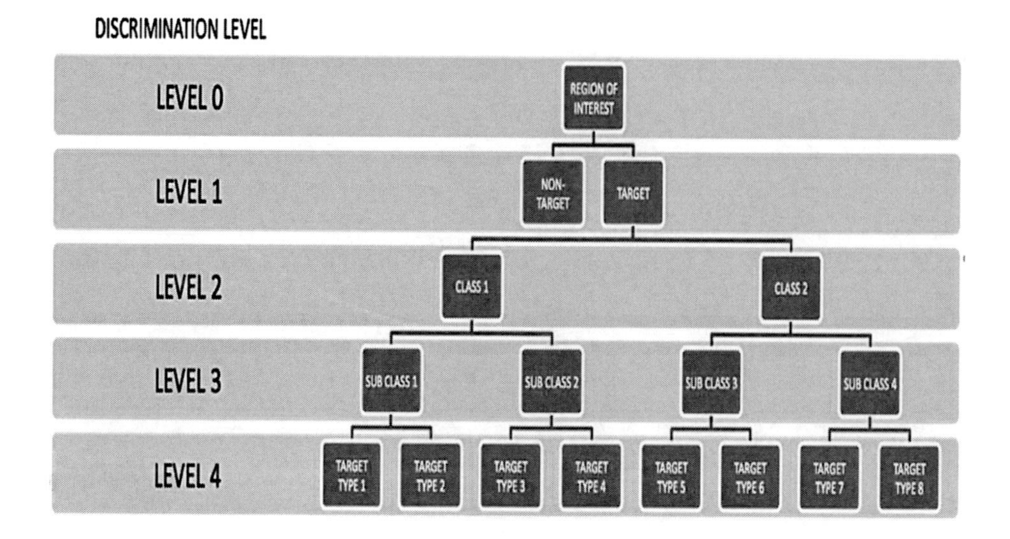

ATR Decision

				TARGETS								Non Target
				CLASS 1				CLASS 2				
				SUB-CLASS 1		SUB-CLASS 2		SUB-CLASS 3		SUB-CLASS 4		
				Target 1	Target 2	Target 3	Target 4	Target 5	Target 6	Target 7	Target 8	
Truth	Targets	CLASS 1	SUB-CLASS 1	Target 1								
				Target 2								
			SUB-CLASS 2	Target 3								
				Target 4								
		CLASS 2	SUB-CLASS 3	Target 5								
				Target 6								
			SUB-CLASS 4	Target 7								
				Target 8								
	Non Target											

Figure 1.16 Multilevel decision tree (top) and corresponding confusion matrix (bottom).

representative of the test data. For results to be statistically justifiable, both training data and test data must be random samples drawn from the same population. If the training data is not representative of the test data, it is not clear what should be expected of the ATR.

***a priori* probability**: The probability estimate prior to receiving new information (e.g., image data). The set of *a priori* probabilities are the priors.

For example, the *a priori* probability of database target classes {tank, truck, APC} would each be 0.333 under the nominal assumption.

The term *confidence* is often applied to an ATR score in a colloquial manner. Better definitions follow.

Confidence: The probability, based on a set of measurements, that the actual value of an event (e.g., target score) is greater than the computed and reported value.

For example, there may be 50% confidence that the actual score of the T-72 is greater than the reported score of 0.8.

Confidence interval: The probability, based on a set of measurements, that the actual value of an event (e.g., target score) resides within a specified interval.

For example, the probability could be 80% that the actual T-72 score falls between 0.7 and 0.9. Note that confidence bounds contradict the interpretation of the ATR score as a probability since a probability must lie within the bounds of [0.0, 1.0] and confidence bounds are typically not so restricted.

Confidence bounds could also be placed around ATR performance results as a whole. For example, the probability could be 80% that the P_{ID} performance of the ATR falls within the bounds of (0.7, 0.9). Although confidence bounds are sometimes provided for ATR performance results, the required rigorous statistical model justifying them is generally lacking. Furthermore, such bounds would only apply to a carefully designed closed experiment, not the infinitely varying real world.

Various competing theories provide ways of measuring confidence. Each of these approaches has major proponents as well as its own terminology and equations. For example, opinions can be said to come in degrees, called degrees of belief or credences.

Degree of belief [bel(A)]: The degree of belief, given to event A, is the sum of all probability masses that support the event A without supporting another event.

Degree of plausibility [pl(A)]: The degree of plausibility quantifies the total amount of belief that might support an event A. The plausibility is the degree of support that could be attributed to A but can also support another event.

Knowledge imprecision of probability estimate: The two quantities bel(A) and pl(A) are often interpreted as a lower and upper bound of an unknown probability measure P on A. The difference pl(A) – bel(A) is an indicator of the degree of knowledge imprecision on $P(A)$.

Pignistic (Latin for betting) probability: Pignistic probability Bet P is the quantification of a set of beliefs into final form for decision making. The value of a pignistic probability falls between that of a belief and a plausibility.

1.6 Experimental Design

An *experimental design* is a blueprint of the procedure used to test the ATR and reach valid conclusions. A good experimental design is critical for understanding the performance of the ATR under all conditions likely to be encountered. The experimental design is *internally valid* if it results in a fair test with no extraneous factors. The experiment is *externally valid* if it is sufficiently well designed for test results to generalize to operational conditions.

Factors jeopardizing **internal validity** of competitive ATR tests include:

- unequal access to training data;
- unequal information on, or ability to negotiate or alter, the experimental design;
- unequal access to *a priori* probabilities for target types or operating conditions;
- unequal access to test data, data very similar to test data, or number of times that the test is taken on the same data; and
- blind test data that is not 100% unseen by some organizations taking the test.

From a scientific viewpoint, it would be nice for all competitive ATR tests to be fair tests. From a business perspective, participants in the test might each seek an advantage.

Internal validity is at the center of all cause–effect inferences that can be drawn from a test. If it is important to determine whether some operating condition (OC) causes some outcome, then it is important to have strong internal validity. Essentially, the objective is to assess the proposition if X, then Y. For example, if the ATR is given data for operating condition X, then the outcome Y occurs.

However, it is not sufficient to show that when the ATR is tested with data from a certain OC, a particular outcome occurs. There may be many other reasons, other than the OC, for the observed outcome. To in fact show that there is a causal relationship, two propositions must be addressed: (1) if X, then Y and (2) if *not* X, then *not* Y.

As an example, X may refer to daytime images and *not* X to night images. Evidence for both of these propositions helps to isolate the cause from all of the other potential causes of the outcome. The conclusion might be that when solar radiation is present, the outcome Y occurs, and when it is not present, the outcome Y doesn't occur. This may be just a first step. To better understand the effect, the ATR could be tested on OC bins for each hour of the diurnal cycle with and without cloud cover.

Factors adversely affecting **external validity** or generalizability of ATR tests include:

- test data that is not representative of true operational sensor data (e.g., not considering dead bugs on window in front of IR sensor, not considering platform vibration or motion);
- very limited set of OCs in training and testing data sets;
- unreasonable hardware size, weight, power, or cost requirements;

- ignoring possibilities of enemy changing tactics, countermeasures, and decoys;
- ancillary information (metadata) provided for test not matching that (in type, update rate, or accuracy) which would be available from a relevant operational system.

If the test set represents the intended mission and scenario along all degrees of freedom (such as range, target type, aspect angles, weather conditions, clutter environment, platform, and target motion), then the performances measured on sufficiently large and representative OC bins, when properly interpreted, will point toward expected mission performance.

1.6.1 Test plan

A test plan provides a tangible description of the experimental design so that an entire Integrated Product Team (government, industry, and occasionally university) can work toward the same goals. A test plan should ideally be carefully developed and agreed on by all stakeholders. It can cover such items as how the test site will be restored after tanks tear up the ground, how and by whom data will be sequestered, and aircraft airworthiness certification. A test plan can be 100 pages long. A good test plan is critical to the success and smooth operation of a field test. The test plan should anticipate equipment failure and suggest workarounds. The author has seen tests where dozens of participants from different organizations had to be sent home because one piece of equipment was broken. The following components make up the test plan:

1. Specification of the product to be tested.
2. Scope of the testing.
3. Safety issues as well as issues concerning security, privacy, ethics, environmental, etc.
4. Test lead/manager and team member responsibilities.
5. Entry and exit criteria.
6. Descriptions of items and features to be tested, and list of items and features not to be tested.
7. Applicable requirements and requirements traceability, and key performance parameters (KPPs).
8. Test procedures and guidelines.
9. Test schedule. (Note: Planning a test of soldiers engaged in specified activities requires almost the precision of choreographing a ballet.)
10. Responsibilities for supplying test resources. Staffing and training needs.
11. Measurement & Analysis: Guidelines for in-progress and post-test analysis and reporting. Pass/fail criteria. Contingency for retest.

Safety concerns are considerable when working around heavy machinery such as helicopters and tanks. Testing with live munitions is obviously more dangerous. Testing is often performed in dangerous locations such as fire-prone California in the summer, tire-wrecking Yuma desert, or Alaska in the winter.

1.6.2 ATR and human subject testing

It may be desirable to compare ATR performance with that of human test subjects. Human subjects can also serve as test targets. The first consideration when using human subjects is whether approval is required from an independent Institutional Review Board (IRB). An IRB is designated to approve, monitor, and review research involving humans. Its function is to protect the rights and welfare of the research subjects. Government contracts often require IRBs when human testing is involved. Human testing can be as innocuous as test subjects looking at a monitor and pointing to targets, with performance recorded. It may be as serious as conditions involving live munitions or active sensors. Whether an IRB is required depends on what is written in a contract, the precise nature of the testing, the relationship between the test subjects and the organization doing the testing, and where the tests take place. Rules about the need for an IRB keep changing.

The IRB must have at least five members. The members must have enough experience, expertise, and diversity to make an informed decision as to whether the research is ethical, informed consent is sufficient, and appropriate safeguards are in place.

IRBs appraise research protocols and related materials (e.g., informed consent documents). The chief objectives of an IRB protocol review is to assess the ethics of the research and its methods, to promote fully informed and voluntary participation by prospective subjects, and to maximize the safety of subjects.

Contracting an IRB, submitting the protocols and other paperwork, and obtaining approval can take up to one year. Keep this in mind when planning a test involving human subjects! If a test is to be conducted at different locations, say different military bases, a separate IRB might be required for each location. A project involving government, industry, and university can potentially have several IRBs, with an agreement needed for one IRB to take the lead.

Some issues that can arise in ATR-related human subject testing include:

- Safety: Will individuals serving in the role of targets be around live munitions, experimental aircraft, hazardous material, or maneuvering vehicles? Is an active sensor used, such as a laser, LADAR, radar, or terahertz camera?
- Privacy: Are faces discernable in a database? Could pictures of minors be inadvertently captured in a data collection? Can the sensor see through clothes? Are the privacy laws of the state in which the test takes place being followed?

- Coercion: Are the test subjects competing against the ATR? Are the test subjects being coerced in the direction of particular results?
- Monetary reward: Are the test subjects, for example, being given a trip to Hawaii to take part in a field test?

The experimental design of a test of human subjects must consider different items compared to a test of an ATR. For example:

- What length of time is the test subject given to make and record a decision?
- What are the many display issues involving type of display: display size, distance from the display, room lighting, ability of test subject to adjust display parameters, etc.?
- Is fatigue an issue?
- Are there noise and other distractions?
- Is the test setup realistic, for example, using a flight simulator or ground station, or less realistic such as using the test subject's own computer?

A typical ATR's software forgets the last set of data that it has been tested on unless programmed otherwise. A human test subject can't help but learn the particulars of a data set during a test and, hence, perform better on future data with similar characteristics.

1.7 Characterizations of ATR Hardware/Software

Several key performance parameters of an ATR hardware/software system are as follows:

- size
- weight
- power requirement
- latency
- cost
- mean time between failure
- security level
- number of source lines of code (SLOC count).

The main cost in ATR system development is often the cost of collecting sufficiently comprehensive training and testing data sets. Air-to-ground data collection costs are measured by the flight hour. Costs include fuel, air crew, and equipment. Setting up an array of ground vehicles might include renting the military test site, renting foreign military vehicles, paying for drivers of the vehicles, and keeping a ground crew in place. Likewise, undersea data collections and testing involve considerable expenses. An ATR is procured and deployed only if there is a budget item for doing so. Once the ATR is deployed, there are costs associated with the logistics trail. If the ATR uses the same electronic boards that are used throughout a platform or series of platforms, the logistics cost is low. Another consideration is the number of years that the chips and other components of the ATR will be available. Some chip and board manufacturers guarantee seven years of availability. If one compares that to the longevity of a platform such as the B-52 bomber, that doesn't seem like a long time. It is common to buy and warehouse all potentially needed spare parts in advance. Keeping counterfeit parts, used parts, and lower-grade parts out of a system are critical issues. Another issue is the maintenance cost of the ATR as target sets, rules of engagement, sensor inputs or computer operating systems keep changing. After the ATR is delivered, who is going to train it on new target types, and where is the budget to do this? Software has to be kept under configuration control. Supposing that the ATR software reports errors, there needs to be a maintenance plan to address these errors as well as an upgrade plan. How will software changes be made when the operating system is no longer supported or when the original development team is disbanded? Many of these issues are not unique to ATR and are well understood by the government and defense contractors.

References

1. T. D. Ross, L. A. Westerkamp, R. L. Dilsavor, and J. C. Mossing, "Performance measures for summarizing confusion matrices: the AFRL COMPASE approach," *Proc. SPIE* **4727**, 310–321 (2002) [doi: 10.1117/12.478692].
2. S. G. Alsing, "The Evaluation of Competing Classifiers," Ph.D. dissertation, Air Force Institute of Technology, Wright-Patterson AFB, Ohio (2000).
3. C. B. Bassham, "Automatic Target Recognition Classification System Evaluation Methodology," Ph.D. dissertation, Air Force Institute of Technology, Wright-Patterson AFB, Ohio (2002).
4. A. Mahalanobis, S. R. F. Sims, and A. Van Nevel, "Signal-to-clutter measure for measuring automatic target recognition performance using complimentary eigenvalue distribution analysis," *Opt. Eng.* **42**(4), 1144–1151 (2003) [doi: 10.1117/1.1556012].
5. K. Fukunaga and W. L. G. Koontz, "Representation of random processes using the finite Karhunen–Loeve transform," *J. Opt. Soc. Am.* **72**(5), 556–564 (1982).
6. M. Self, B. Miller, and D. Dixon, "Acquisition Level Definitions and Observables for Human Targets, Urban Operations, and the Global War on Terrorism," Technical Report No: AMSRD-CER-NV-TR-235, RDECOM CERDEC (2005).
7. A. L. Magnus and M. E. Oxley, "Theory of confusion," *Proc. SPIE* **4479**, 105–116 (2001) [doi: 10.1117/12.448337].

Chapter 2
Target Detection Strategies

2.1 Introduction

An automatic (or aided) target recognizer (ATR) consists of two essential stages: detection and recognition. This chapter covers detection algorithms for literal imagery and ground targets, which are the most basic cases. There are a large number of other cases that aren't covered. Other sensor types (such as vibrometer, high-range-resolution radar, ground-penetrating radar, LADAR, sonar, magnetometer, etc.) and other target types (such as buried landmines, ballistic missiles, aircraft, underground facilities, hidden nuclear material, etc.) require detection algorithms specific to those circumstances. However, the basic strategy covered here still applies.

Several hundred target detection algorithms were evaluated. They were tested on tens of thousands of images of the following types:

- longwave forward-looking infrared (FLIR),
- midwave FLIR,
- Ku-band synthetic aperture radar (SAR), and
- X-band SAR.

Several more-complex algorithms were also designed and tested. Each algorithm is briefly described. The ones that performed best on FLIR imagery are noted. Some insights are shared on target detection with SAR imagery.

Detection approaches with deeper philosophical underpinnings are referred to as *grand paradigms*. Eight grand paradigms are reviewed. In the early days of ATR, there were great debates on the benefits of one paradigm versus another: pattern recognition versus artificial intelligence, model-based versus neural networks, signal processing versus scene analysis, etc. Money flowed to develop the paradigms whose proponents could stimulate the most excitement. Paradigms now generating considerable interest include approaches based on multiscale architectures, biologically inspired designs, and quantum imaging. Each novel paradigm has something to offer. Once the fervor for a new approach dies down, it or its components become additional tools in the ATR system designer's toolbox.

2.1.1 What is target detection?

The most popular ATR architecture is shown in Fig. 2.1(a). The (front-end) target detector visits each image pixel in turn. At each stop, the algorithm computes a test statistic T. The m local maxima of the T-plane image (saliency map) are reported as the m most likely target locations. In particular, for FLIR, where target strength varies with diurnal condition, range, target operating history, sun angle, etc., m is often kept fixed. The detector is then said to operate at a *constant false alarm rate per image*. Fixing m sets bounds on processing requirements. The second stage recognizer operates over regions-of-interest around the m hypothesized target locations. The second stage classifies detected objects as to target/clutter and also type. Thus, the second stage completes the detection process.

Other ATR architectures have been implemented over the years. A popular processor box was sold by several companies in the 1980s, under such names as AutoQ-I, AutoQ-II, GVS-41, and Optivision.[1] This architecture fit the design of Fig. 2.1(b). The first stage used dedicated hardware to obtain connected components by tracking strong edges around the borders of blobs until closure was obtained. This was followed by a detection stage. Detection involved sorting connected components into target-like objects and clutter using various features and rules. This was followed by a third stage classifier.

It is also possible to leave out the detection stage altogether [Fig. 2.1(c)]. A set of finely tuned correlation filters can be applied to the image as a whole, either digitally or optically. This sliding classifier approach can be generalized to any type of classifier, using any type of features, applied to the region about each pixel in the image. The architecture of Fig. 2.1(c) has not won out in either machine or animal vision.

2.1.2 Detection schemes

The main reason for using a target detection stage is to reduce the workload of the classification stage. The front-end detector performs a *prescreening* or *focus-of-attention* function. Two popular approaches to target detection are (1) anomaly detection and (2) correlation.

A target appears anomalous to its immediate background or the image as a whole due to strong contrast, border strength, bright spot, unusual texture, or high variance—just to name a few clues to the presence of a target. All such clues, and hence detectors, can be generalized from single-band to multiband imagery.

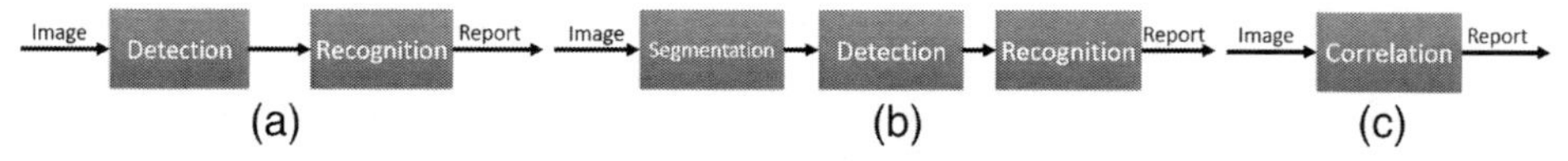

Figure 2.1 Some simplified ATR architectures.

A second approach to detection assumes that targets are well described. Target descriptions are converted to templates. A set of target templates is matched against each area of the image, taking range or scale into account. Correlation is accomplished with raw gray levels or extracted features in the spatial or some transform domain. New types of correlators are invented each year under the general name of *advanced correlation filter*.

Other detection schemes are possible. A detection algorithm can be developed through a training or discovery process. Genetic algorithms and artificial neural networks are currently popular. A trained neural network can be slid across a grayscale or feature image, computing *targetness* at each pixel. It may not be clear how the neural network is making its decisions. Detection schemes are sometimes implemented using pyramids as hierarchical multiscale image representations. Gaussian, Laplacian, Haar and, more generally, wavelet pyramids are popular. Steerable pyramids can be viewed as an over-complete wavelet transform.[2] They provide a multiscale, multi-orientation, image decomposition thought to be similar to that used in visual attention. Another popular approach uses histogram-of-oriented-gradient (HOG) features.[3] This method forms a histogram of the occurrences of gradient orientations within image regions. The histogram is fed into a trained classifier. As with most detection schemes, this method can also be implemented using a multiscale representation.

Mathematical morphology was popular in the early 1990s, leading to such approaches as the rolling ball algorithm[4] and morphological wavelet transform.[5] The rolling ball algorithm treats an image as a 3D surface. A target is a depression in the image into which the rolling ball falls.

Moving target detection requires a different strategy. Multiple frames are required to detect target movement with passive sensor imagery. Moving targets are detected with a radar's *moving target indication* (MTI) mode, not SAR mode. As the target moves, its distance to the radar system changes. The emitted signal is reflected back with a variation in frequency dependent on the speed of the moving object. This shift in frequency is called the Doppler effect. This chapter addresses *stationary target indication* (STI).

Optimality is often claimed in signal processing solutions to target detection. Optimality can only be substantiated under certain precisely defined and restricted conditions that need to be identified and confirmed. In the infinitely varying real world, optimal detectors do not exist. This topic is discussed further in Section 2.5.

FLIR sensors introduce many different types of artifacts and noise that must be accounted for. As the FLIR sensor ages, it produces increasingly noisier imagery until eventually it must be repaired or replaced. Thermal scenes vary widely as a function of current weather and diurnal conditions, past weather conditions, solar loading, climatic zone, background clutter, range, depression angle, and countermeasures. Thermal scenes also vary according to target types and their maintenance, articulations, aspect angles, and operating histories. ATR developers do their best to obtain a comprehensive development set, an open test set, and, most importantly, a blind test set. Performance results over a blind test set are a compelling predictor of future operational performance to the extent that the test data match the future operational location, conditions, and target set. Simple detection algorithms require less testing than finely tuned algorithms.

2.1.3 Scale

Despite often-heard claims to the contrary, all IR target detection algorithms require spatial scale information. It is theoretically possible to bootstrap scale, like biological vision systems viewing long-range targets, but this technology falls into the category of *research topic*. Without scale information, the detector doesn't know whether to search for a target smaller than a pixel, larger than the whole image, or somewhere between.

A SAR ATR receives pre-scaled data. Scale is not a significant issue with SAR, but one shouldn't assume that every SAR sensor produces perfectly scaled imagery. Also, SAR images can be in the slant plane, ground plane, or some proprietary geometry.

An ATR requires high-quality scale information to process passive sensor imagery. The detector algorithm's geometry is computed from range data, sensor field of view, and the sizes and shapes of the targets sought. Under some circumstances, range can be computed with simple geometry. With high depression angle, the range to a ground intersection point can be computed from sensor pointing angle, height above the ground, and digital terrain data or a flat-earth assumption. However, when the sensor is on a ground vehicle or on an aircraft flying a nap-of-the-earth mission, the sensor is essentially looking parallel to the ground. A purely geometric range solution is then highly error prone. Assuming that the targets of interest are ground vehicles, range above the skyline is treated as infinity. What is actually needed is range to the spot where the vehicle's wheels or tracks touch the ground. This is different from the range to the ground surface along the view-ray passing through the center of the vehicle. Range-to-air targets is another matter.

For ATR with passive sensors, the range solution and source of range data are critical system design issues. This highlights a word to the wise: *ATR is a system design problem and not an algorithm design problem.*

2.1.4 Polarity, shadows, and image form

Target polarity is not necessarily known in IR imagery. Consider active and inactive targets separately. The hottest parts of an active target, such as engine and exhaust pipe, are much brighter (i.e., hotter) than the background in white-hot FLIR data. Other parts of the same vehicle may be darker than the background. Consider the case of a heavy armored vehicle such as a main battle tank. The mass of metal cools down during the night and takes time to warm up in the morning, even after the engine is turned on. A good target detector responds to the hot part of a vehicle, even if the average gray level of the vehicle matches that of its background. Target detection sometimes mimics the perceptual quandary discussed in psychology textbooks of detecting a Dalmatian dog against a background of the same average gray level or similar texture. An inactive target can be hotter or cooler than its background. A tank with its engine off might not be considered a threat and hence not a target.

Midwave and longwave IR imagery can have remarkably different character when the sun is shining. Longwave imagery is formed by emitted energy. Midwave imagery is composed of reflected sunlight plus emitted energy. Target appearance is more stable in the longwave band. Solar highlights and self-shadows can appear at different places on a target in the midwave band, depending on relative sun position. A highlight may sometimes aid target detection. More often, solar energy lights up the clutter, littering the scene with bright nuisance objects. There are potential ways of mitigating this problem, such as using one or more subbands within the midwave band or by carefully controlling the sensor pointing angle. However, these kinds of solutions have associated cost and risk.

Targets are brighter (i.e., have higher gray levels) than their backgrounds in SAR imagery. With SAR, when targets are on a mowed grass lawn, their shadows are sharp and clearly darker than the background. Shadow direction is known since the sensor is the illuminator. A solar shadow may exist for a long-parked vehicle in longwave infrared (LWIR) imagery, but it does not move when the vehicle starts moving. Shadow location in midwave infrared (MWIR) imagery is knowable if latitude, longitude, time, date, and sensor-to-scene geometry are known. However, the existence of a target shadow in MWIR imagery depends on whether the vehicle is in the shade of clouds, trees, or buildings.

Consider the point along the image formation chain at which imagery is pulled off for processing. A target detector operates on a FLIR image before annotation overlays are added, and the image is reduced in bits-per-pixel for display. The form of a SAR image is important for target detector design. A detector can be applied to the complex image, magnitude image, log magnitude image, square root image, lin-log image, or magnitude-squared image. DeGraaf has demonstrated that the SAR image formation process has an order-of-magnitude effect on the performance of front-end detection algorithms.[6] Of the more than a dozen ways of forming the SAR image, certain modern spectral estimation techniques are particularly notable for their enhancement of detection through reduction of speckle. Some of these spectral estimation (also known as superresolution) techniques, such as the minimum variance method, produce images with Gaussian statistics. This improves the performance of detection algorithms that are based on Gaussian assumptions.

2.1.5 Methodology for algorithm evaluation

The following discussion is based on extensive testing. Test data consisted of target and clutter-blob ROI images, as well as full-sized images. Targets were tracked and wheeled military vehicles. About 30,000 ROIs were used each for LWIR and MWIR tests. ROIs were given approximately constant spatial scaling, taking into account typical range error. The data was from high-end FLIR sensors of the type used on military helicopters and unmanned air vehicles. LWIR imagery was a mixture of data from second-generation scanning sensors and staring sensors. MWIR data was from staring sensors. Depression angles were from 0 deg to 60 deg, with the majority of data from low grazing angles.

Algorithm evaluation was performed in multiple steps. Two equally weighted evaluation criteria were used: parametric and nonparametric.

We recorded each detector's response to target t and clutter c objects. A simplified T-test was used to measure the distance between the two populations of responses:

$$T = \frac{\overline{X}_t - \overline{X}_c}{\sqrt{\sigma_t^2 + \sigma_c^2}}. \tag{2.1}$$

where $\overline{X}_t - \overline{X}_c$ is the difference of the mean scores on targets and clutter, and $\sigma_t^2 + \sigma_c^2$ is the sum of the variances of the scores.

We also examined the tails of the two distributions. The criterion was the percent of clutter blobs that passed through when 90% of targets were detected. Detection algorithms that survived this initial screening were tested on full-frame images. The final evaluation criterion was the list length necessary to detect 90% of targets.

Sufficient data can't be collected to test algorithms against targets and backgrounds in all of their potential diversity. It is more constructive to report on which algorithms tend to perform well than it is to pick out a most excellent algorithm. Algorithms that tested adequately on FLIR data are enclosed in solid boxes in the text.

2.1.5.1 Evaluation criteria for production systems

Algorithm performance is only one of many considerations in algorithm selection for production systems. The software cost of an algorithm is usually determined by the SLOC count required to implement it. Coding and documentation of the code are quite expensive with a coding standard such as SEI Level 5 [or implementation on a field-programmable gate array (FPGA)]. This weighs in favor of simpler algorithms. One must also consider the sustainability of the ATR long after its original developers have retired or moved on to other endeavors. When an ATR enters production, there may not be a contract in place to keep its original design crew intact, analyzing ATR performance year after year, fixing malfunctions, and adapting the design to new situations. This again suggests simpler, more transparent, algorithms. Another important factor is the processing requirements of the algorithm. Processor choice is generally not under the algorithm designers' control. An algorithm must map well to a particular processor architecture, computer language, and parallelizing scheme. One must also consider the effect of the algorithm on the security classification of the ATR system. Algorithms that use literal templates can turn the ATR into a classified device. This greatly increases logistics cost and difficulty.

2.1.5.2 Target detection: machine versus human

Machine-versus-human detection is an “apples to oranges” comparison. A FLIR ATR system typically receives 14-bit/pixel data. A pilot typically sees 8-bit/pixel video on a very small display in a cockpit full of distractions. The ATR doesn't get distracted or fatigued. An ATR can process 30 different scenes per second from a step-stare sensor mode. A human can't do that, but is better than the ATR at perceiving a scene over an extended time period, such as detecting the person planting a roadside bomb. An ATR receives precise metadata, such as sensor pointing angles, altitude, aircraft velocity, bad pixel list, and some form of range data. The human perceives target size from “scene gist” and comparison with other scene objects. The human receives additional information from past experiences, radio contact, and pre-mission briefings. Current ATRs lack such higher-level understanding. More useful than human-versus-machine tests are assessment of human operators with and without machine cueing.

2.2 Simple Detection Algorithms

Eight categories of simple detectors are reviewed. Their relative performance are noted.

2.2.1 Triple-window filter

Many detectors are defined via two components: the statistical test and the geometry over which the test is performed. The geometry is generally in the form of a hypothesized blob region and its immediate neighborhood. This often takes the shape of a double or, preferably, a triple-window filter (Fig. 2.2).[1]

For a triple-window filter, the rectangular inner window is set slightly smaller than the smallest target sought, taking scale (i.e., pixels per foot) into account. The inner perimeter of the annular outer window is made slightly larger than the largest target. When the detector is centered on the target, much of the target's border falls into the annular middle window. There is no particular requirement for rectangular detection windows. The windows don't necessarily have to be symmetric or co-centric. If processing power and design complexity allow, better or adaptive geometries can be formulated.

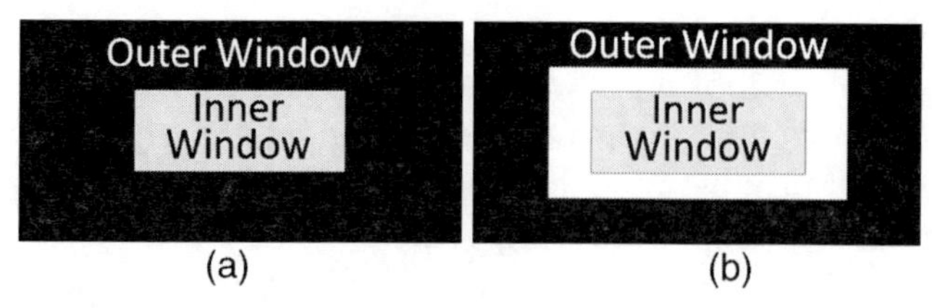

Figure 2.2 (a) Geometry of double-window filter. (b) Geometry of triple-window filter.

Multiple statistical features can be used. However, when a detection algorithm becomes too complex, it starts looking like a classification algorithm. This violates the conventional ATR philosophy of an efficient pre-screener that processes everything, followed by a classification stage applied to selected image regions. Like the triage nurse in an emergency room, the detector assigns a degree of urgency to image regions requiring further assessment and treatment.

Statistical tests are often based on standard tests of the difference of two populations. We evaluated variations of standard tests as well as approaches specifically designed for clutter rejection.

2.2.2 Hypothesis testing as applied to an image

Hypothesis testing is used to determine whether or not to accept a certain statement about two populations. A pertinent statement might be that the pixels momentarily covered by a sliding inner window have characteristics similar to those covered by the surrounding outer window.

Definition 1: A *statistical hypothesis* is an assertion about the distribution of one or more random variables.

Definition 2: A *test* of a statistical hypothesis is a rule that when applied to an image region leads to a decision to accept or reject the hypothesis about the image region.

Definition 3: A function of one or more random variables that does not depend on any unknown parameters is called a *statistic*.

Definition 4: A *test statistic* is a statistic that is used to help make a decision in a hypothesis test.

Definition 5: *Parametric statistical tests* are those whose models specify certain conditions about the parameters of the populations in image regions.

The most powerful tests are those that have the strongest and most extensive assumptions. The T-test, for example, has a variety of strong assumptions underlying its use. When its assumptions are met, the test is the best of its kind. When the underlying assumptions for a test are not even close to the true situation, it is difficult to know how well the test will perform. In some cases it may be desirable to use tests that assume as little as possible about the nature of the imagery.

Definition 6: A *nonparametric statistical test* is a test whose model does not specify conditions about the parameters of the populations in image regions.

Although the assumptions of nonparametric tests are weaker than those of parametric tests, it is clear that even nonparametric tests will never hold for real images. For example, these simple tests, when applied to image data, do not take into account spatial correlation among neighboring pixels, atmospheric attenuation as a function of range, or countermeasures taken by the adversary. In the final analysis, target detectors have to be tested over very large and diverse data sets, rather than chosen on purely theoretical grounds.

Although some detectors are modeled after hypothesis tests, we do not actually perform hypothesis tests in the work described here; instead, we use local maxima of test statistic values to indicate relative target strengths and locations. That is, the detector never tries to reject a hypothesis outright (except when the test statistic value is extremely low). Since we generally leave the number of pixels in each population out of the equations, the test statistics given here are not true to their textbook legacies.

The notation used throughout this chapter is given in Appendix 2.1.

2.2.3 Comparison of two empirically determined means: variations on the T-test

T-test: The data consists of two independent samples, one of size n_1 from the inner window and the other of size n_2 from the outer window. The size of the outer window is set to $n_1 \approx n_2$. For target detection purposes, the outer window can be made slightly larger than the inner window, but if it is too large it can encroach on a nearby target, the skyline, regions of highly different statistics, or the border of the image. The three nominal assumptions are:

1. Inner- and outer-window samples are independent random samples from their respective populations.
2. The two samples are mutually independent.
3. The inner-window pixels fit a normal distribution $N(\mu_1, \sigma_1^2)$, and the outer-window pixels fit a normal distribution $N(\mu_2, \sigma_2^2)$, where the means μ_1 and μ_2 and variance are unknown.

The simplified T-test statistic is as follows for the one-sided and two-sided tests, respectively. They measure the difference between inner- and outer-window sample means, normalized by the square root of the sum of inner- and outer-window samples variances:

$$T_1 = \frac{\overline{X}_1 - \overline{X}_2}{\sqrt{s_1^2 + s_2^2}}, \qquad T_2 = \frac{|\overline{X}_1 - \overline{X}_2|}{\sqrt{s_1^2 + s_2^2}}.$$

Several points are in order. This test is remarkably robust against departures from normality. The test statistic is invariant to linear transformations of gray level. It has a high value if the inner- and outer-window means differ by a large amount. However, the test will also have a high value if the sample variances are very low. This can happen for s_2 if the background is bland (e.g., sky in FLIR). These occurrences are not necessary indicative of a target being present (rejection of the null hypothesis). Therefore, two variations of a test are employed. One variation adds a small constant to a standard deviation whenever it is used in the denominator. Another variation clamps a standard deviation to a typical value, determined from a large set of images, whenever it falls below this typical value, and it is used in the denominator. The best performing of these three variations will be the reported rating for each detector that uses variance in the denominator.

The T-test is a very powerful statistical test. So, why not just choose it as a detection statistic and be done with the analysis? One reason is that high, not low, inner-window variance sometimes indicates presence of a target. Also, as previously stated, image data does not match the strict assumptions of the T-test.

We again note that the notation is explained in this chapter's first appendix, labeled as Appendix 2.1.

After we describe a test statistic, we will list several variations that were also tested. No claim is made that any of these variations have sound statistical pedigree. For example, consider

$$T_3 = \frac{(\overline{X}_1 - \overline{X}_2)}{(k_1 + k_2)^{1/4}}, \qquad T_4 = \frac{|\overline{X}_1 - \overline{X}_2|}{(k_1 + k_2)^{1/4}}.$$

The inner-window variance might have no bearing on whether a target is present and can be ignored to yield two common statistics used in traditional constant false alarm rate (CFAR) detectors (see Section 2.5):

$$\boxed{T_5 = \frac{\overline{X}_1 - \overline{X}_2}{s_2}}, \qquad T_6 = \frac{|\overline{X}_1 - \overline{X}_2|}{s_2}.$$

Remember that tests that perform particularly well over FLIR imagery are enclosed in solid boxes in this chapter. Similarly,

$$T_7 = \frac{\overline{X}_1 - \overline{X}_2}{k_2^{1/4}}, \qquad T_8 = \frac{|\overline{X}_1 - \overline{X}_2|}{k_2^{1/4}}, \qquad T_9 = \frac{\overline{X}_1 - \overline{X}_2}{s_i}.$$

Let subscript i denote full-frame image statistics. However, there are some caveats to using global image statistics in detection algorithms. With forward-looking imagery and ground targets, it is preferable to ignore sky pixels when computing global image statistics. In very wide field-of-view images, the statistics of a region near the detector, smaller than the image as a whole, should be used in place of global image statistics. The region should be wider than high since background statistics vary with range:

$$T_{10} = \frac{|\overline{X}_1 - \overline{X}_2|}{s_i}, \qquad T_{11} = \frac{\overline{X}_1 - \overline{X}_2}{MD_2}, \qquad T_{12} = \frac{|\overline{X}_1 - \overline{X}_2|}{MD_2},$$

$$T_{13} = \frac{\overline{X}_1 - \overline{X}_2}{k_2}, \qquad \boxed{T_{14} = \frac{\overline{X}_1 - \overline{X}_i}{\sqrt{S_1^2 + S_i^2}}}, \qquad T_{15} = \frac{\overline{X}_1 - \overline{X}_2}{s_1},$$

$$T_{16} = \frac{|\overline{X}_1 - \overline{X}_2|}{k_2}, \qquad T_{17} = \frac{\overline{X}_1 - \max(\overline{X}_2, \overline{X}_i)}{\sqrt{s_1^2 + \max(s_2^2, s_i^2)}}, \qquad T_{18} = \left|\frac{\overline{X}_1 - \max(\overline{X}_2, \overline{X}_i)}{\sqrt{s_1^2 + \max(s_2^2, s_i^2)}}\right|,$$

$$T_{19} = \frac{|\overline{X}_1 - \overline{X}_2|}{s_1}, \qquad T_{20} = \frac{(\overline{X}_1 - \overline{X}_i)s_1^2}{\max(s_2^2, s_i^2)}, \qquad T_{21} = \left|\frac{(\overline{X}_1 - \overline{X}_i)s_1^2}{\max(s_2^2, s_i^2)}\right|.$$

To reduce computations, simple tests can be designed based only on sample means. In FLIR, these are called tests of temperature difference. Two tests of this type are given followed by several variations. Note that test T_{22} would not have been a top performer if the test database had a higher percentage of cold targets against warmer backgrounds. Test results must be viewed in relation to operational considerations. Is a cold vehicle a *target*? So, although test T_{23} performed worse than T_{22} on this test set, it is more robust to target polarity and should be considered if operational considerations so warrant. Also note that a 2D difference of Gaussian (DoG) or 2D symmetric Gabor filter can be viewed as a smoothed version of triple-window filter T_{22}:

$$\boxed{T_{22} = \overline{X}_1 - \overline{X}_2}, \quad T_{23} = |\overline{X}_1 - \overline{X}_2|, \quad T_{24} = \frac{(\overline{X}_1 - \overline{X}_2)}{\overline{X}_i}, \quad T_{25} = \frac{|\overline{X}_1 - \overline{X}_2|}{\overline{X}_i},$$

$$T_{26} = \frac{\overline{X}_1 - \overline{X}_2}{\overline{X}_1}, \qquad T_{27} = \frac{(\overline{X}_1 - \overline{X}_i)S_1^2}{S_i^2}, \qquad T_{28} = \frac{|\overline{X}_1 - \overline{X}_2|}{\overline{X}_1}.$$

To compare the behavior in the central portion of two independent samples of equal size, the difference of the arithmetic means is divided by the inner- and outer-window grayscale ranges R_1 and R_2, respectively. This provides two test statistics similar to the T-test. They are attributed to F. M. Lord (see Ref. 7, pp. 276–277). We also provide four variations:

$$T_{29} = \frac{\overline{X}_1 - \overline{X}_2}{R_1 + R_2}, \qquad T_{30} = \frac{|\overline{X}_1 - \overline{X}_2|}{R_1 + R_2}, \qquad T_{31} = \frac{\overline{X}_1 - \overline{X}_2}{R_2},$$

$$T_{32} = \frac{|\overline{X}_1 - \overline{X}_2|}{R_2}, \qquad T_{33} = \frac{[\overline{X}_1 - \max(\overline{X}_2, \overline{X}_i)]s_1^2}{\max(s_2^2, s_i^2)},$$

$$T_{34} = \left| \frac{[\overline{X}_1 - \max(\overline{X}_2, \overline{X})]s_1^2}{\max(s_2^2, s_i^2)} \right|.$$

2.2.4 Tests involving variance, variation, and dispersion

Assume that inner- and outer-window pixels are two independently drawn random samples of approximately equal size from a common normally distributed population. The F-test statistic is used to test inner- and outer-window sample variances for equality. In contrast with the T-test, the F-test is very sensitive to deviations from the normal distribution. The F-test is followed by five variations:

$$T_{35} = \frac{s_1^2}{s_2^2}, \qquad T_{36} = \frac{k_1}{k_2}, \qquad T_{37} = \frac{MD_1}{MD_2},$$

$$T_{38} = \frac{q_1}{q_2}, \qquad T_{39} = \frac{d_1}{d_2}, \qquad T_{40} = \frac{s_1^2}{s_i^2}.$$

Cacoullos (see Ref. 7, p. 261) provides another version of this test, given as T_{41}. Another form of this test intended for very large sample sizes is provided by Sachs, given here as T_{42} (see Ref. 7, p. 265). These tests are followed by 10 variations:

$$T_{41} = \frac{s_1^2 - s_2^2}{s_1^2 s_2^2}, \qquad \boxed{T_{42} = \frac{s_1 - s_2}{\sqrt{s_1^2 + s_2^2}}}, \qquad T_{43} = \frac{|s_1^2 - s_2^2|}{s_1^2 s_2^2},$$

$$T_{44} = \frac{s_1^2}{\max(s_2^2, s_i^2)}, \qquad T_{45} = \frac{s_1}{\sqrt{s_1^2 + s_2^2}}, \qquad T_{46} = \frac{s_1 - s_2}{\sqrt{s_1^2 + \max(s_2^2, s_i^2)}},$$

$$T_{47} = \frac{|s_1 - s_2|}{\sqrt{s_1^2 + s_2^2}}, \qquad T_{48} = \frac{|MD_1 - MD_2|}{\sqrt{MD_1^2 + MD_2^2}}, \qquad T_{49} = \frac{|d_1 - d_2|}{\sqrt{d_1^2 + d_2^2}},$$

$$T_{50} = \frac{|q_1 - q_2|}{(q_1 + q_2)^{1/3}}, \qquad T_{51} = \frac{|k_1 - k_2|}{(k_1 + k_2)^{1/3}}, \qquad T_{52} = \frac{s_1 - s_2}{s_1^2 + s_2^2}.$$

The simplest measure of target variance is just the inner-window sample variance. This test is followed by nine variations.

$$T_{53} = s_1^2, \qquad T_{54} = q_1, \qquad T_{55} = q_2, \qquad T_{56} = \frac{s_1}{s_2}, \qquad T_{57} = \frac{s_1}{s_i},$$

$$T_{58} = \overline{X}_1, \qquad T_{59} = \overline{X}_2, \qquad T_{60} = s_2^2, \qquad T_{61} = k_2, \qquad T_{62} = k_1.$$

Dispersion and variation measure properties similar to variance. The dispersions of two independent samples can be compared by means of the grayscale ranges R_1 and R_2. In analogy to the F-test, the ratio of the grayscale ranges is used (see Ref. 7 p. 275). This test is followed by 14 variations. Although T_{63} performed well, it is not robust to noise, such as bad pixels.

$$\boxed{T_{63} = R_1/R_2}, \qquad T_{64} = R_1 - R_2, \qquad T_{65} = \frac{R_1 - R_2}{s_2}, \qquad T_{66} = \frac{R_1 - R_2}{s_i},$$

$$T_{67} = \frac{K_1 - K_2}{s_i}, \quad T_{68} = \frac{(K_1 - K_2)s_1}{s_i}, \quad T_{69} = \frac{(R_1 - R_2)s_1}{s_i}, \quad T_{70} = \frac{(R_1 - R_2)s_1}{s_2},$$

$$T_{71} = \frac{d_1}{d_2}, \qquad T_{72} = \frac{I_1}{I_2}, \qquad T_{73} = d_1 - d_2, \qquad \mathrm{T}_{74} = \frac{|d_1 - d_2|}{\sqrt{s_1^2 + s_2^2}},$$

$$T_{75} = I_1 - I_2, \qquad T_{76} = \frac{d_1 - d_2}{s_2}, \qquad T_{77} = \frac{(d_1 - d_2)s_1}{s_2}.$$

The ratio of the standard deviation to the mean value is called the coefficient of variation, or, occasionally the coefficient of variability, denoted by V. The coefficient of variation is a dimensionless relative measure of dispersion with the mean value as unit.[7] The test statistic for comparing two coefficients of variation is given by Sachs (see Ref. 7 p. 275). This test is followed by four variations. Let

$$V = \frac{s}{\overline{X}} \tag{2.2}$$

and

$$W = \frac{MD}{\overline{X}}. \tag{2.3}$$

$$T_{78} = \frac{|V_1 - V_2|}{\sqrt{V_1^2 + V_2^2}}, \qquad T_{79} = \frac{V_1 - V_2}{\sqrt{V_1^2 + V_2^2}}, \qquad T_{80} = \frac{MD_1 - MD_2}{\sqrt{MD_1^2 + MD_2^2}},$$

$$T_{81} = \frac{W_1 - W_2}{\sqrt{W_1^2 + W_2^2}}, \qquad T_{82} = \frac{d_1 - d_2}{\sqrt{d_1^2 + d_2^2}}.$$

2.2.5 Tests for significance of hot spot

What would be hotter in a typical scene than an operating engine or exhaust pipe? Under some conditions the "hottest" point on a target might be the brightest point in a FLIR image. Under other conditions the target's hottest spot is considerably hotter than any other point on the target or in the local surround. But, this is not always the case due to atmospheric effects, target aspect angle, and other manmade objects in the scene. (When a test is applied to SAR, the concept of hottest spot is replaced by that of strongest scatterer.)

Tests can be devised to detect small hot spots. A unimodal distribution is skewed if considerably more probability mass lies on one side of the mean than the other.[7] A very hot spot on the target causes skewness of the distribution function. Another way to test the significance of the hot spot is to compare the difference between inner-window hottest point $\hat{X}_1$ and outer-window mean $\overline{X}_2$. Population minimum $\hat{\hat{X}}$ is occasionally used in these tests.

Although skewness can be determined exactly from the moments, the following simpler measure based on the difference of the sample mean and median sometimes proves satisfactory:

$$\text{Skewness}: \frac{\overline{X} - \tilde{X}}{s}. \tag{2.4}$$

From this we derive the following test statistic using the outer-window standard deviation. This test is followed by four variations:

$$T_{83} = \frac{\overline{X}_1 - \tilde{X}_1}{s_2}, \qquad T_{84} = \frac{|\overline{X}_1 - \tilde{X}_1|}{s_2}, \qquad T_{85} = \frac{\overline{X}_1 - \tilde{X}_1}{s_1},$$

$$T_{86} = \frac{\overline{X}_1 - \tilde{X}_1}{s_i}, \qquad T_{87} = \frac{|\overline{X}_1 - \tilde{X}_1|}{s_i}.$$

To determine the significance of the blob's hot spot, we also compare the inner-window maximum gray level to the outer-window mean gray level using several tests and their variations. Although some of these tests performed quite well, they are not robust against sensor noise, such as bad pixels. These tests and variations are given in Appendix 2.2.

2.2.6 Nonparametric tests

One logical way to compare two populations is with their distribution functions F. The simplest way to measure the discrepancy between two distribution functions is the largest vertical distance between their two graphs. The Kolmogorov–Smirnoff test statistic is given by Conover (see Ref. 8, p. 428):

$$T_{124} = \sup_x |F_1(x) - F_2(x)|.$$

The Cramer–von Mises test statistic is related to the area between the two distribution plots, summed over discrete graylevel values, followed by six variations (see Ref. 8, p. 423):

$$T_{125} = \sum [F_1(x) - F_2(x)]^2, \quad T_{126} = \sum |F_1(x) - F_2(x)|,$$

$$T_{127} = \sum |f_1(x) - f_2(x)|, \quad T_{128} = E_1, \quad T_{129} = \sum |F_1(x) - F_2(x)|^4,$$

$$T_{130} = \sum |F_1(x) - F_2(x)|^2, \quad T_{131} = |E_1 - E_2|.$$

A Mann–Whitney-like test uses ranks (see Ref. 7, p. 293). First, a rank is assigned to each pixel in the ROI based on gray levels. The ROI is then partitioned by a triple-window filter. The test statistic is the sum of the ranks $r(x)$ over the pixels within the inner window. Several inner-window sizes are used, and the highest test statistic value is chosen. This best inner-window size defines a best fit box b around the blob:

$$T_{132} = \sum_{\text{inner window}} r(x).$$

A popular SAR test uses the standard deviation within the best-fit box about the object, denoted by subscript b. This test was also applied to the FLIR data and is followed by four variations.

$$T_{133} = s_b, \qquad T_{134} = \sum_{\text{inner window}} r^2(x), \qquad T_{135} = \frac{s_b}{s_i},$$

$$T_{136} = \frac{\overline{X}_b s_b}{s_i}, \qquad \boxed{T_{137} = \frac{\overline{X}_b}{s_i}}.$$

2.2.6.1 Percent-bright tests

The weighted-rank fill ratio measures the percentage of the total energy contained in the brightest pixels of an object.[9] With our implementation, it is the sum of the gray levels of the k brightest (i.e., hottest) pixels in the inner window normalized by the sum of the gray levels in the area covered by the triple-window filter. The brightest pixels under the entire filter are marked corresponding to either a 1% or 2% level. T_{140} and T_{143} percent-bright statistics are at the 1% and 2% levels, respectively. T_{138} and T_{139} simply compute the percent of the pixels in the inner window that are marked as bright, at a 1% level.

For k corresponding to the 1% of brightest pixels in the filter area,

$$T_{138} = \frac{\sum_{k\text{ brightest pixels}} n_{1i}}{n}, \qquad T_{139} = \frac{\sum_{k\text{ brightest pixels}} n_{1i}}{n_1},$$

$$\boxed{T_{140} = \frac{\sum_{k\text{ brightest pixels}} x_{1i}}{\sum_{ROI} x_i}}.$$

For k corresponding the 2% of the brightest pixels in the filter area, we get the same type of equations denoted by T_{141}, T_{142}, and $\boxed{T_{143}}$, respectively. Similarly,

$$T_{144} = \frac{(1/k) \sum_{k\text{ brightest pixels}} x_{1i} - \overline{X}_2}{s_2}, \qquad T_{145} = \frac{(1/k) \sum_{k\text{ brightest pixels}} x_{1i} - \overline{X}_2}{(s_2/s_1)}.$$

2.2.7 Tests involving textures and fractals

Except for tree trunks, buildings, and poles, most of the edges in a FLIR scene viewed from a low grazing angle are horizontal. Line-to-line discontinuity in the image caused by a second-generation (scanning) FLIR's sensing elements introduce horizontal, but not vertical noise edges. Interlaced staring sensors have the same problem. The circuitry of non-interlaced staring FLIR sensors can introduce horizontal streaks, vertical streaks, or both. Nevertheless, targets can sometimes be detected by strong vertical edges. We tested a number of detectors based on total variation.[10] Results are given in Appendix 2.3. Similarly, tests can be derived from average window gradient values $\overline{G}$. Results are given in Appendix 2.4. Several detectors based on total variation and gradient performed quite well. Substituting corner features for gradients also worked well.

Fractal dimensionality is sometimes measured as

$$T_{223} = \frac{\log[M_1] - \log[M_2]}{\log[2]},$$

where M_1 is the number of 1-pixel × 1-pixel boxes needed to cover a thresholded blob image, and M_2 is the number of 2-pixel × 2-pixel boxes needed to cover the blob image.

2.2.8 Tests involving blob edge strength

Again, construct a triple-window filter (Fig. 2.3). The inner window is made smaller than the smallest target sought. The inner perimeter of the outer window is made slightly larger than the largest target sought. Thus, if a target is fairly well centered in the filter, most of its border should fall in the annular middle window. The basic spoke filter approach operates as follows. Construct eight spokes from the center to the outer window and compute the log of the

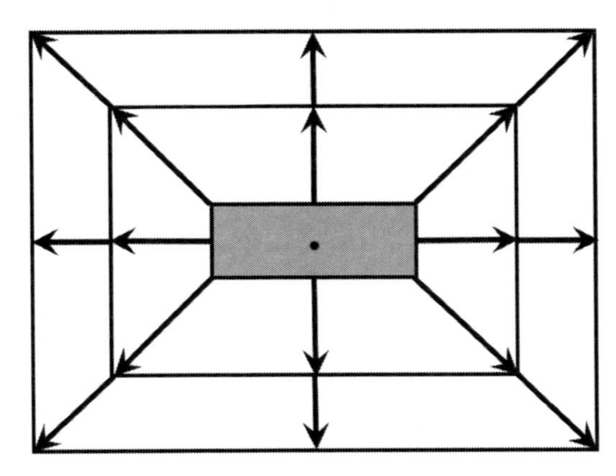

Figure 2.3 Spoke geometry #1.

absolute value of the strongest Sobel edge crossing each of the spokes. This is done separately for the middle and outer window (an elaboration of the traditional spoke filter of Minor and Sklansky).[11]

Let $\overline{e}_1$ denote the average of the logs of the maximum absolute edge strengths in the middle window perpendicular to the eight spokes, and $\overline{e}_2$ the corresponding value for the outer window. Let σ_1 and σ_2 denote the corresponding sample standard deviations of edge strengths. Tests using edge strength are direct analogs of those described previously using gray levels. Results are given in Appendix 2.5. None of these test types worked well.

Half of the spokes in the eight-spoke filter are in the diagonal directions. Military vehicles viewed from a low grazing angle are rectangular in appearance. Diagonal edges are few in number and weak in strength. The spoke filter test was repeated with other filter geometries shown in Fig. 2.4. Each was tested with the same test statistics as in the eight-spoke case. Thus, $27 \times 12 = 324$ variations were tested. Let α denote the spoke geometry using the numbers (2–13) given in Fig. 2.4, and let β denote the statistical test. The detectors that performed best, denoted by $T_{\alpha,\beta}$ are: $T_{5,229}$, $T_{5,230}$, $T_{5,228}$, $T_{8,230}$, $T_{8,228}$, $T_{8,61}$, $T_{9,228}$, $T_{9,229}$, $T_{9,230}$, $T_{10,224}$, $T_{10,225}$, $T_{13,224}$, $T_{13,230}$, and $T_{13,228}$.

All of the spoke tests were repeated without using logs when summing edge strength. Results were similar. In general, for detection of vehicles in FLIR imagery, a rectangular spoke geometry works best, with bottom edges de-emphasized compared to top or side edges. This is because target bottoms are often hidden by grass, bushes, and the roll of terrain when viewed from a low grazing angle. The best spoke filters outperformed other simple detectors for daytime LWIR and MWIR imagery. Spoke filters were outperformed by various kinds of tests detecting hot regions in nighttime FLIR imagery.

2.2.9 Hybrid tests

We tested a large number of detectors that were the joint statistics of two types of tests. The hybrid tests are highly correlated in performance with the two

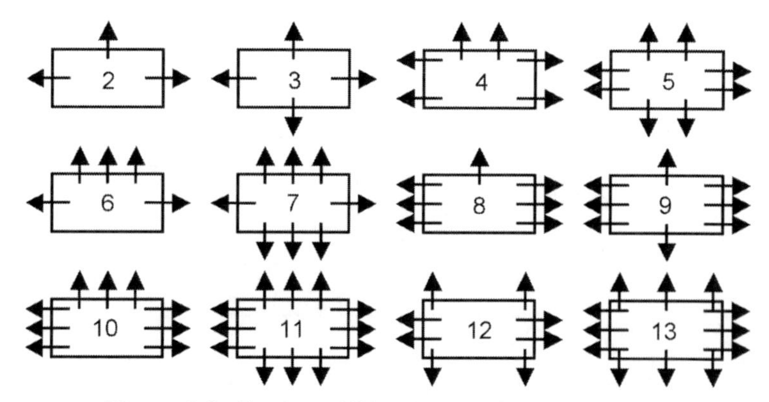

Figure 2.4 Twelve additional spoke filter geometries.

types of tests compounded to form them. None of the hybrid tests was an outstanding performer. Hybrid tests are listed in Appendix 2.6.

2.2.10 Triple-window filters using several inner-window geometries

One problem with the triple-window filter is that the single inner window does not provide a good match to the shapes of each target, at each aspect angle, for the ensemble of targets sought. For example, with forward-looking imagery, a long-bed truck viewed from the side is short and wide. Viewed from the front it is short and narrow. An air defense unit is taller and wider than a truck when viewed from the front, but narrower than the long-bed truck when viewed from the side. For this test, we applied four triple-window filters to each pixel visited. The filters differ only in inner-window shape. Note that this test only works well when the inner windows are roughly equal in area.

The reported test statistic value for each filter location was the maximum of the four values. Thirteen of the better tests were repeated with this variation. In nearly all cases, performance was better with multiple inner windows. We also tried varying the shape of the outer window to fit that of the inner window. That did not help. Another variation of this test allows an octagonal inner window to shrink or expand to fit the hypothesized blob area. Its performance was similar to that of the version with four rectangular inner windows.

2.3 More-Complex Detectors

As processing power increases, one can now consider using detection algorithms that appeared too computationally demanding in the past. Several of these are described.

2.3.1 Neural network detectors

Several kinds of neural network detection algorithms were tested. Consider one that worked fairly well on both SAR and *down-looking* IR imagery. This detector is composed of 16 parts. Each part consists of a five-window filter feeding features to a separate neural network. Each part is tuned to a different target aspect angle, 22.5 deg apart. The geometry of each part consists of a center rectangle, surrounded by a set of concentric annular rectangles, all within a larger circular region (Fig. 2.5). As a detector visits each pixel, first- through fourth-order statistics are computed for each of its five regions. All of these features are fed into a neural network trained on like data. A separate network is used for each of the 16 aspect angles. In test mode, the T-plane image is constructed from the maximum of the 16 network scores at each pixel.

The advantage of this neural network detector is its nonlinear mapping of features to detection score. The main disadvantages are the training

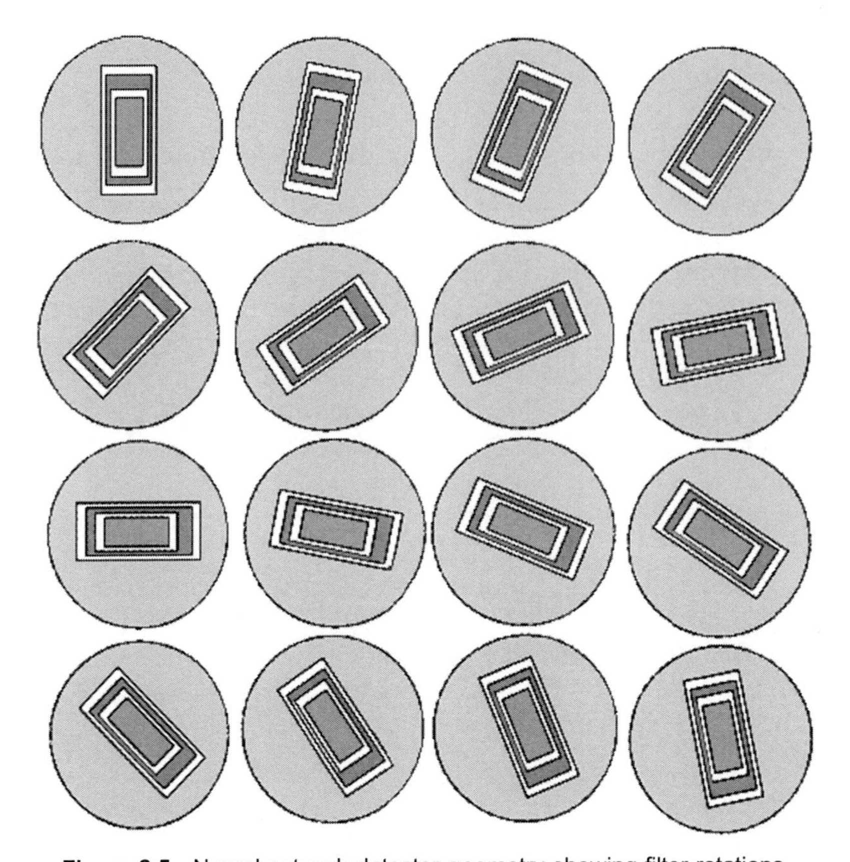

Figure 2.5 Neural network detector geometry showing filter rotations.

requirement and the specificity of the detector to a particular set of target types and training data.

2.3.2 Discriminant functions

When we tested the simple detectors, we computed the correlation of the responses of each detector to each other. There were no surprises. Detectors with a similar design had highly correlated output. The objective is to combine the least correlated good detectors. Any type of classifier can be used to do the combining, treating detector outputs as features. A satisfactory alternative is to separately apply several good detectors to each image and choose several strongest, non-redundant, detection points from each.

One can also choose a set of detectors following a narrative. For example, one might observe the following:

- Targets tend to have strong borders.
- Targets have contrast to their backgrounds.
- Targets sometimes contain a small bright spot.

Then a set of good detectors is chosen to model the observations.

A combination of simple detectors performs better than any single detector. However, the point of diminishing returns is rapidly reached. A combination of two detectors yields significant improvement. Three detectors works marginally better. Adding a fourth detector rarely if ever helps.

2.3.3 Deformable templates

An edge-vector-based template matcher was tested as a front-end detector. The template matcher was specifically devised to have robust performance against FLIR imagery. The templates were designed to automatically adjust their scale and also deform slightly to match detected object size and shape. Regions of each template were unequally weighted in computing the template's match, since the bottoms of targets are often compromised by grass, bushes, and terrain when viewed from a low grazing angle.

This template matcher performed better than the best simple detector. The disadvantages are processing requirements, the cost of template ensemble design, and specificity of the templates to a particular set of targets.

2.4 Grand Paradigms

The target detectors presented thus far are quite basic and simple to implement. Detection approaches with deeper philosophical underpinnings will be called *grand paradigms*. In the early days of ATR, there were great debates (often at conferences) on the benefits of one paradigm versus another: pattern recognition versus artificial intelligence, model-based versus neural networks, signal processing versus scene analysis, optical versus digital implementation, etc. Current debates concern open source solutions versus proprietary algorithms, qualia versus representationalism, and local processing versus "out in the cloud." The popularity of a novel paradigm generally follows the plot shown in Fig. 2.6. As the hype builds for a celebrated new paradigm, the money flows toward its development. Eventually, the hype dies down, and the new paradigm becomes just another tool in the engineer's toolbox. With equal amounts of talent and effort applied to each paradigm, it is questionable whether one methodology will consistently outperform all others in real-world conditions. However, this remains to be seen. Each innovative paradigm has something to offer, if only just a different way of viewing the problem. New paradigms provide new tools. Various tools from the toolbox are mixed and matched to build a system meeting particular requirements. As Professor Laveen Kanal noted, ATR engineers all have essentially the same tricks in their bag of tricks. It is which tricks are pulled out of the bag and how they are combined that is important. Eight grand paradigms are described.

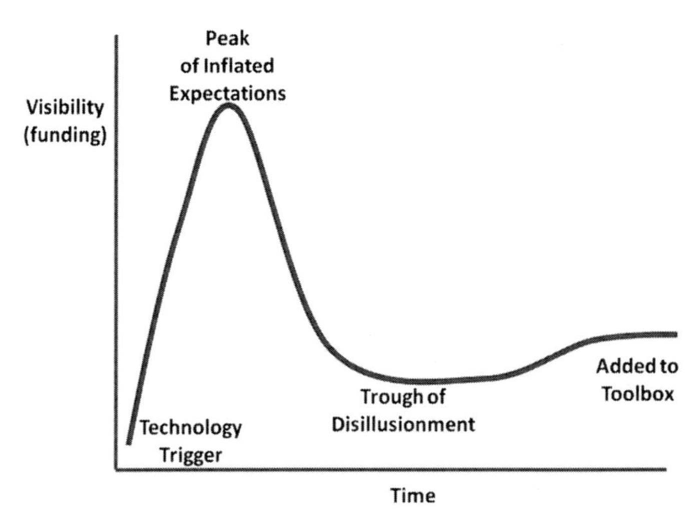

Figure 2.6 The popularity of and funding for ATR (and detector stage) paradigms typically follows the hype cycle. Hype cycles characterize the over-enthusiasm and subsequent disappointment that typically follows the introduction of new technology. (Adapted from Ref. 12.)

2.4.1 Geometrical and cultural intelligence

To make maximum use of scene geometry, the sensor's location and pointing angles must be known precisely. LADAR, laser, or radar range data are often available. The ground surface is modeled with digital terrain elevation data. The locations of large manmade and natural objects such as buildings, roads, and tree lines are available from a "cultural" database or determined, for example, from LADAR data. When the geometry of the scene is precisely known, the detector is adjusted for target size and expected atmospheric effects as it slides over the image. The detector utilizes location, time of day, and day of year, together which indicate sun angle and suggest something about weather conditions and shadow locations. Climatic zone, suggesting expected clutter and target conditions, is determined from latitude and longitude. In the visible band, climatic zone indicates target paint color, e.g., desert beige, forest green, or snow white.

A smart search considers the mission and ConOps. A target is an object of military interest. A *force structure* describes how opposing military personnel, weapons, and equipment are organized for the operation, missions, and tasks per doctrine or as dictated by the ground environment. A trained image analyst makes heavy use of context and force structure. The analyst detects a target in a likely location: by a bridge, near a road, inserted into a tree line or at a mountain pass. An algorithm utilizes information of this type via digital terrain and cultural data. Terrain and cultural information are used to limit the search areas for certain targets. A tank is unlikely to appear in the middle of a lake, on a steep slope, or on the roof of a large industrial building. Vehicles travel in groups according to known doctrine.

Most notable for this approach is the Semi-Automated Image Intelligence Processing (SAIP) system.[13] SAIP utilizes several-false alarm mitigation techniques of these types, including digital terrain analysis and area delimitation, cultural clutter identification, and recognition of force structure (e.g., maneuvering battalions).

2.4.2 Neuromorphic paradigm

Biomimicry is a discipline that studies nature's best ideas and then imitates these designs and processes to solve difficult problems. This suggests modeling target detection after biological processes.[14] Recent advances in modeling visual processes result from f-MRI brain imaging experiments with humans and the use of more-invasive techniques with monkeys. Recognition is not impossible without the prescreening operation (detection) taking place in the dorsal stream of the visual cortex. It is just that covert spotlights-of-attention speed up the recognition process as compared to overtly shifting gaze from spot to spot in a cluttered scene. Thus, the human vision system separates detection from recognition as in the model given in Fig. 2.1(a). This is necessitated by the limited processing capacity of the visual brain and the evolutionary survival requisite of quick decisions.

Visual target prescreening is a parallel operation, particularly for targets that are roughly at the same range. Attention is not purely a bottom-up, saliency-map-forming process, but emerges from mechanisms of feedback and competition. Attention is goal directed, using scene context, expectations, and higher-level knowledge. The initial coarse category guess and 3D location resulting from processing in the dorsal stream are fed back and integrated with the bottom-up analysis along the slower ventral parvocellular pathway of the visual cortex. Object recognition takes place serially along the ventral pathway at the spotlighted location.

Both the target detection and recognition stages of an ATR can be closely modeled after biological vision—to the extent that it is understood. There is a vast body of literature on this subject, e.g., see Refs. 15 and 16. One advantage of a strategy of biomimicry is that understanding of biological vision is advancing at a steady pace at no cost to ATR designers.

2.4.3 Learning-on-the-fly

Learning-on-the-fly paradigms tightly couple the human-in-the-loop with the machine. The machine performs detection, and the human either accepts or rejects each detected object. Regions-of-interest about the machine-detected objects are displayed to the human for final decision. Each corresponding decision is fed back to the machine to adjust the training of the detection/clutter-rejection algorithm (or algorithms). Then, during a mission, the algorithm, and possibly also feature set, adapt to the current local conditions and target set. This presumably combines the best of both worlds: workload reduction by means of electronics and insightful decisions by an experienced human teacher.

However, as was determined by extensive testing, there are several limitations to this seemingly perfect combination of man and machine, including these three:

1. The human-in-the-loop is not making decisions based on the exact same set of pixels or features used by the algorithm.
2. The human-in-the-loop makes much broader use of context, past experience, and pre-mission briefing information than is now possible with an algorithm. The decision fed back to the algorithm is based on more than just the information content of the pixels processed by it.
3. The hardware/software system must have met all of its key performance parameters to have been deployed. Each deployed system will have exactly the same performance at the time of deployment. During on-the-fly training, the performance of each system will drift from its original design in ways that are not predictable. It will not even be known whether a system still meets its minimum performance requirements. Several errors by a distracted human-in-the-loop could make performance worse.

2.4.4 Integrated sensing and processing

Let us consider some research topics related to target detection. *Computational imaging* involves any technique for which computation plays an integral role in the image formation process. Computational imaging considers an image as not something for viewing, but rather as a mathematical representation of targets and background. It incorporates signal processing goals, such as target detection, into the sensor system's design. We use the term computational imaging to exclude image processing techniques applied to formed images. A SAR system can be said to perform computational imaging to form an image. A detector applied to the formed SAR image is not doing computational imaging. A detection algorithm embedded into the image formation process, where imaging and detection are jointly optimized, is doing computational imaging.

Lensless imaging systems use multiple coded-aperture masks to generate image components. The image is decodable by image processing. This approach is particularly useful for x-ray and gamma-ray imaging for which lenses and mirrors do not exist. The approach can also be applied to FLIR and visible-band imaging. A system can be designed so that only image components necessary for target detection are produced.

Compressive imaging systems use coded apertures to construct a compressed image directly, without first producing a normal array of intensity values.[17] Compressive imaging is motivated by the fact that most natural images can be compressed by 10:1 or more and then reconstructed with little loss of fidelity. This is particularly true of often-murky IR images. Targets are outliers from the natural background in the compressed space. Image decompression is not necessary to do the target detection. Target detection is performed directly in the low-dimensional space. Although compressed sensing typically exploits spatial correlations, correlations also exist spectrally and temporally. Compressive sensing ideas are beginning to be explored for target detection.[18,19]

One can make the case that not much has changed in sensor-aided detection since Galileo detected the moons of Jupiter with his telescope. At a most basic level, EO/IR imaging sensors and target detection follow the same model used by Galileo. But now, advances in quantum mechanics—understanding the very nature of light—promise revolutionary breakthroughs in sensor design and target detection. Progress may require abandoning the 400-year-old paradigm.

A photon is the basic unit of light. Photons exhibit wave–particle duality. Like all elementary particles, photons are best explained by quantum mechanics. Much research is devoted to the application of photons: quantum optics, quantum computers, quantum teleportation, nonlinear optical processes, quantum cryptography, ghost imaging, quantum communication, quantum lasers, circularly polarized light detection, etc. A photon is ordinarily treated as a wave having a particular wavelength. With normal cameras, the information content of photons is not fully exploitable for target detection. Photons carry such information as frequency, polarization, entanglement, phase, arrival time, orbital angular momentum, linear momentum, etc. For the purpose of target detection, frequency is exploited by multiband sensors. Polarization diversity is rarely used to aid target detection. Other properties of photons have never been exploited for target detection.

Many fruitful areas of research remain to be explored in designing novel sensors, integrating sensors with processing, and exploiting the properties of photons.

2.4.5 Bayesian surprise

Consider a soldier scanning the battlespace with a pair of binoculars. Nothing much is happening. Nothing much is changing. Then, suddenly, he is surprised by the appearance of an enemy vehicle.

A surprise occurs when expectations aren't met. A spatial and/or temporal anomaly in imagery may be referred to as a "surprise." Surprises are located under the following Bayesian framework, as conceived and formulated by Baldi and Itti.[20,21] Background information is captured by a prior probability distribution

$$\{P(M)\}_{M \in \mathrm{M}} \tag{2.5}$$

over the space of models M. New data refers to a region of an image as would be encountered when scanning a single image, or a new frame's data of a particular image region (in stabilized video). A new piece of data D changes the prior distribution for all models in the model space into the posterior distribution $\{P(M|D)\}$ via Bayes' theorem:

$$\forall M \in \mathrm{M}, \quad P(M|D) = \frac{P(D|M)}{P(D)} P(M). \tag{2.6}$$

The new data D carries no surprise if it leaves beliefs unaffected; that is, if it looks much like the old data. D is surprising if the posterior distribution resulting from observing D significantly differs from the prior distribution. Surprise S is quantified as the distance between the posterior and prior distributions:

$$S(D, M) = \mathrm{dist}[P(M), P(M|D)]. \tag{2.7}$$

The distance can be computed as Kullback–Leibler (KL) divergence:

$$S(D, M) = KL[P(M|D), P(M)] = \int_M P(M|D) \log \frac{P(M|D)}{P(M)} dM. \tag{2.8}$$

This paradigm is put into practice using standard features, such as local contrast, edge orientation, and motion.

2.4.6 Modeling and simulation

The IR target detection problem can be attacked from a thermal modeling perspective. NVTherm (updated as NVThermIP) is the Army Night Vision and Electronic Sensor Directorate's software-based model for thermal imaging systems.[22] The model has been extensively verified by laboratory measurements and field tests. This validation is focused on predicting human recognition performance, rather than automated search. The target contrast metric used by NVTherm is called RSS (root sum of squares). RSS is determined by the first- and second-order statistics of target (*tgt*) and background (*bkg*). The outer window of a double-window design is $\sqrt{2}$ the dimensions (maximum width and height) of the target:

$$RSS = [(\mu_{tgt} - \mu_{bkg})^2 + \sigma_{tgt}^2]^{1/2}. \tag{2.9}$$

A variation is

$$C_{tgt} = \frac{[(\mu_{tgt} - \mu_{bkg})^2 + \sigma_{tgt}^2]^{1/2}}{2\mu_{scene}}, \tag{2.10}$$

where μ_{scene} is the sample mean in the vicinity of the target.[23,24] Either of these equations can be used as a metric for a double- or triple-window filter.

Xpatch® is a set of prediction codes and analysis tools to predict and simulate radar signatures from 3D target models. Xpatch can be integrated into a model-based ATR system.

Modeling and simulation is useful for developing, training, and testing algorithms when sufficient real data can't be collected due to cost or unavailability of certain target types or conditions. Modeling the entire image formation chain helps in designing detectors for adverse atmospheric conditions such as dust, snow, fog, turbulence, and rain.

2.4.7 SIFT and SURF

Lowe's patented Scale-Invariant Feature Transform (SIFT) algorithm is quite popular for target detection, recognition, and also image registration.[25] A similar approach is Speeded-Up Robust Feature (SURF).[26]

Let $I(x)$ denote the input image. Gaussian G scale images and DoG D scale images are generated as

$$G(x,\sigma) = g_\sigma(x) * I(x), \tag{2.11}$$

$$D[x,\sigma(s,i)] = G[x,\sigma(s+1,i)] - G[x,\sigma(s,i)], \tag{2.12}$$

where g_σ is the Gaussian kernel of width σ, $\sigma = \sigma_0 2^{i+s/S}$, $\sigma_0 = (1.6) \cdot 2^{1/S}$, with octave scale index i, $s = 1, 2, \ldots, S$.

SIFT keypoints (x, σ) are taken at the local maxima/minima in the (D) images that occur at multiple scales in the scale pyramid. Each pixel in a D image is compared to its eight neighbors at its level of the pyramid as well as its nine corresponding neighbors in the levels above and below. If the pixel is the max or min of its 26 neighbors in scale space, it is an aspirant keypoint. Weak keypoints are cast off. Each surviving keypoint is assigned a dominant orientation and is further described by an $8 \times 4 \times 4 = 128$ element feature vector, representing a 4×4 neighborhood, each neighborhood point associated with an 8-element directional gradient histogram. A support vector machine trained on like data is commonly used to classify a group of SIFT keypoints as to target or no target.[27]

2.4.8 Detector designed to operational scenario

To treat target detection merely as a problem in computer vision, neuromorphics, statistics, or thermal modeling divorces it from the intended application. The alternative is to develop and evaluate target detectors under a paradigm true to the operational scenario. The goal is for the detection process to be fully automated as a workload reduction strategy. A sensor may operate in a wide-field-of-view, wide-area-search mode for initial target detection. Once a potential target is found, the sensor system switches to a narrow field of view. This could be a spotlight or object-following mode. Recognition and clutter rejection then occur by human, machine, or, preferably, a combination of the two. The human-in-the-loop may not be on the same platform or even in the same country as the sensor. Once a false detection is rejected, the system leaves the spotlight mode and re-enters the wide-area-search mode. If a target is found, it is only after a series of false alarms. Variations on this theme are common. The sensor might also have an ultra-narrow field of view for final object confirmation. The detection and confirmation sensors do not have to be of the same type or even located on the same platform.

There are several points worth noting with this operational scenario as it affects target detection algorithms. A bright target with high signal-to-noise ratio at near-zero spatial frequency but low signal-to-noise ratio at higher spatial frequencies is detectable but not recognizable. A target detection algorithm should not be given extra credit for detecting true targets that cannot be verified by the human-in-the-loop, per rules-of-engagement. Too many detections that can't be confirmed waste the operator's time. In FLIR imagery, the algorithm generally has 14 bits/pixel to work with, but the human gets to see only 8 bits/pixel on a display. If the dynamic range of the local scene about the target is less than 8 bits, then this difference is immaterial. With higher dynamic range in the target area, a highly skewed distribution function might not affect detection but will affect the human's ability to recognize the target.

Size, weight, and power (SWaP) are always key system design parameters. Even though processing chip technology is advancing at a remarkable rate, so is the military requirement for smaller, lighter, less power hungry, and cheaper hardware. It adds to both human and machine workload when targets are detected that can't be verified within the strict guidelines imposed by rules-of-engagement. This suggests the use of less elaborate algorithms for the detection stage of an ATR. Less elaborate algorithms require less hardware to implement.

2.5 Traditional SAR and Hyperspectral Target Detectors

Finding a target in a large volume of imagery is like finding a needle in a haystack. The debate between Thomas Edison and Nikola Tesla illustrates two ways to attack the problem.

> "If Edison had a needle to find in a haystack, he would proceed at once with the diligence of the bee to examine straw after straw until he found the object of his search. I was a sorry witness to such doings, knowing that a little theory and calculation would have saved him ninety percent of his labor." — Nikola Tesla

The detection algorithms presented thus far in this chapter follow Edison's approach: try everything and see what works. Why not? Detection algorithms are easy to code, and computer time is cheap. However, as Tesla suggests, there is nothing wrong with using "a little theory" to guide one's search for a best solution. Many technical papers on target detection in SAR and hyperspectral imagery delve deeply into the theory underlying their formulation. However, theoretical justification only goes so far in the infinitely variable real world. With limited time and money, it is not obvious how much to spend on theory as compared to large-scale testing of many alternatives. The author's preference is to emphasize testing. However, the more theoretical approach to target detection, as discussed in the next two subsections, still remains popular.

2.5.1 Target detection in SAR imagery

A SAR system is used on an aircraft or spacecraft to detect and recognize targets at long range. The radar transmits and receives successive pulses of radio waves. The flight path of the radar simulates a large antenna aperture. After considerable processing, the received echoes form an image of targets and background. The image resolution can be made independent of range to target, hence such characterizations as 1-ft SAR or 10-ft SAR. A smooth surface reflects the incident radar pulse like a mirror. Corner reflectors produce very bright spots in the image. Target and target-like objects are dominated by specular point scattering. Natural backgrounds are dominated by diffuse return. Speckle noise results from coherent summation of scatterers distributed randomly within a pixel. For high-resolution SAR imagery (1-ft resolution or better), target and clutter statistics are scene dependent and difficult to model. When hundreds of detection algorithms are tested, the winners are not often what the theoretical models predict. It may not even be clear why they are working, such as in the case of neural networks. However, with some humility to Tesla's point of view, some of the algorithms tested should be based on reasonable physical and statistical models.

The signal processing approach to SAR target detection emphasizes modeling the probability density functions (pdf) of just clutter (as for the simplest anomaly detectors), or targets and clutter. A threshold is chosen for the detection response to provide a constant probability of false alarm. At least, that is the objective. This results in what is popularly referred to as a CFAR detector. By contrast, the methods described earlier in this chapter don't use hard thresholds and instead simply feed the m strongest detections per image to the back-end classifier.

The traditional SAR detection approach is based on the standard sliding triple-window filter [Fig. 2.2(b)]. With low-resolution SAR, the center window consists of a single pixel. With high-resolution SAR imagery such as what we are addressing, the center window consists of a group of pixels about the size of the smallest target sought. The basic one-parameter CFAR test declares a target if

$$\frac{\overline{X}_1}{\overline{X}_2} > \tau, \tag{2.13}$$

where $\overline{X}_1$ is the mean value of the pixels in the inner window, $\overline{X}_2$ is the mean value of the pixels in the outer window, τ is a hard detection threshold, and pixels are in the magnitude domain. So, if the inner-window mean is greater than a constant times the outer-window mean, a target is declared at that spot on the image as the detector visits each image pixel in turn.

A basic parametric two-parameter CFAR test is given by

$$\frac{\overline{X}_1 - \overline{X}_2}{\sigma_2} > \tau, \tag{2.14}$$

where σ_2 is the sample standard deviation of the pixels in the outer window.

More-complex assumptions result in more-complex algorithms of this general type. Although traditional SAR target detection follows a signal processing or Bayesian philosophy, the resulting algorithms look very much like some of those previously discussed. The survey paper by El-Darymli et al.[28] covers the derivations and variations of CFAR tests for SAR imagery in the magnitude domain ($A = \sqrt{I^2 + Q^2}$), power domain ($P = A^2$), and log domain [$L = \log(A^2)$]; where a pixel in a complex-valued SAR image is denoted by $I + jQ, j = \sqrt{-1}$.

For high-resolution SAR targets, the problems with the signal processing approach to target detection are the usual ones: (1) the target set consists of targets of varying sizes, shapes and strengths; (2) clutter pixels aren't independent and identically distributed; (3) targets are rare, while difficult clutter blobs are common and target-like; (4) weak targets result from partial occlusion and camouflage; (5) *a priori* probabilities of targets and clutter are unknown; and (6) *probability of false alarm* is a misnomer since we don't know anything about real-world probabilities.

2.5.2 Target detection in hyperspectral imagery

A hyperspectral sensor forms images in which each pixel is a vector. Each element of the vector represents a different spectral band. Hyperspectral imaging differs from multispectral and visible color imaging in that hyperspectral images are generally composed of more bands wherein each band covers a very narrow spectral slice. Countermeasures against hyperspectral sensors are difficult when they cover a very large portion of the spectrum, e.g., from visible to LWIR. Although hyperspectral imagery has very high spectral resolution, it generally has very low spatial resolution. To put it another way, spatial resolution is traded for spectral resolution.

Consider a target about a pixel in size. Hyperspectral anomaly detectors often use the standard triple-window filter [Fig. 1.2(b)], where the center window nominally covers a single pixel.[29–31] The background is often modeled by a mean vector and covariance matrix. As the filter visits each pixel, the anomalousness of that pixel's spectral signature is given by the Mahalonobis distance between hypothesized target and background. A target is declared if

$$(\mathbf{X}_1 - \overline{\mathbf{X}}_2)\Sigma_2^{-1}(\mathbf{X}_1 - \overline{\mathbf{X}}_2) > \tau, \tag{2.15}$$

where $\mathbf{X}_1$ is the observed spectrum at the center pixel, $\overline{\mathbf{X}}_2$ is the mean vector of the pixels in the outer window, and Σ_2^{-1} is the covariance of the pixels in the outer window. This equation is the basis of the benchmark Reed–Xialo (RX) algorithm.[32] The same equation can be used for higher-resolution multispectral sensors by replacing the center pixel spectrum by the inner-window mean vector. Approaches also exist for finding subpixel-sized targets.

Hyperspectral imagery tends to be of lower spatial resolution than FLIR or three-band color imagery. One problem with the RX algorithm is that the outer window of a triple-window filter covers too few pixels to reliably estimate the mean vector and covariance matrix of the pixels falling within it. In some implementations, the outer window of the filter is considered to cover the entire image outside the inner and middle window. This approach is called global RX. This provides more pixels to work with. Then, the problem is that the outer window covers different regions with different characteristics and hence different statistics. Segmentation-based methods segment the image into N regions and apply the RX algorithm to each region using that region's background statistics.

Shadows pose another problem to the implementation of the RX algorithm, particularly in the visible to near-IR band. This problem is sometimes dealt with by using the square root of the pixel values as a way of reducing dynamic range and making uniform noisy regions more Gaussian.[33]

Alternatively, if the hyperspectral signatures of the targets are known, a spectral matching approach can be used. The spectral signature of the unknown is matched against target signatures stored in a library.

Any of the detection algorithms described for single-band imagery can be adapted for hyperspectral imagery. This includes detection algorithms that are trainable binary classifiers. Many alternatives to the RX algorithm have been developed and will be developed each year. Some are quite complex. In an invited conference paper titled "Is there a best hyperspectral detection algorithm?" Manolakis et al.[29] end their paper with the following advice:

> "...any small performance gains attained by more sophisticated detectors are irrelevant in practical applications because of the limitations and the uncertainties about many aspects of the situation in which the detector will be deployed."[29]

2.6 Conclusions and Future Direction

Various simple target detection algorithms were described. Each was tested on thousands of images produced by different types of sensors. We pointed out those that performed well with mid- and longwave FLIR imagery. A number of more complex algorithms were described. They tend to perform better than the simpler algorithms but are more expensive to implement and are potentially less robust.

Although testing was done over very large databases, the databases were neither large enough nor varied enough to determine absolute winners. Performance results are never definitive because performance is always a function of the artifacts, in particular sensor imagery and the uniqueness of target sets and conditions in an infinitely varying world.

Major research projects have been underway to answer fundamental questions in physics, biology, and imaging, such as:

- What is the nature of light?
- How does biological vision work?
- Can thermal and SAR imaging be modeled so as to predict detection and recognition performance?
- How best can man and machine work together?

Since target detection is such a vital problem, it is often used as a funding imperative and test case for developing technology. Eight "grand paradigms," each with profound philosophical foundation, were described in relation to target detector design.

The general conclusion is that simple target detectors, in the form of triple-window filters, can be designed using variations on standard test statistics and standard image processing features. A combination of two or three simple detectors suffice for the majority of ATR systems. A number of reasons were provided for favoring simple detectors. More elaborate detectors can be designed for achieving better performance at the expense of size, weight, power, cost, complexity, and potential lack of robustness. The eight grand paradigms provide starkly different ways of viewing the problem and utterly different tools for achieving a solution. It is too early to say which approach will win out in the long term.

References

1. B. J. Schachter, "A survey and evaluation of FLIR target detection/segmentation algorithms," *Proc. of Image Understanding Workshop*, 49–57 (Sept. 1982).
2. E. P. Simoncelli and W. T. Freeman, "The steerable pyramid: A flexible architecture for multi-scale derivative computation," *Second IEEE Int. Conf. on Image Processing*, Washington, D.C., 444–447 (Oct. 1995).
3. N. Dalal and B. Triggs, "Histogram of oriented gradients for human detection," *IEEE Computer Science Conference on Computer Vision and Pattern Recognition*, 886–893 (2005).
4. ATRWG Technology Committee, "Mathematical Morphology in ATR Algorithm Development," Joint U.S. Department of Defense-Industry Working Group (Oct. 27, 1992; revised June 1993).
5. A. Ye and D. Casasent, "Morphological and wavelet transforms for object detection and image processing," *Applied Optics* **33**(35), 8226–8239 (1994).
6. S. R. DeGraaf and B. J. Schachter, "Adaptive SAR imaging and its impact on ATD/R and image exploitation," *Tri Service Radar Symposium* (June 1998).
7. L. Sachs, *Applied Statistics: A Handbook of Techniques, Second Edition*, Springer Verlag, New York (1984).
8. W. J. Conover, *Practical Nonparametric Statistics, Third Edition*, John Wiley & Sons, New York (1999).
9. L. M. Novak, G. J. Owirka, and C. M. Netishen, "Performance of a high-resolution polarimetric SAR automatic target recognition system," *Lincoln Lab Journal—Special Issue on Automatic Target Recognition* **6**(1), 11–23 (1993).
10. N. Ahuja and B. Schachter, *Pattern Models*, John Wiley & Sons, New York (1983).
11. L. G. Minor and J. Sklansky, "The detection and segmentation of blobs in infrared images," *IEEE Trans. Sys. Man, and Cyber.* **11**(3), 194–201 (1981).
12. Gartner Group, Hype Cycle Research Methodology, WWW.gartner.com/technology/research/methodologies/hype-cycle.jsp, 2012, (accessed 20 August 2012).
13. L. M. Novak, G. J. Owirka, W. S. Brower, and A. L. Weaver, "The automatic target-recognition system in SAIP," *Lincoln Lab Journal* **10**(2), 187–202 (1997).
14. B. J. Schachter, "Closed-loop neuromorphic target cuer," *Aerospace and Electronic Systems Magazine* **28**(8), 10–17 (2013).
15. R. Miikkulainen, R. Bednar, J. A. Choe, and J. Sirosh, *Computational Maps in the Visual Cortex*, Springer, New York (2005).

16. J. Braun, C. Koch, and J. L. Davis, Eds., *Visual Attention and Cortical Circuits*, The MIT Press, Cambridge, Massachusetts (2001).
17. J. Ke, "Architectures for Compressive Imaging with Applications in Sensor Networks, Adaptive Object Recognition and Motion Detection," Ph.D. thesis, The Univ. of Arizona (2010).
18. K. Krishnamurthy, R. Willett, and M. Raginsky, "Target detection performance bounds in compressive imaging," *EURASIP Journal on Advances in Signal Processing* **1**, 1–19 (2012).
19. C. F. Hester and K. K. Dobson, "Using compressive imaging as a fast class formation method in automatic target acquisition," *Proc. SPIE* **7696**, 76960P (2010) [doi: 10.1117/12.849717].
20. P. Baldi and L. Itti, "Of bits and wows: A Bayesian theory of surprise with applications to attention," *Neural Networks* **23**, 649–656 (2010).
21. L. Itti and P. Baldi, "Bayesian surprise attracts human attention," *Vision Research* **49**(10), 1295–1306 (2009).
22. *Night Vision Thermal Imaging Systems Performance Model: User's Manual and Reference Guide*, U.S. Army Night Vision and Electronics Sensors Directorate, Fort Belvoir, Virginia (2001).
23. R. H. Vollmerhausen, E. Jacobs, and R. G. Driggers, "New metric for predicting target acquisition performance," *Optical Eng.* **43**(11), 2806–2818 (2004).
24. R. H. Vollmerhausen, E. Jacobs, J. Hixson, and M. Friedman, "The targeting task performance (TTP) metric: A new model for predicting target acquistion performance," U.S. Army CERDEC, Night Vision and Electronic Sensors Directorate, Tech Report AMSEL-NV-TR-230 (Jan. 2006).
25. D. Lowe, "Distinctive image features from scale-invariant keypoints," *International Journal of Computer Vision* **60**(2), 91–110 (2004).
26. H. Bay, T. Tuytelaars, and L. V. Gool, "SURF: Speeded up robust features," *Proc. of the Ninth European Conference of Computer Vision*, 404–417 (May 2006).
27. S. Sahli, Y. Quyang, Y. Sheng, and D. A. Lavigne, "Robust vehicle detection in low-resolution aerial imagery," *Proc. SPIE* **7668**, 76680G (2010) [doi: 10.1117/12.850387].
28. K. El-Darymil, P. McGuire, D. Power, and C. Moloney, "Target detection in synthetic aperture radar imagery: a state-of-the-art survey," *Journal of Appl. Remote Sensing* **7**(1), 071598 (2013) [doi: 10.1117/1.JRS.7.071798].
29. D. Manolakis, R. Lockwood, T. Cooley, and J. Jacobson, "Is there a best hyperspectral detection algorithm?" *Proc. SPIE* **7334**, 733402 (2009) [doi: 10.1117/12/816917].

30. D. Manolakis, D. Mardin, and G. A. Shaw, "Hyperspectral image processing for automatic target detection application," *Lincoln Laboratory Journal* **14**(1), 79–116 (2003).
31. D. Manolakis, E. Truslow, M. Pieper, T. Cooley, and M. Brueggeman, "Detection algorithms in hyperspectral imaging systems," *IEEE Signal Processing Magazine* **31**(1), 24–33 (2014).
32. I. S. Reed and X. Yu, "Adaptive multi-band CFAR detection of an optical pattern with unknown spectral distribution," *IEEE Trans. On Acoustics, Speech and Signal Processing* **38**, 1760–1770 (1990).
33. D. Borghys, I. Kasen, V. Archard, and C. Perneel, "Hyperspectral anomaly detection: comparative evaluation in scenes with diverse complexity," *Journal of Electrical and Computer Engineering* **5**, 1–16 (2012).

Appendices

Appendix 2.1 Notation used for inner-window (window 1) and annular outer-window (window 2) sample statistics

Sample means:	$\overline{X}_1 = \frac{1}{n_1}\sum_{i=1}^{n_1} x_{1i}$	$\overline{X}_2 = \frac{1}{n_2}\sum_{i=1}^{n_2} x_{2i}$	Eq. (2.16)
Sample variances:	$s_1^2 = \overline{(X_1^2)} - (\overline{X}_1)^2$	$s_2^2 = \overline{(X_2^2)} - (\overline{X}_2)^2$	Eq. (2.17)
Mean absolute deviation:	$MD_1 = \frac{1}{n_1}\sum_{i=1}^{n_1} \lvert x_{1i} - \overline{X}_1\rvert$	$MD_2 = \frac{1}{n_2}\sum_{i=1}^{n_2} \lvert x_{2i} - \overline{X}_2\rvert$	Eq. (2.18)
Third-order statistic:	$q_1 = \frac{1}{n_1}\sum_{i=1}^{n_1} (x_{1i} - \overline{X}_1)^3$	$q_2 = \frac{1}{n_2}\sum_{i=1}^{n_2} (x_{2i} - \overline{X}_2)^3$	Eq. (2.19)
Fourth-order statistic:	$k_1 = \frac{1}{n_1}\sum_{i=1}^{n_1} (x_{1i} - \overline{X}_1)^4$	$k_2 = \frac{1}{n_2}\sum_{i=1}^{n_2} (x_{2i} - \overline{X}_2)^4$	Eq. (2.20)
Mean deviation from kurtosis of a normal Distribution:	$d_1 = \left\lvert \frac{MD_1}{s_1} - 0.7979 \right\rvert$	$d_2 = \left\lvert \frac{MD_2}{s_2} - 0.7979 \right\rvert$	Eq. (2.21)
Sample gray ranges:	$R_1 = \max_i(x_{1i}) - \min_i(x_{1i})$	$R_2 = \max_i(x_{2i}) - \min_i(x_{2i})$	Eq. (2.22)
Horizontal total variation:	$HTV_1 = \sum_{\text{inner window}} \lvert x_{i,j} - x_{i,j+1}\rvert$	$HTV_2 = \sum_{\text{outer window}} \lvert x_{i,j} - x_{i,j+1}\rvert$	Eq. (2.23)
Vertical total variation:	$VTV_1 = \sum_{\text{inner window}} \lvert x_{i,j} - x_{i+1,j}\rvert$	$VTV_2 = \sum_{\text{outer window}} \lvert x_{i,j} - x_{i+1,j}\rvert$	Eq. (2.24)
Interdecile range (80%)—a dispersion statistic:		$I = \text{DZ9} - \text{DZ1}$	Eq. (2.25)

In the table above, DZ1 and DZ9 are the first and ninth deciles, respectively; thus, $I = \text{I80}$ encompasses 80% of the sample distribution.

Let Image$_i$1 through Image$_i$9 denote image region i binarily thresholded at deciles 1 through 9, respectively; $i = 1, \ldots, 4$. Let:

Maximum horizontal total variation over binarily thresholded deciles:	$\text{HTV\_MAX} = \max\left\{ \begin{array}{l} \text{HTV(Image1)}, \\ \ldots, \text{HTV(Image9)} \end{array} \right\}$	Eq. (2.26)
Horizontal total variation over image region thresholded at average gray level:	$\text{HTV\_AVG} = \text{HTV(Image5)}$	Eq. (2.27)

For HTV_MAX$_1$ and HTV_MAX$_2$, deciles are computed and thresholding is done for inner and outer windows separately. For HTV_MAX$_3$ and HTV_MAX$_4$, deciles are computed and thresholding is done for the entire area under the filter, where subscripts 3 and 4 refer to the total variation being measured in inner and outer windows, respectively. HTV_MIN, VTV_MAX, VTV_AVG, and VTV_MIN are defined analogously.

Entropy:	$E = -\sum_i f_i \log(f_i)$ or $E \approx -\sum_i f_i \log(f_i + 1)$	Eq. (2.28)
Sample size:	n_1 = number of pixels in inner window $\approx n_2$ = number of pixels in outer window	Eq. (2.29)

With this assumption of sample size, the number of pixels in the windows are generally left out of the equations. Another reason for leaving the sizes of pixel populations out of the usual equations is to avoid biasing the detector from the top to the bottom of an image as the detector changes size with approximated range-to-target. We occasionally use n to denote the number of the pixels in the inner plus middle plus outer window, that is, the total pixel count under the triple-window filter.

Appendix 2.2 Other tests for significance of the maximum value of a sample

$$\boxed{T_{88} = \frac{\hat{X}_1 - \overline{X}_2}{\sqrt{s_1^2 + s_2^2}}} \qquad T_{89} = \frac{\hat{X}_1 - \overline{X}_2}{s_2} \qquad T_{90} = \hat{X}_1 - \overline{X}_2$$

$$\boxed{T_{91} = \hat{X}_1} \qquad T_{92} = \frac{\hat{X}_1 - \hat{X}_2}{s_2} \qquad T_{93} = \frac{\hat{X}_1 - \overline{X}_2}{k_2}$$

$$T_{94} = \frac{\hat{X}_1 - \overline{X}_i}{s_i} \qquad T_{95} = \frac{\hat{X}_1 - \overline{X}_2}{s_i} \qquad T_{96} = \hat{X}_1 - \overline{X}_i$$

$$T_{97} = \frac{\hat{X}_1 - \hat{X}_2}{s_i} \qquad \boxed{T_{98} = \hat{X}_1 - \hat{\hat{X}}_2} \qquad \begin{aligned} T_{99} = {} & \max(\hat{X}_1, \hat{X}_2) \\ & - \min(\hat{\hat{X}}_1 - \hat{\hat{X}}_2) \end{aligned}$$

$$T_{100} = \frac{\hat{X}_1}{s_i} \qquad T_{102} = \hat{X}_2 \qquad T_{103} = \hat{\hat{X}}_1 \qquad T_{104} = \hat{\hat{X}}_2$$

And variations:

$$T_{105} = \frac{\hat{X}_1 - \overline{X}_2 - s_2\sqrt{n_2/4}}{s_2} \qquad T_{106} = \frac{(n/n-1)\overline{X}_1 - (1/n-1)\hat{X}_1 - \overline{X}_2}{s_2}$$

$$T_{107} = \frac{|(n/n-1)\overline{X}_1 - (1/n-1)\hat{X}_1 - \overline{X}_2|}{s_2} \qquad T_{108} = \frac{\overline{X}_1 - \hat{X}_1 - 2\overline{X}_2}{s_2}$$

$$T_{109} = \frac{\overline{X}_1 - \hat{X}_1 - 2\overline{X}_2}{s_i} \qquad T_{110} = \frac{\hat{X}_1 - \overline{X}_1 - s_2\sqrt{n_2/4}}{s_i}$$

$$T_{111} = \frac{\hat{X}_1 - \overline{X}_1 - s_2\sqrt{n_2/4}}{k_2} \qquad T_{112} = \frac{n}{n-1}\left(\frac{\overline{X}_1 - \hat{X}_1}{s_i}\right)$$

$$T_{113} = \frac{\left|\frac{n}{n-1}\left(\frac{\overline{X}_1 - \hat{X}_1}{s_i}\right)\right|}{s_i} \qquad T_{114} = \frac{\overline{X}_1 + \hat{X}_1 - 2\overline{X}_2}{s_2}$$

$$T_{115} = \frac{\overline{X}_1 + \hat{X}_1 - 2\overline{X}_2}{k_2} \qquad T_{116} = \frac{\overline{X}_1 + \hat{X}_1 - 2\overline{X}_2}{s_i}$$

$$T_{117} = \frac{\overline{X}_1 + \hat{X}_1 - \overline{X}_2 - \overline{X}_i}{s_i} \qquad \boxed{T_{118} = \frac{\overline{X}_1 + \hat{X}_1 - \overline{X}_2 - \overline{X}_i}{s_2 + s_i}}$$

$$\boxed{T_{119} = \frac{\overline{X}_1 + \hat{X}_1 - \overline{X}_2 - \overline{X}_i}{s_2}} \qquad \boxed{T_{120} = \frac{\overline{X}_1 + \hat{X}_1 - \overline{X}_2 - \overline{X}_i}{k_2}}$$

$$T_{121} = \frac{(\overline{X}_1 + \hat{X}_1 - \overline{X}_2 - \overline{X}_i)s_1}{k_2} \qquad T_{122} = \frac{(\overline{X}_1 + \hat{X}_1 - \overline{X}_2 - \overline{X})s_1}{s_2}$$

$$T_{123} = \frac{(\overline{X}_1 + \hat{X}_1 - \overline{X}_2 - \overline{X})s_1}{s_i}.$$

Appendix 2.3 Tests based on total variation

$$T_{146} = \frac{|HTV_1 - HTV_2|}{\sqrt{(HTV_1)^2 + (HTV_2)^2}} \qquad T_{147} = \frac{HTV_1}{VTV_1}$$

$$T_{148} = HTV_1 \qquad T_{149} = \frac{HTV_1 + VTV_1}{HTV_2 + VTV_2}$$

$$T_{150} = \frac{HTV\_MAX_1 - HTV\_MAX_2}{HTV\_MAX_1^2 + HTV\_MAX_2^2} \qquad T_{151} = \frac{|HTV_1 - HTV_2|}{s_2}$$

$$T_{152} = \frac{|HTV_1 - HTV_2|}{\sqrt{(VTV_1)^2 + (VTV_2)^2}} \qquad T_{153} = \frac{HTV\_MAX_3 - HTV\_MAX_4}{HTV\_MAX_3^2 + HTV\_MAX_4^2}$$

$$T_{154} = \frac{|HTV_1 - HTV_2|}{s_i} \qquad \boxed{T_{155} = \frac{|HTV_1 + VTV_1|}{\sqrt{(VTV_1)^2 + (VTV_2)^2}}}$$

$$T_{156} = \frac{HTV\_MAX_1 + VTV\_MAX_1}{HTV\_MAX_2 + VTV\_MAX_2} \qquad T_{157} = \frac{|HTV_1 - HTV_2|}{k_2}$$

$$T_{158} = \frac{VTV_1 + |HTV_1 - HTV_2|}{\sqrt{(HTV_1)^2 + (HTV_2)^2}} \qquad T_{159} = \frac{HTV\_MIN_1 + VTV\_MIN_1}{HTV\_MIN_2 + VTV\_MIN_2}$$

$$T_{160} = \frac{HTV_1 + VTV_1}{k_2} \qquad \boxed{T_{161} = \frac{|VTV_1 - VTV_2|}{\sqrt{(VTV_1)^2 + (VTV_2)^2}}}$$

$$T_{162} = \frac{HTV\_MAX_3 + VTV\_MAX_3}{HTV\_MAX_4 + VTV\_MAX_4} \qquad T_{163} = \frac{HTV_1 + VTV_1}{s_2}$$

$$T_{164} = \frac{(HTV_1 - HTV_2)}{VTV_2} \qquad T_{165} = \frac{HTV\_MIN_3 + VTV\_MIN_3}{HTV\_MIN_4 + VTV\_MIN_4}$$

$$T_{166} = \frac{HTV_1 + VTV_1}{s_i} \qquad T_{167} = \frac{(HTV_1)(\overline{X}_1 - \overline{X}_2)}{s_2}$$

$$T_{168} = \frac{(HTV_1)(\overline{X}_1 - \overline{X}_2)(\overline{G}_1 - \overline{G}_2)}{s_2} \qquad T_{169} = \frac{HTV_1}{VTV_2}$$

$$T_{170} = \frac{(HTV_1)(\overline{X}_1 - \overline{X}_2)}{k_2} \qquad T_{171} = \frac{(HTV\_MAX_1 - HTV\_MAX_2)}{VTV\_MAX_2}$$

$$T_{172} = \frac{|VTV_1 - VTV_2|}{s_2} \qquad T_{173} = \frac{(HTV_1)(\overline{X}_1 - \overline{X}_2)}{\overline{G}_2}$$

$$T_{174} = \frac{(HTV\_MAX_3 - HTV\_MAX_4)}{VTV\_MAX_4}$$

$$T_{175} = \frac{HTV_1(\overline{X}_1 - \overline{X}_2)}{S_i} \qquad T_{176} = \frac{|HTV_1 + VTV_1|}{\sqrt{(HTV_1)^2 + (HTV_2)^2}}$$

$$T_{177} = \frac{(HTV_1)(\overline{X}_1 - \overline{X}_2)(\overline{G}_1 - \overline{G}_2)}{s_i} \qquad T_{178} = \frac{|VTV_1 - VTV_2|}{s_i}$$

$$T_{179} = \frac{HTV_1 + |VTV_1 - VTV_2|}{\sqrt{(HTV_1)^2 + (HTV_2)^2}} \qquad \boxed{T_{180} = \frac{HTV_1 + VTV_1}{HTV_1 + VTV_1 + HTV_2 + VTV_2}}$$

$$T_{181} = VTV_1 \qquad \boxed{T_{182} = \frac{|HTV_1 + VTV_1|}{\sqrt{(HTV_2)^2 + (VTV_2)^2}}}$$

$$T_{183} = HTV\_MAX_1 - HTV\_MIN_1 \qquad T_{184} = \frac{VTV_1}{HTV_2}$$

$$T_{185} = \frac{|HTV_1 + VTV_1|}{\sqrt{(VTV_2)^2 + (VTV_1)^2}} \qquad T_{186} = VTV\_MAX_1 - VTV\_MIN_1$$

$$T_{187} = \frac{HTV_1 + VTV_1}{VTV_2} \qquad T_{188} = \frac{VTV_1}{\sqrt{(VTV_2)^2 + (VTV_1)^2}}$$

$$T_{189} = HTV\_MAX_3 - HTV\_MIN_3 \qquad T_{190} = \frac{HTV_1}{s_i}$$

$$T_{191} = \frac{|HTV\_MAX_1 - HTV\_MAX_2|}{\sqrt{(HTV\_MAX_1)^2 + (HTV\_MAX_2)^2}} \qquad T_{192} = HTV\_MAX_1$$

$$T_{193} = VTV\_MAX_1 \qquad T_{194} = \frac{|HTV\_MAX_1 - HTV\_MAX_2|}{\sqrt{(VTV\_MAX_1)^2 + (VTV\_MAX_2)^2}}$$

$$T_{195} = HTV\_MIN_1 \qquad T_{196} = VTV\_MIN_1$$

$$T_{197} = \frac{|HTV\_MAX_3 - HTV\_MAX_4|}{\sqrt{(VTV\_MAX_3)^2 + (VTV\_MAX_4)^2}} \qquad T_{198} = HTV\_AVG_1$$

$$T_{199} = VTV\_AVG_1 \qquad T_{200} = \frac{|HTV\_MIN_1 + VTV\_MIN_1|}{\sqrt{(VTV\_MIN_1)^2 + (VTV\_MIN_2)^2}}$$

$$T_{201} = HTV\_MAX_3 \qquad T_{202} = VTV\_MAX_3$$

$$T_{203} = \frac{|HTV\_MAX_1 + VTV\_MAX_1|}{\sqrt{(VTV\_MAX_1)^2 + (VTV\_MAX_2)^2}} \qquad T_{204} = HTV\_MIN_3$$

$$T_{205} = VTV\_MIN_3 \qquad T_{206} = \frac{|HTV\_MAX_3 + VTV\_MAX_3|}{\sqrt{(VTV\_MAX_3)^2 + (VTV\_MAX_4)^2}}$$

$$T_{207} = HTV\_AVG_3 \qquad T_{208} = VTV\_AVG_3$$

$$T_{209} = \frac{|HTV\_MIN_3 + VTV\_MIN_3|}{\sqrt{(VTV\_MIN_3)^2 + (VTV\_MIN_4)^2}} \qquad T_{210} = VTV\_MAX_3 - VTV\_MIN_3$$

Appendix 2.4 Tests based on gradients

$$T_{211} = \overline{G}_1 - \overline{G}_2 \qquad T_{212} = \overline{G}_1 \qquad T_{213} = \frac{\overline{G}_1 - \overline{G}_2}{s_2}$$

$$T_{214} = \frac{\overline{G}_1 - \overline{G}_2}{s_i} \qquad T_{215} = \frac{\overline{G}_1 - \overline{G}_2}{k_2} \qquad \boxed{T_{216} = \frac{\overline{G}_1 - \overline{G}_2}{\sqrt{s_1^2 + s_2^2}}}$$

$$T_{217} = \frac{(\overline{G}_1 - \overline{G}_2)(\overline{X}_1 - \overline{X}_2)}{\sqrt{s_1^2 + s_2^2}} \qquad T_{218} = (\overline{G}_1 - \overline{G}_2)(\overline{X}_1 - \overline{X}_2)$$

$$T_{219} = (\overline{G}_1/\overline{G}_2)(\overline{X}_1 - \overline{X}_2)$$

$$T_{220} = \frac{|\overline{G}_1 - \overline{G}_2|}{s_i} \qquad T_{221} = \frac{\overline{G}_1 - \overline{G}_2}{\sqrt{s_1^2 + \max(s_2^2, s_i^2)}} \qquad T_{222} = \overline{G}_1/s_i$$

Appendix 2.5 Tests involving blob edge strength

$$T_{224} = \frac{\overline{e}_1 - \overline{e}_2}{\sqrt{\sigma_1^2 + \sigma_2^2}} \qquad T_{225} = \frac{|\overline{e}_1 - \overline{e}_2|}{\sqrt{\sigma_1^2 + \sigma_2^2}} \qquad T_{226} = \frac{\overline{e}_1 - \overline{e}_2}{\sigma_2} \qquad T_{227} = \frac{|\overline{e}_1 - \overline{e}_2|}{\sigma_2}$$

$$T_{228} = \frac{\overline{e}_1}{\overline{e}_2} \qquad T_{229} = |\overline{e}_1 - \overline{e}_2| \qquad T_{230} = \overline{e}_1 - \overline{e}_2 \qquad T_{231} = \frac{\overline{e}_1 - \overline{e}_2}{R_2}$$

$$T_{232} = \frac{|\overline{e}_1 - \overline{e}_2|}{R_2} \qquad T_{233} = \overline{e}_1 \qquad T_{234} = \frac{\sigma_1 - \sigma_2}{\sqrt{\sigma_1^2 + \sigma_2^2}} \qquad T_{235} = \frac{\hat{e}_1 - \overline{e}_2}{\sqrt{\sigma_1^2 + \sigma_2^2}}$$

$$T_{236} = \hat{e}_1 - \overline{e}_2 \qquad T_{237} = [\overline{e}_1 - \overline{e}_2]^2 \qquad T_{238} = \overline{e}_1\overline{e}_2 \qquad T_{239} = \frac{\overline{e}_1 - \overline{e}_2}{\sqrt{s_1^2 + s_2^2}}$$

$$T_{240} = \frac{\overline{e}_1 - \overline{e}_2}{(k_1 + k_2)^{1/4}} \qquad T_{241} = \frac{(\overline{e}_1 - \overline{e}_2)k_1}{k_2} \qquad T_{242} = \frac{(\overline{e}_1 - \overline{e}_2)k_1}{s_i k_2} \qquad T_{243} = \overline{e}_1/s_2$$

$$T_{244} = \frac{(\overline{e}_1 - \overline{e}_2)}{\overline{G}_2} \qquad T_{245} = \overline{e}_1/k_2 \qquad T_{246} = \overline{e}_1/s_i$$

Appendix 2.6 Hybrid tests

$$T_{247} = \frac{(\overline{X}_1 - \overline{X}_2)(s_1 - s_2)}{s_1^2 + s_2^2} \qquad T_{248} = \frac{(\overline{X}_1 - \overline{X}_2)(s_1 - s_2)}{\sqrt{s_1^2 + s_2^2}} \qquad T_{249} = \frac{(\overline{X}_1 - \overline{X}_2)s_1}{s_2}$$

$$T_{250} = \frac{(\overline{X}_1 - \overline{X}_2)s_1}{s_2^4} \qquad T_{251} = \frac{(R_1 - R_2)(s_1 - s_2)}{\sqrt{s_1^2 + s_2^2}} \qquad T_{252} = \frac{(\overline{X}_1 - \overline{X}_2)(\overline{e}_1 - \overline{e}_2)}{s_2}$$

$$T_{253} = \frac{(\overline{X}_1 - \overline{X}_2)s_1}{s_i} \qquad T_{254} = \frac{(\overline{X}_1 - \overline{X}_2)s_1}{s_i^2} \qquad T_{255} = \frac{(\overline{X}_1 - \overline{X}_2)(k_1 - k_2)}{(k_1 + k_2)^{1/4}}$$

$$T_{256} = \frac{(V_1 - V_2)(s_1 - s_2)}{V_1^2 + V_2^2} \qquad T_{257} = \frac{(\overline{X}_1 - \overline{X}_2)s_1}{s_2^8} \qquad T_{258} = \frac{(\overline{X}_1 - \overline{X}_2)s_1}{\sqrt{s_i}}$$

$$T_{259} = \frac{(\overline{X}_1 - \overline{X}_2)(k_1 - k_2)}{k_1 + k_2} \qquad T_{260} = \frac{(V_1 - V_2)(k_1 - k_2)}{V_1^2 + V_2^2} \qquad T_{261} = \frac{(\overline{X}_1 - \overline{X}_2)^2 s_1}{s_2^8}$$

$$T_{262} = \frac{(\overline{X}_1 - \overline{X}_2)^2 s_1}{s_i^2} \qquad T_{263} = \frac{(\overline{X}_1 - \overline{X}_2)(\overline{e}_1 - \overline{e}_2)s_1}{s_2} \qquad T_{264} = \frac{(\overline{X}_1 - \overline{X}_2)(\overline{e}_1 - \overline{e}_2)}{s_i}$$

$$T_{265} = \frac{(\overline{X}_1 - \overline{X}_2)^2 k_1 s_1}{s_2^8} \qquad T_{266} = \frac{(\overline{X}_1 - \overline{X}_2)^2 k_1 s_1}{s_i^2} \qquad T_{267} = \frac{(\overline{X}_1 - \overline{X}_2)(\overline{e}_1 - \overline{e}_2)s_1}{s_i}$$

Chapter 3
Target Classifier Strategies

3.1 Introduction

A target classifier receives image or signal data about a detection point. It infers the category of the object portrayed by the data. The classification decision can benefit from a host of other available information; the more information the better.

ATR often involves a client–contractor relationship. The contractor is committed to providing a quality product to the customer. Yet, target classification is sometimes viewed in a naïve fashion. The customer throws data "over the fence." The contractor is asked to classify the "targets." Little thought is given to the breadth and scope of the problem. The usual "solution" involves showing that the contractor's favorite classifier outperforms several alternatives.

However, the true nature of the target classification problem is more complex. Ironically, choice of a classification paradigm may be the least important aspect of target classification. We will outline the issues involved in target classification. This will be followed by a review of a number of different types of classifiers.

3.1.1 Parables and paradoxes

If no prior assumptions are made about the exact nature of the classification problem, is any reasonable classifier superior to any other? The answer is NO according to the No Free Lunch theorem.[1] Self-deception results from choosing a favorite classifier *a priori* or with limited testing, without a deep understanding of the problem and a well vetted test plan.

In the absence of encompassing assumptions, is there a best set of features to use for target classification? The answer is NO according to the Ugly Duckling theorem.[2] A good set of features results from understanding the true nature of the problem. Choice of features always biases classifier decisions.

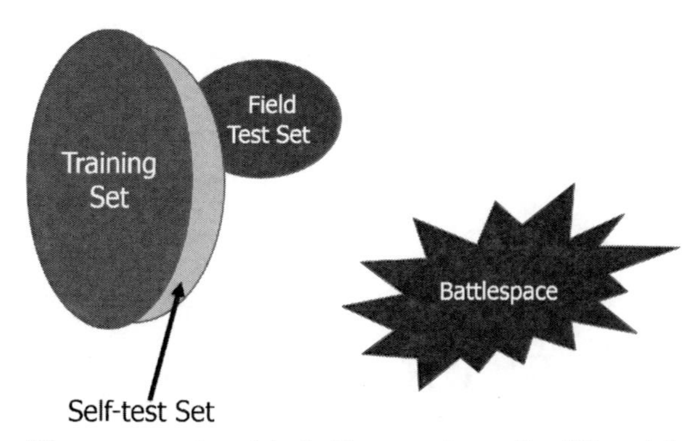

Figure 3.1 Conditions encountered in battle may be quite different from training data, laboratory blind test data, or field test data.

Is the best model always the one that performs best on the training or validation data? The answer is NO according to Occam's razor (see Fig. 3.1). When the problem is not well defined in a statistical sense, Occam's razor prefers simple solutions over more complex ones. One should proceed with simple models until simplicity can be traded for greater explanatory power. However, it is not quite as easy as saying that the simplest model seeming to meet performance requirements is always best. Besides, there are various ways of thinking of simplicity:

- model described by less complex equations,
- model with fewest assumptions,
- model with simplest architecture,
- model requiring fewest lines of code,
- model requiring least processing power,
- model easiest to analyze, explain and repair,
- model requiring least training data,
- model easiest to map to real-time hardware (e.g., FPGA), and
- model least affected by possible future changes to sensor and situation.

No Free Lunch Theorem – D. H. Wolpert and W. G. Macready[1]

"A general-purpose universal optimization strategy is impossible. The only way that one strategy can outperform another is if specialized to the structure of the specific problem under consideration."[3] There is no black box classifier that performs best on all problems. There is a danger in choosing a classification algorithm based on its performance on a small test set. It is important to incorporate problem-specific knowledge in designing the classification stage of an ATR, including a deep understanding of sensors, metadata sources, platform, and concept of operations. The classifier must be tested on a sufficiently comprehensive test set using a well-devised test plan.

The Ugly Duckling Theorem – S. Watanabe[2]
The list of possible classification features is infinite. To say that two objects are similar because they have similar features, some features have to be judged *salient*. As long as all possible features characterizing objects are ascribed equal relevance, the ugly duckling will be as similar to an arbitrary normal duckling as any two ducklings are similar to each other. Certain features must be considered pertinent to rate one particular duckling as ugly.

Occam's razor – William of Occam – 1300s, modern interpretation
Given two models of the data with all other things being equal, the simpler model is preferable. Among competing hypotheses that predict equally well, the one with the fewest assumptions is preferable. Simplicity and comprehensibility are goals in and of themselves. Simpler models also happen to be less expensive to implement. This does not necessarily mean that the simplest model will lead to better generalization or greater accuracy. There are an infinite number of possible and more complex models. There is a substantial burden of proof on those proposing a complex model. In a particular context, with extensive testing and fine tuning, a more complex model might ultimately provide better predictions than the simpler model.

Feature vectors extracted from training data samples can be envisioned as points in a high-dimensional feature space. Each point has an associated class label. Training often involves partitioning the feature space into regions. Each region is assigned a class label derived from the labels of the preponderance of the points within the region. Inference involves determining in which region an unlabeled feature vector resides. Compared to a simple boundary, a complex boundary between classes always performs better over the training set. This is because a region with a complex boundary can contain a higher percentage of points of a single class. When existing test data is used over and over again, it becomes training data. This is because region boundaries are tweaked before each test. Results reported in the technical literature and presented at conferences are often the end product of algorithms repeatedly tested on, and tuned to, a test set.

When choosing a "best" classifier, one needs to have some idea of how well the existing training data will match future battlespace data. A classifier is said to *overfit* the training data if it gets good performance on the known data sets but could provide bad performance under yet unseen conditions (Fig. 3.1). But, how do we know that any method for preventing overfitting in the past will prevent a bad fit to future data? There is no covenant that prevents nature and the enemy from breaking the "rules." The best that one can do is try to understand the engineering problem better than anyone else, judiciously choose classifiers for evaluation, develop as sufficiently comprehensive training set as can be afforded, subject the seemingly best few classifiers to blind testing or testing "in the field," and then hope that once the

ATR gets into production the character of the classification problem will not change. A simpler model is easier to analyze in the sense that there will be more chance to find flaws in the design—flaws that can spawn catastrophic prediction failures.

3.2 Main Issues to Consider in Target Classification

ATR is a system design problem, not an algorithm design problem. There must be a clear understanding of how the entire system will be used. This understanding should be across an entire Integrated Product Team, including customer, prime contractor, sensor supplier, processor supplier, lead integrator, ATR algorithm/software supplier, and end-user community. ATR design works best when the ATR designers participate in the sensor design, development, and test, rather than designing the ATR for a sensor of uncertain virtue that will only exist after the ATR is finalized and delivered.

ATR is an interdisciplinary problem. If the ATR is to fly on a helicopter, it is a good idea for the algorithm designers to fly onboard the same type of helicopter to get a feeling of how difficult it is for the pilot to follow maneuvering ground targets under conditions of changing wind direction. The algorithm designers need to talk to those programming the processor box to learn how small changes to the algorithms may reduce the difficulty of mapping the algorithms to the processor. The algorithm designers' assumptions about latency might not be achievable in the chosen hardware. The algorithm designers need to talk to the drivers of foreign military vehicles used in a data collection. What can be turned on/off, rotated, or added/subtracted on the vehicle (e.g., headlights, air conditioner, heater, turret, skirts, fuel drums)? How do these changes affect a target's signature? The algorithm designers need to talk to the sensor designers to determine what types of functions and parameters can be adjusted within the sensor and turned on or off, and how these adjustments affect the digital scene. Can the ATR control any of these settings? The algorithm designers need to talk to the subject matter experts (SMEs) in the government labs to get their point of view, and determine what databases exist and whether any data collections are planned. It is much easier and cheaper to piggy-back on someone else's data collection than to run your own. The algorithm designers need to talk to the end-users. What will really help them do their job better? The algorithm designers need to talk to other relevant experts such as statisticians, system designers, those running human perception labs and simulators, and those designing similar systems whether in government, industry, or academia. Algorithm developers need to work closely with a business development team. What does the budget look like? What is the upgrade cycle for a particular platform? Is there a good working relationship with the customer? How likely is a program to survive? Large amounts of money have gone into ATR development for programs that haven't survived, such as the Comanche helicopter and Future Combat Systems. Can the developing technology be transitioned to other platforms?

Success is achievable by paying close attention to the items listed in the following subsections. Each complication can be dealt with by thorough analysis, risk mitigation strategies, and extensive testing.

3.2.1 Issue 1: Concept of operations

A concept of operations (ConOps) is a document describing the characteristics of a proposed system from the perspective of the intended users of the system. It conveys the qualitative and quantitative attributes of the system to all stakeholders. ConOps evolve from a vision of how certain capabilities could be utilized to achieve a collection of military objectives.

A ConOps document includes the following elements:

- statement of the goals and objectives of the system,
- strategies, policies, and constraints affecting the system,
- organization, authority, responsibilities, and interactions among the stakeholders,
- operational processes for deploying the system, and
- key performance parameters (KPPs).

Proper ATR design flows out of the ConOps. Without a ConOps, ATR design is unbounded. It will not be clear who the stakeholders are and what expectations they have. Expectations regarding target classification are irrepressibly unrestrained without an explicit ConOps. Customers will always ask why the ATR doesn't work on data that it wasn't designed to operate over.

3.2.2 Issue 2: Inputs and outputs

The ATR can be viewed as a black box that transforms its inputs to desired outputs. The inputs include data from one or more sensors operating in one or more modes. Inputs also include various forms of metadata, mission-specific data, and possibly information passed from other platforms or by radio contact with the operating base. Metadata provides context to the classification problem and can include some form of range information, digital terrain elevation data, bad pixel list, sensor elevation and pointing angles, latitude, longitude, and platform motion. Metadata must be time-tagged or synchronized with image data. The operational system might be able to collect additional metadata or feature data to resolve classification ambiguities, but collection of each additional piece of data has associated costs.

For each detection, the ATR commonly generates a list of target classes, at some level of specificity, with a probability estimate apportioned to each class. These probability estimates are predictions or inferences of what the detected object may be. The ATR could instead output its decisions and probability estimates for each level of a decision tree. The output of the ATR can include other information, such as the activity in which targets are engaged. The output often also includes a scaled ROI or video clip about each high-probability target type. The ATR output is commonly in the form of a track file, even for cases when detected targets are stationary.

A target classifier, if designed properly, will benefit from additional statistical information that can aid the classification process. If *a priori* class probabilities are known, even approximately, this information should be provided to the ATR. For example, it may be known that the enemy has ten times as many T-62 tanks as T-72 tanks. The distinction between T-62 and T-72 might be important. Or, it may be immaterial as to whether a tank is a T-62 or T-72, so the two classes can be merged into one.

Risk is fundamental to all military operations. The risk of not classifying a high-value target correctly must be balanced against the risk of mistakenly classifying the wrong object as a high-value target. The military cost or risk associated with each type of class error should be provided to the ATR. The cost of incorrectly classifying a Scud launcher might be 100 times the cost of incorrectly classifying a fuel truck.

It must be understood what will be done with the output of the ATR. Will ROI images of high-probability targets be shown to an aircrew for visual analysis and action? Is the classifier then just a suggestion of what the target may be—but the human observers are the real decision makers? Or is automatic action taken based on the classifier's output? What is the role of the ATR in its interactions with humans-in-the-loop? Where are these humans? In the same aircraft as the ATR, or sitting at a ground station in the same country; or is the data being sent to the cloud for later action? Or is the data piling up in some mega-storage warehouse, the vast majority of which will never be used for anything?

3.2.3 Issue 3: Target classes

Target types and classes cover a wide range of possibilities. Traditional targets are military vehicles, including ground vehicles, ships, drug-running boats, submarine periscopes, mobile refineries, aircraft, etc. Another category is enemy soldiers and irregular forces. Dismounts might only be considered targets if they are carrying large weapons. Targets can also include mines, improvised explosive devices, incoming missiles or munitions, rocket launchers, tiny drones, pick-up trucks with large guns (called "technicals"), tunnels, shipping containers with nuclear material, or anything else of military interest. Targets can also be associated with rescue operations, such as life vests floating on the water, small lifeboats, flares, downed aircraft, etc. There are many video analytic, homeland security, and commercial problems comparable to ATR problems such as locating icebergs or forest fires from aircraft, airport security, and face recognition.

Target classifiers distinguish a target of one class from that of other classes at some level of specificity. Levels of specificity can be labeled as detection, classification, recognition, identification, friend from foe, or fingerprinting; for example. Is the detected object:

- target versus clutter,
- large vehicle versus small vehicle,
- tank versus truck versus APC,
- T-72 tank versus T-82 tank versus M60 tank versus M1 tank,
- friendly T-72 tank versus enemy T-72 tank,
- enemy T-72 tank with its engine running versus cold enemy T-72 tank, or
- bomb-damaged enemy T-72 tank versus intact enemy T-72 tank?

An important question is whether the required level of classification is possible given the input data. It may not be possible to tell a friendly T-72 from an enemy T-72 with the available data. Do other types of data or systems exist to help make the decision, such as from a Blue Force Tracking System or intelligence information?

A very important issue is whether target classes are actually known at the time of ATR design and will remain fixed into the distant future; or more likely, whether relevant target classes are country and mission dependent. Then the issue is: How does the classifier train on the classes that are mission dependent when the mission changes day by day? There are not likely to be engineers in the theater of operation to retrain and validate the ATR. There must be a mechanism in place for providing the ATR with a list of targets of interest. The classifier stage of the ATR must be reactionary to the target list within the bounds of its design. What happens when the pilot gets a command over the radio to search for a different target type?

3.2.4 Issue 4: Target variations

Many military vehicles have articulated parts (Fig. 3.2). A tank's turret and guns can move. Hatches can be in the open or closed position. Soldiers might stick their heads out of hatches. Grass, bushes, and the roll of terrain obscure the bottoms of vehicles. Vehicles can have attachments such as fuel drums, chains, or objects draped over them. Open-bed trucks carry various types of cargo. Vehicles are purposely placed into cul de sacs, along tree lines, or alongside buildings. With certain sensor types, a target's appearance changes as a function of depression angle, aspect angle, range, sun angle, and under different weather and diurnal conditions. A hyperspectral sensor may detect a vehicle by the spectral reflection from its special military paint. Is it known for certain that the tank's crew didn't repaint their rusted vehicle with house paint?

Figure 3.2 Many target types have articulated parts. An ATR must recognize them in all of their variations. The Scud launcher shown here is most dangerous when its missile is in the launch position. (Photo from defense.gov.)

Military vehicles often share components with other similar vehicles or sometimes quite different vehicles. Some vehicle types use the exact same top structure, but the bottom chassis is completely different (e.g., tracked versus wheeled). More commonly, the bottom chassis is the same for a large number of vehicle types, but the top structure is different. In these cases, it is impossible to distinguish the vehicle types when only the common part of the vehicle is observable. Some military aircraft have commercial counterparts.

In the thermal IR part of the spectrum, various parts of a vehicle can appear quite different depending on which parts are turned on or recently used, such as engine, exhaust pipe, bogey wheels, driveshaft, internal heaters, or lights.

The critical issue in classifying a target with EO/IR imagery is scale. Without accurate scale information, it is not known whether the target is smaller than a single pixel or larger than the whole image. Is it a hummingbird or a helicopter? What are the sources of scale or equivalently range information?

Classifiers must be robust to target variation and varying appearance under different conditions. It needs to be understood that under some conditions and aspect angles, certain vehicle types cannot be discriminated from each other (Fig. 3.3).

3.2.5 Issue 5: Platform issues

The nature of the platform in which the ATR resides affects the target classification problem. Platform vibration is a major issue. Vibration may be

Figure 3.3 Target types are more easily distinguished from broadside and directly overhead views and are less easily distinguished from front, back, and diagonal views. (FLIR images are from rdl.train.army.mil.)

dampened by the sensor system, but there is always residual vibration. Vibration is much worse under some circumstances, such as after a missile is fired. Forward velocity introduces motion parallax, making it difficult to register one EO/IR image to the next. Platform rotation introduces scene rotation, which can be corrected with data from an inertial measurement unit (IMU). Depending on how the sensor is mounted and its pointing direction, part of the platform might appear in the image. With a helicopter-mast-mounted IR sensor, the helicopter blades whirl over the background video. With a remotely operated undersea vehicle using visible-band sensors, robot arms and own-vehicle shadow may appear in the image.

3.2.6 Issue 6: Under what conditions does a sensor supply useful data?

It is obvious that a visible-band sensor is more useful in day than night. Even in daytime, low-reflection camouflage paint and patterns, solar highlights, shadows, concealments, and decoys make recognition difficult. A visible-band sensor is not effective when the target appears in the image next to the sun. An IR sensor is more useful against hot active targets in clear weather and is less useful against cold-soaked (i.e., ambient temperature) vehicles, after and during rain, in a sandstorm, or when the battlefield is full of smoke and fire. A long-range, laser-illuminated, range-gated SWIR sensor is highly affected by atmospheric turbulence. A visible-band hyperspectral sensor is affected by atmospheric conditions and target paint. SAR is best against stationary targets. A high-range-resolution (HRR) radar mode works well against moving targets. An undersea visible-band sensor is affected by marine snow and attenuation of the red band. An acoustic sensor is useless if there are helicopters and other sources of loud noise in the area. Active sensors such as radar and LADAR might need to be turned off at times to avoid enemy detection of their transmissions.

3.2.7 Issue 7: Sensor issues

Will there be sufficient information to classify a target? It is often the case that the ATR is under design and test while the operational sensor is being designed. The sensor's exact characteristics might not yet be determined. Or the ATR might be delivered for integration onto a platform, and the sensor changed a few years later.

Consider IR sensors. There are a dozen types of potential defects in IR imagery. These include: bad pixels, nonuniformity, clipped histogram, periodic noise on data lines, flicker; bugs, dirt, water drops, optical distortion, focus issues, and scratches on the lens or window in front of the sensor. Some FLIR sensors are interlaced. Each frame is composed of separate fields imaged in succession. If the platform or target is moving, the fields might not properly fit together. Then it may be necessary to use a single field, with corresponding reduction in resolution. Certain FLIR cameras use both time-delay integration scanning and interlacing, resulting in a number of geometric nonuniformity issues. With an IR camera, stray reflections or temperature variations from within the camera or window in front of the camera can distort the image. This is called *narcissus*. A FLIR camera may have a useful lifetime of a dozen years, with its imagery progressively worse toward its end of life.

Analog, color National Television System Committee (NTSC) standard cameras are still in military use. Most of these are now actually digital cameras with analog output. Much of the chromatic information is lost during formation of the analog signal. Interlacing, vignetting, over and under saturation, and image artifacts are common.

Image data is sometimes stored or sent to the ATR in a compressed form. Even the best modern codecs introduce image artifacts at high compression rates. These may show up in the form of ringing around bright spots or periodic blockiness. Image compression corrupts extracted features.

Image formation can be described as a chain of operations, some of which take place within the sensor system. A display viewed by an operator can be at the end of the chain. Annotation or graphics are sometimes embedded into imagery as a step in the image formation chain, rather than being stored as a separate overlay. It is important to pull off image data for ATR processing before embedded symbology, bit-depth reduction, and compression spoil image integrity.

3.2.8 Issue 8: Processor

Some military vehicles have been in operation for many years. There aren't many new manned aircraft under development. Military systems don't get updated as often as cell phone models, for example. Adding an ATR to an existing platform can mean making use of the meager processing resources on that platform. Even if a board slot is available, adding a new board type to an existing platform might be prohibited by logistics, outdated backplane design, or weight or power limits. Adding a unique processor chip can also be prohibited if there is no guarantee that the chip will be available for a number of years into the future. Startup companies offering very high-performance, massively parallel processors often go out of business after their venture funds are exhausted. There is a long list of such companies.

In the long run, as processors become smaller and more powerful, ATR might not be viewed as a separate processor box, but rather as just another mode of a sensor. That is, military sensors might be sold with image exploitation built into them, as is currently done with handheld and cell phone cameras. The processing and algorithms required for ATR are also very similar to those required for self-driving cars. ATR designers shouldn't ignore technology advances in related fields. The traditional concept of ATR in a large, rugged enclosure (air transport rack) will fade away when ATR can be implemented on a single chip or very small module. As an aside, the standard Air Transport Rack is commonly referred to as an "ATR," not to be confused with an automatic target recognizer.

3.2.9 Issue 9: Conveying classification results to the human-in-the-loop

Despite dire warnings in the popular press of intelligent machines becoming our overlords, trained humans far outperform machine classification in tasks requiring intuition, judgment, flexibility, common sense, creativity, verbal consultation, understanding human culture, and scene gist. Except for fully autonomous systems such as cruise missiles, a human is normally in the loop to make the final decision on target class. The human decides the action to be taken. The ATR is no more than a workload reducer for the human operator. Design considerations include understanding the human's role in interpreting ATR conclusions. Human vision has its limits in terms of time to make a decision and number of decisions that can be made before fatigue sets in. This suggests keeping the false alarm rate very low. Target detection and classification probabilities must be conveyed to the human in an understandable form. This can be in the form of a report. More likely, information will be conveyed via a graphical user interface (GUI). Targets can be circled on an image, with a color code indicating probability. However, it is difficult to display a low-dynamic-range target embedded in a high-dynamic-range image without losing target detail. One solution is to display targets as ROIs separate from the larger image. Individual ROIs might have lower dynamic range, making it easier to map them to 8-bits/pixel. Should the operator be given the ability to adjust the contrast and brightness of each ROI or switch to false color? What additional information should be conveyed to the operator, such as range-to-target or target location on a map? How should hyperspectral, complex-valued SAR (with real and imaginary terms), 3D LADAR, or HRR radar signals be presented to a human operator?

How should the ATR's conclusions be displayed to the operator? Suppose that the ATR's conclusions are as follows for a detection:

- 95% tracked vehicle,
- 90% tank, and
- 35% T-72, 36% M1, 7% T-62, 12% tank of unknown type.

Then the question is: At what level of specificity should these ambiguous conclusions be shown to an operator who is under time pressure with myriad distractions? Should these mixed inferences be conveyed graphically, in numbers, in color codes, or with synthetic speech? What if, as the target is tracked, the ATR's conclusions change every frame time?

The human operator requires additional controls to interact with the ATR. He should be able to point to a detection, indicating not to show him anything like that again, and point to another detection and tell the ATR to show him more things like that. The operator might want to dial down the number of detections or only be shown highest-probability decisions—taking into account the mission and rules of engagement. What overburdened pilots hate the most is false alarms. They also hate inconclusiveness. A machine that makes mistakes, even the types of mistakes that a human makes, is liable to be shut off or ignored.

A cockpit with displays is referred to as a *glass cockpit*. Not all military cockpits have displays. Those that do usually have quite small displays. There are times when a pilot (or driver) might be wearing night vision goggles and can only see the displays in grainy monochrome. Helmet-mounted displays (HMDs) are becoming more common. HMDs have severe design constraints that need to be understood when used to convey ATR results to an operator. Some HMD displays show a sensor image to one eye, allowing the pilot's other eye to look out the cockpit window.

3.2.10 Issue 10: Feasibility

Target classification can cover a wide range of sensors, platforms, target types, depression angles, and performance requirements. Problem difficulty ranges from that of a single target type with high-resolution imagery against a benign background to that of a multitude of ill-defined targets, some moving, some stationary, some partially obscured, in a complex cluttered cityscape imaged at low resolution. The classification problem can be attacked with a wide range of paradigms and tools, including pattern recognition, signal processing, expert systems, artificial intelligence, biomimicry, evolutionary algorithms, 3D geometry, and modeling and simulation. How do we determine which paradigm is best without competitively testing every paradigm? How do we determine if a particular target classification problem is solvable? Solvable is defined as meeting the key performance requirements called for by contract.

Consider hard evidence. If a trained image analyst can recognize the targets, you have evidence that the problem is solvable in theory. But we cannot download the analyst's neural code. Therefore, there is no guarantee that algorithms can be formulated to solve the problem. If a target is more discernable to a human with a feature image than with the raw data, this is a clue that the feature image may be useful to the ATR. If a project is long-term and well-funded, has dozens of engineers working it with a fleet of data collection aircraft at their disposal, and has extensive government test and evaluation planned for varying conditions and locals, it can be assumed that the project has credibility. A university or small company report of fantastic target recognition performance over limited data with self-test and an unclear relationship between training and test data should be taken less seriously. Press releases by commercial companies claiming "breakthroughs" in artificial intelligence, computer vision, and quantum computers are not hard evidence.

Reproducibility is the ability of an entire experiment to be reproduced under similar conditions. It is a foundation of the scientific method. However, replicating an experiment conducted by others will not get a researcher a publication in a refereed journal or a large research contract. An extensive research study resulting in failure is unlikely to be published. Our culture rewards the reporting of positive data. Innovation and claimed breakthroughs are highly compensated—not cautiousness. In recent years it has been determined that the majority of studies in biomedical research and psychology cannot be replicated. ATR results reported in the engineering literature are often not reproducible. They are not generalizable to diverse sets of target types, conditions, viewing geometries, and missions. Many published papers on target classification represent results for favorite classifiers tuned to the test data and competing classifiers tested with little tweaking.

An ATR that has met all of its performance requirements, is in production, and has been proven in battle should be taken seriously. For example:

- The AN/APG-78 fire control radar system allows the Apache attack helicopter to detect, classify, and prioritize ground targets during daytime and night-time and in poor weather and obscured conditions.

- Identification Friend or Foe (IFF) systems are used to positively identify friendly forces. This is a form of cooperative recognition, at least from the point of view of friendly forces. Such systems have been successfully used since the 1940s.
- Israel's Iron Dome air defense system has successfully intercepted and destroyed short-range rockets and artillery shells. The radar system detects the rocket's launch and tracks its trajectory. The missile-firing unit launches an interceptor missile equipped with electro-optical sensors. It distinguishes rockets deemed threats from those that will not land in designated areas.
- Northrop Grumman's Airborne Laser Mine Detection System (ALMDS) is mounted on the MH-60S helicopter. Flying over sea lanes, it finds and geolocates mine-like objects with its pulsed laser light and streak tube receivers by imaging the near surface of the ocean in 3D, day or night.
- Naval Air Warfare Weapons Division (China Lake) personnel developed algorithms/software for the helicopter-mounted Automatic Radar Periscope Detection and Discrimination (ARPDD) system. The system achieved initial operational capability in 2013.
- The Joint Surveillance and Attack Radar System (Joint STARS) is an advanced airborne command, control, intelligence, surveillance and reconnaissance system used by the U.S. Air Force. It provides all-weather surveillance and targeting of moving and stationary targets.

Once a system is deployed, discussion of it usually disappears from the literature on target recognition.

Some realism needs to be introduced into the analysis of the problem. Suppose that the classification problem is determining whether a detected person is a farmer holding a rake/hoe versus a terrorist holding a rifle. Is there sufficient resolution to solve the problem? Suppose that the problem is detecting (metal, plastic, and ceramic) land mines that have been buried for several years. Grass and bushes have grown over the area. Is there any combination of sensors that can provide adequate signal-to-noise ratio (SNR) to solve the problem? If a trained dog (Fig. 3.4) or pig can solve the problem, that is a hint that the problem is potentially solvable, say by an artificial nose. But, it may take 100 years of research to develop an electronic nose as good as a bloodhound's nose. Suppose that the problem is finding suspicious activity using data from a video-rate gigapixel sensor on a UAV. Is sufficient processing power available to solve the problem on the UAV, or is sufficient bandwidth available to transmit the data to the ground? Suppose that the problem is finding the one person in the crowded marketplace wearing the explosive vest. Is the clutter-to-target ratio too high to solve the problem? That is, are there just too many people, and is there too much movement, too wide an area, and too weak a signal? Suppose that the problem is finding

Figure 3.4 If a trained dog can find an IED, this is a hint that the problem is potentially solvable by an electronic device. However, artificial noses can't yet match a dog's sniffer. Photo is Pfc John Casey with his partner Roxy at Forward Operating Base Sharana in Afghanistan. (Photo by Sgt. Morales from army.mil.)

tunnels 50 m underground. Is there a gravity mapper or any known sensor that has the sensitivity to detect the tunnels?

If the ATR is working well and enemy forces are being clobbered, how easy is it for the enemy to change tactics? Is the enemy going to park its tanks next to schools, hospitals, playgrounds, museums, religious buildings, historical sites, behind cover—or leave them uncovered in the open as sitting ducks?

Does the problem involve so many target types, say >100, that reliable identification is impossible? Or is it just too costly to collect a sufficiently comprehensive training set? Just as cars are changed from one model run to the next, so are military vehicles. However, military vehicles tend to have longer production runs than cars. Are there too many versions or variations of the same target type to reliably identify it? There may be a number of factories in different countries producing variations of the same military vehicle. Skirts can be placed over the bogey wheels of tanks, fuel drums can be added, and objects can be draped over vehicles. The appearance of a target also depends on what is around it. When a vehicle is standing in place, its exhaust heats up the ground around it. At night the upper surfaces of vehicles radiate heat into a clear atmosphere. Wheels and treads leave hot tracks. Windshields and optics reflect sunlight. Some vehicles exhaust diesel smoke. Vehicles moving through the desert kick up considerable amounts of dust. A vehicle's shadow is very distinct and distinguishing in SAR imagery when the background is a neatly mowed lawn. But the shadow breaks up with a rougher background. It is difficult, if not impossible, to neatly segment a highly textured vehicle from a highly textured background. This is frequently the case in thermal imagery, in which targets often have bright hot spots but indistinct borders. Camouflage patterns make segmentation difficult in the visible band. Therefore, target classification is often encumbered by features that are a confounding mixture of target and background.

Suppose that an ATR is getting reasonable performance on a small set of data. The next step is to perform extensive test and evaluation on a wider set of relevant data. This usually involves field trials or laboratory blind tests. Test and evaluation needs to be done by independent organizations using well-defined test plans and evaluation equations. Without a good test plan, performance results are advertising rather than science. Competitive tests by several leading contenders set a bound on what is achievable. It is also interesting to compare human performance to machine performance, as in the story of John Henry "the steel driving man." If the goal performance level is not achievable, then a clever new idea is needed, or more development is called for (possibly decades worth), or the required performance level needs to be lowered, or the required level of classification needs to be reduced.

3.3 Feature Extraction

Feature extraction is one of the most important steps in target recognition. It refers to the extraction of a set of descriptors, attributes, or relevant information from a target image (or signal). The input is a ROI containing a detected object. The output is a feature vector (or 2D or 3D feature image). Features must be extracted carefully so that the feature set contains as much relevant information as possible with little irrelevant information or noise. Feature extraction teases out useful information from difficult data. Good features disentangle the underlying factors of variation in the data. When the input image is too large or redundant, particularly when compared to the amount of training data available, feature extraction serves as a special type of dimensionality reduction.

One can list several desirable attributes of a good feature set, but there is no dictate that a feature set must possess any of these attributes:

- In-class versus cross-class variance: Intraclass variance should be small, while interclass separation should be large. Thus, the set of features derived from within the same class should be similar from sample to sample, while features derived from different classes should vary significantly. This broad concept of class, for example {tank, truck,

APC}, only makes sense if it is understood that a single broad class should actually be treated as multiple subclasses.

- Robustness to noise and distortion: Features need to be robust against noise in image data such as "uncorrected" bad pixels, residual nonuniformity, interlacing, under- or oversaturation, optical distortion, electronic noise, dust, turbulence, atmospheric attenuation, platform motion; or with active sensors, weak signal due to range, etc.
- Computational efficiency: There should be a balance in the processing requirements of the stages of an ATR, including feature extraction. Processing capacity is affected by size, weight, power, logistics, and cost limits.
- Controllable invariances: Features generally require invariance to certain types of variations, but only within limits. For example, there are often known bounds on range-to-target error.
- Sparseness, dimensionality: Sparse codes reduce redundancy and are necessary for certain types of classifiers. Sparse binary codes are quite popular. High-dimensional feature vectors can often be mapped to an embedded lower-dimensional manifold without loss of discriminating power. Raw data rarely is the optimal choice for a feature set, but is regaining popularity with deep-learning techniques. However, remember that the eyes don't feed raw video data to the brain.

Feature extraction rises to the level of feature detection when the extracted features can be verified using ground or image truth. Then we can talk about false features, missed features, and probability of feature detection versus false alarms. Features that can be verified with proper instrumentation include vibration, velocity, length, aspect angle, and temperature.

Suppose that the objective is to recognize a person. Eyes, nose, mouth, ears, and hair color are good features. However, some features might be missing due to sunglasses, earmuffs, hat, or the person sipping a drink. The ATR is often faced with a situation where the full set of features is not available. This can be due to long range, obscuration, weather, sensor defects, target velocity, target construction, or sensor mode. In the worst case, features are missing or corrupted due to actions by the adversary. A challenging adversary will remove or modify features through the use of camouflage, concealment, alteration, or deception. In some cases the absence of a feature can be a critical feature in and of itself. Consider the guy in the police lineup with the missing right pinky finger.

Features can be classified into different categories based on various criteria:

1. Pixel Level, Local, Global
   a. Pixel level: Features that are calculated at each pixel, such as hue, saturation, brightness, grayscale. (However, with most color

cameras, only one color filter is used at each photodiode. A pixel is actually assigned a three-color vector by borrowing color information from its neighbors.)

b. Local: Features that are computed within a small neighborhood about the pixel, such as edges and corners. A sparse set of features is reported with thresholding or by suppressing all but the local maxima.

c. Global: Features calculated over an entire image or a regular subdivision of an image, such as grayscale histogram, Fourier magnitude, histogram of oriented gradients, optical flow field, or n^{th}-order statistics.

2. Domain or Sensor Specific

   a. Domain: Different kinds of features are used for different recognition problems such as fingerprint, iris, face, mammogram, road network, handwritten characters, or speech. In a military context, different features are used for the recognition of different types of targets such as aircraft, buried landmines, undersea mines, nuclear material, ground vehicles, ships, tunnels, incoming missiles, IEDs, micro-UAVs, or shooters. For example, jet engine modulation (JEM) features characterize a jet engine from its spectrum when illuminated by a radar.

   b. Sensor: Different types of features are appropriate for different types of sensor data, such as LADAR, FLIR, interferometric SAR (IFSAR), hyperspectral, gravitometer, acoustic, vibrometer, fully polarimetric HRR radar, sonar, terahertz sensor, stereo camera, ground-penetrating radar, etc.

3. Raw or Preprocessed Data

   a. Raw: Features can be derived from raw sensor data. These include histogram, edge vectors, and n^{th}-order statistics.

   b. Preprocessed: Preprocessing includes super-resolution, color correction, video stabilization, scaling based on estimated range to target, histogram equalization, image mosaicking, radar autofocus, sharpening, and noise reduction.

4. Low Level versus High Level

   a. Low-level features are generally simple features extracted from raw or lightly processed data. These include edges, corners, radar cross-section, optical flow vectors, and target velocity.

   b. High-level features are formed from combinations of low-level features. These include target shape, number of wheels, human gait, or soldier's uniform.

5. Whole ROI versus Segmented Target

   a. Whole ROI: Features can be extracted from a rectangular ROI about a detected target. Some of the features then contain a puzzling mixture of foreground target and background clutter. This is not so bad if the target is a hawk against a clear sky rather than a leopard in the jungle.

   b. Segmented target: If a target can be segmented from its background, then some features, such as a histogram, will contain only target data. Other features such as target shape or center can then be more easily computed.

6. Single-Scale, Multiscale, and Multilook, etc.
   a. Single-scale: Features are derived directly from individual images.
   b. Multiscale: A feature set is derived from a hierarchy of image representations. Examples are wavelet features and features derived from a pyramidal representation.
   c. Multilook: Spatiotemporal features are derived from the target's appearance in successive frames or scans, preferably taken from different viewpoints. Multiple 2D views can be combined to construct a view-invariant 3D representation.
   d. Multisensor: Features combine data from two or more sensors. A common example is a stereo pair of cameras. Data can also be combined from different sensor types, such as a FLIR camera and a HRR radar. In both of these examples, a better 3D depiction will emerge.
   e. Multiplatform: Features are derived from sensors on more than one platform.
7. Supervised versus Unsupervised Feature Extraction
   a. Supervised: The ATR designer determines the types of features to extract. When problem specific, these are called "hand crafted" features.
   b. Unsupervised: The recognition process trains on raw data and determines what features to extract. This is done by auto-encoders and the early layers of deep-learning algorithms.[4]
8. Mathematical versus Semantic versus Biologically Inspired
   a. Mathematical: Features derived with complex code that have obscure semantic meaning. Examples are wavelet and Fourier features.
   b. Semantic: Features that are readily understood by a person. 2D features with semantic meaning can be viewed on a monitor. For example, an edge image can be observed and understood. This allows the ATR developer to determine how well the feature extraction process is working. Target velocity and length have clear meaning.
   c. Biologically inspired: Features based on an understanding of the types of features extracted by biological sensors and processing, including the retina and visual cortex. Many different kinds of

creatures have survived evolutionary challenges. The human isn't necessarily the best model. Common biological models include fly, shrimp, eagle, bat, jumping spider, deep-sea fish, and horseshoe crab. For undersea sensing, the hammerhead shark is a more appropriate model than human vision because human vision is "designed" for imaging in the atmosphere.

9. Physics-based

   Features that take advantage of physical phenomena or physical laws, such as polarization, Doppler, Planck's law, and quantum spin Hall effect. Buried landmine and IED detection usually have deep physical underpinning. Thermal neutron activation (TNA) and quadruple resonance (QR) devices directly detect the explosive signatures of landmines and IEDs. TNA detects neutron-activated gamma rays from nitrogen in the explosive material. QR utilizes radio frequency magnetic field pulse tuned to explosive compounds. Ground-penetrating radar systems detect dielectric discontinuities below the surface.

   Certain multiband thermal systems detect the fine quartz particles resting on the surface of disturbed earth after the larger particles have sunk into the ground. This Restrahlen effect is a quartz double-reflection feature centered at 8.5-μm and 8.9-μm wavelengths. In addition, vegetation often changes color around disturbed earth.

10. Model-based
    a. 3D Models: Objects to be recognized come from a library of 3D models. Recognition involves finding transforms and correspondences that best overlay 3D features onto each 3D model, or 2D features onto 2D projections of the 3D models.
    b. Target parts: At a component level, features are target parts. Recognition is a true hierarchical process that includes a parts-based phase. For example, for face recognition, the parts are eyes, nose, and mouth. For tanks, the parts are turret, cannon barrel, chassis, tread, and bogey wheels.
    c. Relational features: Features are the relationship of one entity to another. Examples are distance of vehicle to road, distance between vehicles in a possible convoy (Fig. 3.5), gun-slinger following man with his arms raised, arrangement of vehicles at a launch site, and angle of weapon to body. Features can be related temporally, such as elapsed time between observed flash and acoustic blast.

3.4 Feature Selection

The feature extraction process produces a large number of potential features. Some of these features are redundant, irrelevant, unstable, or noisy. Feature selection chooses a preferred subset of the extracted features. The benefits are

Figure 3.5 Not all features are from a target's physical properties. Relational features such as distance between vehicles are useful for recognizing a convoy. Photo from www.ng.mil.

reduced classification error, shorter training time, better generalization, clearer explanation of why classification should work, and fewer on-line computations.

Too many features and too few training samples result in the "curse of dimensionality." A feature vector with thousands or hundreds of thousands of elements is overkill when equal or better performance could be achieved with a properly chosen subset of features. However, no feature selection process will compensate for misconstruing the statistics or ConOps if, for example, the training set is small and battlespace conditions don't match conditions under which training data was collected.

A feature is redundant if it has high mutual information with other features. A redundant feature adds no supplementary evidence beyond that contained in the equivalent original features. An example of redundant features in a color image are {cyan(x), magenta(x), yellow(x)} and {hue(x), saturation(x), brightness(x)}, wherein the original features are {red(x), green(x), blue(x)}. However, classifiers don't make perfect use of the information contained in features. Some redundancy is tolerable. It may be that {hue(x), saturation(x), brightness(x)} better disentangles the scene information for some classification task than {red(x), green(x), blue(x)}.

A feature is irrelevant if it has low mutual information with the target class. An irrelevant feature provides no useful information for the classification task. Examples are time-of-day for SAR (but not for FLIR), color when the task is character recognition, or mean gray-level when the task is fingerprint recognition.

Examples of noisy features are the highest frequency terms in a Fourier or wavelet domain, edges extracted from interlaced video, compressed/decompressed SAR imagery, optical flow vectors extracted from FLIR imagery before non-uniformity correction, and EO/IR features obtained under conditions of atmospheric shimmering.

Consider several feature selection strategies:

- Filter methods choose individual or groups of features based on their relevance (discriminating power). These methods are based on the mutual information between features and target classes, or they use statistical tests such as the T-test or F-test. With a mutual information approach, a feature is judged good if it is highly correlated to a class but is not highly correlated to any other feature. Another approach is to find a minimum number of features that separate classes as consistently as the full set of features.
- Wrapper methods use a specific classification algorithm to determine the relative merits of alternative feature sets. The virtuousness of a set of features is judged by the performance of the chosen classifier with those features. Performance is measured over the training set. Some wrapper methods are explained:
  - A cross-validation (jackknife) method partitions the training data into N subsets. Training takes place over $N - 1$ subsets, with the Nth holdout subset used for testing. This process is repeated N times, with a different holdout subset used each time. Feature performance is averaged over the N holdout subsets. This is a weak approach for military data, particularly if the data is all collected at the same time and place, and over the same targets.
  - A bootstrap method uses a randomly chosen subset of data samples for training, and the remainder for testing. This method differs from the jackknife method in that there is redundancy in the training set from one trial to the next.
  - Boosting procedures can be interpreted as greedy feature selection processes. Many weak classifiers are combined to form a committee machine. In the training stage, training samples are prioritized according to training error. The weak classifiers, trained after the stronger classifiers, are forced to focus on the more difficult training samples. Boosting procedures can be used for feature selection. An information gain criterion is used for choosing features.[4]
  - With recursive feature elimination, for example with a support vector machine (SVM), during training, one recursively removes the feature with the least weight magnitude in the SVM solution. SVM is retrained on all remaining features after each successive feature is eliminated.

- Fully embedded methods bury the feature selection process into the trainable classifier. Deep-learning algorithms are a popular example of this concept.[5] This approach makes sense with lots of training data and little understanding of which feature types not only have high discriminating power but are likely to be robust to differences between training and operational data.

Insight and intuition facilitate both feature extraction and selection. It helps if features have physical interpretation or semantic meaning. If the task is distinguishing a T-72 tank from a T-62 tank, it helps to ask professional image analysts how they do it and choose features accordingly. The analysts might use one small spot on the target to make the decision. Features resulting from target data can be more useful than features resulting from a mixture of target and background—if it is possible to reliably get such features. If the sensor produces occasional bad pixels pegged at maximum and minimum values, then features highly affected by the pegged pixels aren't good choices. It doesn't make sense to use a feature whose meaning is distorted by the distribution of target types in the training data. For example, a hot engine is not a good IR feature if all of the vehicle types in a training set are active except for one vehicle whose engine is broken. This is a surprisingly common problem. To distinguish a person carrying a weapon from a person carrying a farm implement, it helps to know the positions and poses in which weapons are carried and fired.

One further step can be taken. Feature vectors represent points in a high-dimensional feature space. Feature vectors might map well to a lower-dimensional manifold embedded within the higher-dimensional space. Classes (or subclasses) are separated by sparsely populated regions on the manifold. Linear or nonlinear mapping techniques can be used to map high-dimensional feature vectors to lower-dimensional spaces while retaining discriminating power. Popular techniques of this type include principle components analysis, self-organizing maps, auto-encoders, and newer manifold learning techniques. Once a manifold is formed, the distance between an unknown and a known point on the manifold is measured along the surface of the manifold. This is analogous to measuring the distance between points on a cloth tape measure.

There are many different feature extraction and selection algorithms. It is not always possible to develop a sufficiently comprehensive training set to select an optimal set of features. For ATR purposes, feature sets should all be considered suboptimal. For the same data, many different kinds and subsets of features can produce similar performance. Features that work well for one set of conditions might perform poorly under different conditions. The experience of the ATR design team is paramount.

3.5 Examples of Feature Types

Simple operators are often applied to a ROI to produce a feature image. Summary statistics are computed for the ROI image as a whole, or the image is partitioned into overlapping blocks with summary statistics computed for each block. Several examples are shown in Figs. 3.6 and 3.7.

Features derived from combinations of moments were originally proposed for pattern recognition by Hu in 1962.[6] The basic moments m_{pq} and central moments μ_{pq} of order $p + q$ for image region or segmented blob $f(x, y)$ are defined as

$$\begin{aligned} m_{pq} &= \sum_x \sum_y x^p y^q f(x, y), \\ \mu_{pq} &= \sum_x \sum_y (x - \bar{x})^p (y - \bar{y})^q f(x, y); \qquad p, q = 0, 1, 2, \ldots, \end{aligned} \tag{3.1}$$

where $\bar{x} = \frac{m_{10}}{m_{00}}, \bar{y} = \frac{m_{01}}{m_{00}}$, and $(\bar{x}, \bar{y})$ is the center of grayscale mass of the region.

Features can be formed from combinations of moments.[6] Some of these features are insensitive to translation, rotation, and affine transforms. Other types of moments have been proposed, including ridgelet, Zernike, Gaussian–Hermite, Legendre, Fourier–Mellin, geometrical, and complex.

Features can be obtained by partitioning an image into overlapping blocks and then extracting features local to each block. Popular features of this type include histogram of oriented gradients (HOG) for single images and histogram of optical flow (HOF) vectors for video snippets. These will be discussed in the following subsections.

Popular descriptive features for target recognition go by such acronyms as SIFT, SURF, RIFT, PCA-SIFT, and GLOH. These feature types emphasize invariance, such as to scale and illumination.

Different kinds and operating modes of coherent imaging systems produce different kinds of information, resulting in different types of features for use

Figure 3.6 Examples of several feature types for an IR image of a jeep and a visible image of a Predator-B UAV. Feature types from left to right are raw gray scales, edge image, Laplacian image, histogram, and Fourier magnitude. (Jeep image from NVESD.Army.mil. UAV image from Grandforks.af.mil.)

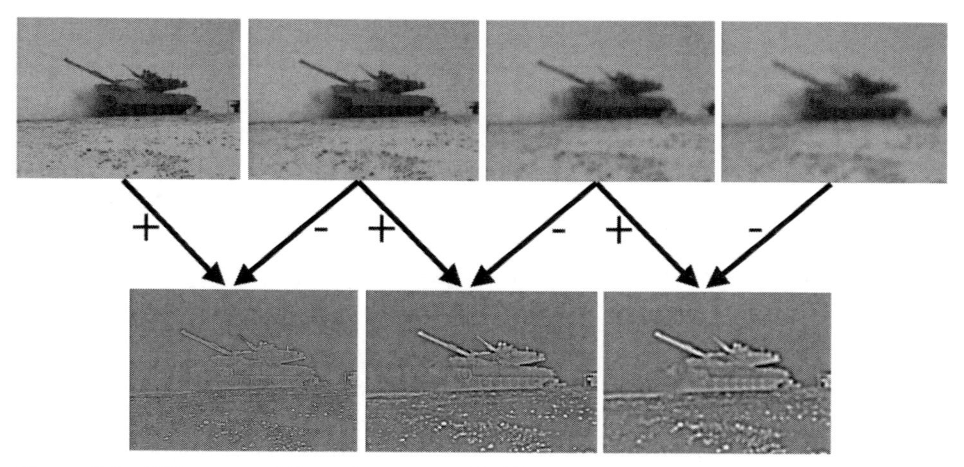

Figure 3.7 Difference-of-Gaussians pyramid. (Tank photo by Sgt. Chad Menegay, WWW. Army.mil/NewsArchives.)

by an ATR. Synthetic aperture radar, holography, and sonar produce information that is inherently complex valued. SAR image formation involves Fourier transforming in-phase I and quadrature phase Q returned components, and projecting 3D data onto a chosen 2D plane to produce a complex-valued 2D image. A complex-valued SAR pixel thus consists of a magnitude part $[M = (I^2 + Q^2)^{1/2}]$ and a phase part $[P = \tan^{-1}(Q/I)]$. The phase part is ignored for display. SAR magnitude backscatter features form an image of bright spots representing strong scatterers (e.g., from target corners). Complex-valued raw *phase history* refers to data that has not been Fourier transformed and cannot be viewed as an image. Moving targets are recognized with 1D HRR profiles. Some radar systems have the ability to send and receive energy with different polarizations. Micro-Doppler signatures result from frequency modulation of the returned radar signal resulting from a target's micro-motion.

3.5.1 Histogram of oriented gradients

The first flyable ATR, called AUTO-Q, used dedicated electronics to convert a stream of video images into a stream of gradient images.[7] Detection, segmentation, and recognition were all done using gradient vectors. Gradient vectors are once again popular using the HOG approach.[8] The image is partitioned into overlapping blocks. The gradient vectors from each block are mapped to a histogram. Histogram bins correspond to gradient directions. A concatenation of the individual histograms is used as a feature vector. As with all such approaches, variations are common. A more detailed explanation follows, focusing on how the approach can be used in an ATR.

Step 1. An image is normalized and noise-cleaned in some manner. For forward-looking imagery, the region above the skyline is normalized differently from the way the region below the skyline is normalized.
Step 2. A potential target is detected within the image using some detection algorithm.
Step 3. An appropriately sized, shaped, and scaled ROI is extracted about the detection point. The ROI should fit tightly around the detected object.
Step 4. The ROI is partitioned into N overlapping blocks.
Step 5. Each block is partitioned into M cells, where each cell is composed of an array of A pixels.
Step 6. Gradient vectors (having magnitude and direction) are computed at each pixel location within a cell. The magnitude-weighted gradients are mapped to a histogram of H bins. That is, each gradient vector maps to the histogram bin determined by its angular direction. Its contribution to the histogram is determined by its magnitude (or magnitude squared, or thresholded magnitude). Each bin corresponds to a gradient direction, e.g., {0 deg, ±22.5 deg, ±45 deg, ±67.5 deg, ±90 deg, ±112.5 deg, ±135 deg, ±157.5 deg, 180 deg} or {0 deg, 22.5 deg, 45 deg, 67.5 deg, 90 deg, 112.5 deg, 135 deg, 157.5 deg} for unsigned angles.
Step 7. The M histograms arising from the M cells within a block are concatenated to form a single histogram, which is then treated as a vector $\mathbf{v}$.
Step 8. For each block, compute the normalization factor f and use it to normalize the elements of $\mathbf{v}$:

$$f = \frac{\mathbf{v}}{\sqrt{\|\mathbf{v}\|_2^2 + \epsilon}}.$$

Step 9. Concatenate the N normalized histograms from the N blocks to form a single feature vector $\mathbf{V}$. Vector $\mathbf{V}$ feeds the ATR's classifier.

Typical values include:

- Spatially scaled ROI size: 64 pixels wide × 128 pixels high (if the candidate target is a suspected dismount). 128 pixels wide × 64 pixels high (if the candidate target is a suspected vehicle).
  - Note: For military targets at range, the spatially scaled ROI can, by design, consist of much fewer pixels than this nominal value.
- Number of overlapping blocks per ROI: $N = 8 \times 16$ or $16 \times 8 = 128$.
- Number of cells per block: $M = 4$.
- Number of pixels per cell: $A = 8 \times 8 = 64$.
- Number of histogram bins per cell: $H = 8$ or 16.
- Length of final feature vector $\mathbf{V}$: 8 bins/cell × 4 cells/block × 8 × 16 blocks/ROI = 4096 bins = 4096 elements in the feature vector.

3.5.2 Histogram of optical flow feature vector

When applied to spatiotemporal data, HOG features can be obtained using spatiotemporal gradients. Alternatively, optical flow vectors can be used in place of gradient vectors, forming the HOF method.[9,10] The HOG and HOF features are often combined to form HOG-HOF features. These features are commonly extracted at a hierarchy of scales.[10] Instead of extracting these features densely over blocks covering the entire ROI, the HOG-HOF features are sometimes extracted locally about spatiotemporal interest points (STIPs).

3.6 Examples of Classifiers

The paradigm of target classification is as follows, regardless of whether the technique is called template matching, neural, or statistical (Bayesian or frequentist), etc. Image (or signal) data plus temporally synchronized metadata is collected. A potential target is detected in the (e.g., image) data. This target is sometimes referred to as a *blob* or an *unknown*. The blob image is transformed in some manner such as centered, spatially scaled, segmented, or rotated, using the available metadata (e.g., range, roll, pitch). A feature vector (or feature image) is assembled from features extracted from the blob image. Irrelevant and redundant features are discarded. The resultant feature vector $\mathbf{X}$ represents a candidate target. To know which target $\mathbf{X}$ represents requires further processing, which is called *classification*. The candidate target is then assigned to one of r subclasses $\theta_1, \ldots, \theta_r$, where each subclass is a member of a broader class $\theta_i \in \{C_1, C_2, \ldots, C_q\}$ according to the evidence, i.e., the vector of features $\mathbf{X} = \{x_1, \ldots, x_d\}$. For example, a subclass may be T-72 tank at ~45-deg aspect angle, the broader class being T-72 tank.

Supervised classifiers require pairs of input data $\{\mathbf{X}_i, Y_i\}$ where $\mathbf{X}_i$ are the feature vectors, and Y_i are the corresponding labels for the feature vectors. The Y_i can also be expressed as a vector $\mathbf{Y}_i$ with a nonzero entry in the nth position to indicate the nth class type. Unsupervised classifiers require only the set $\{\mathbf{X}_i\}$. They map each unknown $\mathbf{X}$ to a cluster but do not assign a meaningful class label to the cluster. The samples assigned to a particular cluster have common characteristics.

Basic classification algorithms that we will cover next do not do justice to the complexity of the military target classification problem. There are many other issues. What metadata is available, and what is done with it? What errors are inherent in the metadata? How well is the metadata synched to the image data? What *a priori* information is available? How does the sensor used to collect the training data relate to the operational sensor? What are the operating modes of the sensor? Can modes (e.g., radar mode, camera field-of-view) be switched, or can other sensors be called up to get additional information when needed? Can the ATR control the platform or sensor mode to get a better look at the target? What are the roles of the classifier and the human-in-the-loop? How does the system report its results? What performance is required to successfully complete a mission? How is the mission list of target classes loaded into the system? How does the system know if the sensors are working correctly or if the weather is degrading the data? Does the data in the classifier turn the system into a classified military device? How does the system secure itself if captured?

3.6.1 Simple classifiers

There is nothing wrong with using a simple classifier. Simple classifiers are robust to unforeseen circumstances, and easy to design, test, implement, and maintain. Only when a simple classifier can't meet performance requirements should a more complex classifier be used. Even then, a simple classifier can help determine the added benefit versus cost of a more complex approach.

3.6.1.1 One-class classifiers

A one-class classifier distinguishes targets from non-targets. For a particular mission, everything but the target of interest can be considered a distraction or clutter. The assumption of the one-class classifier is that in the infinitely varying real world clutter blobs are not well described. This is especially true at high resolution rather than for point-like objects and noise-like clutter. Many target detectors can be viewed as one-class classifiers. The one-class classifier cannot provide the posterior probability of targets because information on non-targets is neither available nor convincingly assumed. That is our underlying assumption as ATR developers exploiting a rich set of features. Radar engineers often make the opposite assumption with limited target descriptors or low-resolution data.

3.6.1.2 Two-class linear classifiers

A linear classifier computes a single hyperplane to separate one class from another. Suppose that a magical long-range sensor can determine the height and weight of animals. Suppose that 13 animals are imaged. Five of them are elephants, and eight are giraffes. The 13 animals result in a training set of size $m = 13$; $D = \{(\mathbf{X}_1, Y_1), \ldots, (\mathbf{X}_{13}, Y_{13})\}$, $Y_i \in \{\text{elephant, giraffe}\}$. Each training sample is represented by a two-element feature vector; $\mathbf{X}_i = (w_i, h_i)$, where w denotes animal weight, and h denotes height. Figure 3.8(a) shows a plot of the 13 vectors represented as points in the 2D feature space. The two classes are linearly separable since any of a number of straight lines can be drawn separating them, as shown in Fig. 3.8(b). A separating line can be expressed as $aw + bh = c$. There are many possible solutions for a, b, and c, each resulting in a different separating line. Which solution is deemed optimal depends on the chosen definition of optimal. For example, linear regression defines optimality in terms of means and variances, and uses all 13 points to obtain a

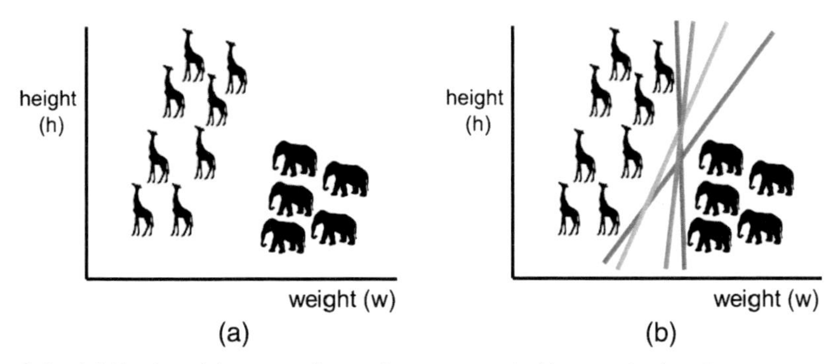

Figure 3.8 (a) Each training sample can be represented by a point in a feature space. This example is for a two-class problem, where training vectors each have two elements. (b) The two classes in this example are linearly separable by any of a number of straight lines.

solution. As we shall see shortly, an approach called support vector machine uses a different definition of optimality.

Thus, two sets of points are linearly separable in 2D space if they can be completely separated by a single line. This single line is not necessarily unique. Two sets of points are linearly separable in higher-dimensional feature space if they can be separated by a single hyperplane, where a hyperplane is just the multidimensional counterpart of a line.

Let **X** be the feature vector of an unknown object, where unknown means not yet assigned a class label. A weight vector **W** and threshold b are learned from the labeled training data D. Function f converts the biased dot product of the two vectors into the desired output. The output score of the classifier is

$$out = f(\mathbf{W} \cdot \mathbf{X} - b). \tag{3.2}$$

All output values above zero can be assigned to one class, and all values below zero can be assigned to the other class. When the bias term b is left out of this type of equation, it is assumed to be the last element of the weight vector with a 1 placed into an additional element of the feature vector.

3.6.1.3 Support vector machine

A support vector machine (SVM) constructs a hyperplane separating two classes of linearly separable points.[11] The thesis of the approach is that not all points in feature space are equally important in constructing the separating plane. The points closest to the decision surface are called support vectors. These points are circled in bold in the example of Fig. 3.9(a). They are the most difficult to classify because they are closest to points of the other class. Although many lines can separate the black and green points in Fig. 3.9(a), SVM maximizes the margin around the separating hyperplane, as shown in Fig. 3.9(b). The decision function depends only on the support vectors.

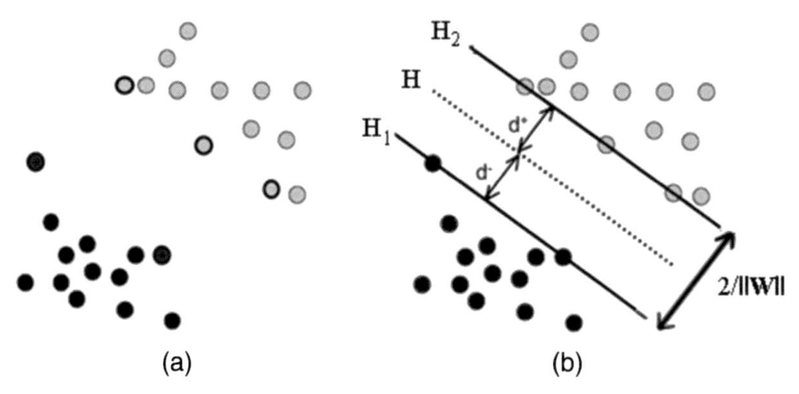

Figure 3.9 (a) Feature vectors from two linearly separable classes. The five support vectors have dark perimeters. (b) The best separating hyperplane, according to SVM, is the one with the most margin, shown here as a dashed line.

The training data set D consists of n points $\mathbf{X}_i$, such that

$$D = \{(\mathbf{X}_i, \mathrm{Y}_i)|\mathbf{X}_i \in R^d, Y_i \in \{-1, +1\}\}, \tag{3.3}$$

where Y_i is either -1 or $+1$, indicating the class of $\mathbf{X}_i$.

In Fig. 3.9(b), the points on the planes H_1 and H_2 are support vectors. The H_1 and H_2 planes are described by the equations

$$\begin{aligned} H_1 &: \mathbf{X} \cdot \mathbf{W} + b = +1, \\ H_2 &: \mathbf{X} \cdot \mathbf{W} + b = -1. \end{aligned} \tag{3.4}$$

According to the SVM method, the hyperplane H provides the best separation between the two sets of points. The distance between hyperplanes H_1 and H_2 is $2/\|\mathbf{W}\|$. Minimizing $\|\mathbf{W}\|$ maximizes the margin. The following constraint is added:

$$\begin{aligned} &\mathbf{X}_i \cdot \mathbf{W} + b \geq +1 \text{ for } \mathbf{X}_i \text{ of the first class } Y_i = +1, \\ &\mathbf{X}_i \cdot \mathbf{W} + b \leq -1 \text{ for } \mathbf{X}_i \text{ of the second class } Y_i = -1. \end{aligned} \tag{3.5}$$

The two equations can be combined to yield

$$\mathbf{Y}_i(\mathbf{X}_i \cdot \mathbf{W} + b) \geq 1 \text{ for all } i, \text{ where } Y_i \in \{-1, +1\}. \tag{3.6}$$

This is a constrained optimization problem that can be solved by the Lagrangian multiplier method. The objective is to find the hyperplanes that maximize the margin by minimizing $\|\mathbf{W}\|^2$, such that the discrimination boundary is conformed:

$$\text{Minimize } \frac{1}{2\|\mathbf{W}\|^2} \text{ such that} \\ Y_i(\mathbf{X}_i \cdot \mathbf{W}) + b = 1. \tag{3.7}$$

The problem is formulated in a dual form as

$$\text{Maximize in } \eta_i: \\ L(\eta) = \sum_{i=1}^{n} \eta_i - \frac{1}{2}\sum_{i,j} \eta_i \eta_j Y_i Y_j \mathbf{X}_i^T \cdot \mathbf{X}_j \\ \text{subject to:} \\ \eta_i \geq 0, \text{and to the constraint in } b \\ \sum_{i=1}^{n} \eta_i Y_i = 0. \tag{3.8}$$

In this formulation, the input feature vectors only appear inside a dot product. There are two extensions of SVM. Nonlinear SVM modifies the training vectors using a kernel function $k(\mathbf{X}_i, \mathbf{X}_j) = \phi(\mathbf{X}_i) \cdot \phi(\mathbf{X}_j)$ instead of $\mathbf{X}_i \cdot \mathbf{X}_j$. With good selection of a kernel function, nonlinearly separable points can be separated. However, finding such a kernel is not guaranteed. With the kernel method, the actual values of $\phi(\mathbf{X})$ need not be known as long as the dot product is known. Common inner product kernels are given in Table 3.1.

Basic SVM is a binary classifier. Multiclass SVMs are usually implemented by combining several two-class SVMs, such as with a decision tree or ensemble of binary SVM classifiers, each trained to recognize a single class or subclass.

SVMs are often compared to artificial neural networks (ANNs). There are many types of ANNs and several variations of SVM. Basic two-class SVM has a strong foundation in optimization theory, reaches a global minimum, and isn't subject to different performance each time it is re-trained on the same data. SVM code is readily available. SVM doesn't come in a bewildering number of varieties as do ANNs. A basic ANN has some good features, too. It constructs decision surfaces for multiple classes through training and then outputs a posterior probability estimate for each class. It is well suited for very large training sets and is relatively insensitive to noise and mislabeled training

Table 3.1 Several inner product kernels that can be used with SVM.[12]

Type of SVM	Inner product kernel $k(\mathbf{X}_i, \mathbf{X}_j)$	Note
Hyperbolic tangent (sigmoid)	$\tanh(\alpha \mathbf{X}_i \cdot \mathbf{X}_j + c)$	Equivalent to a two-layer perceptron network.
Polynomial (homogeneous)	$(\mathbf{X}_i \cdot \mathbf{X}_j)^p$	Power p must be chosen.
Polynomial (inhomogeneous)	$(\mathbf{X}_i \cdot \mathbf{X}_j + 1)^p$	Power p must be chosen.
Gaussian radial basis function (RBF)	$e^{-\frac{1}{2\sigma^2}\|(\mathbf{X}_i - \mathbf{X}_j)\|^2}$	Gaussian width σ must be chosen.

vectors. ANNs can be constructed to handle very large multiclass problems (>1000 classes). There are dozens, if not hundreds, of types of ANNs. As noted in Table 3.1, some versions of the SVM are a lot like some ANNs. In conclusion, which one works best depends on the nature of the data and skill of the ATR designer. With both SVMs and ANNs, it is difficult to understand exactly how and why the classifier is making its decision. It is always better to be paradigm neutral: competitively test several classifier types rather than choosing one as a leap of faith.

> "It is a capital mistake to theorize before one has data. Insensibly, one begins to twist facts to suit theories, instead of theories to suit facts." – Sherlock Holmes (Arthur Conan Doyle, 1888)

3.6.2 Basic classifiers

3.6.2.1 Single-nearest-neighbor classifier

One of the oldest classifier types is the single-nearest-neighbor (1NN) classifier. The training database D consists of pairs

$$D = \{(\mathbf{X}_1, Y_1), (\mathbf{X}_2, Y_2), \ldots, (\mathbf{X}_m, Y_m)\},$$

where $(\mathbf{X}_i, Y_i)$ is the ith input feature vector and its subclass label, $Y_i \in \{\theta_1, \theta_2, \ldots, \theta_q\}$, and $\theta_i \in \{C_1, C_2, \ldots, C_q\}$, where C_i denote broader classes. Elements of the feature vectors are assumed to be properly normalized, for example, to all fall in the range [0,1]. An r-dimensional feature vector represents a point in an r-dimensional space. The nearest-neighbor classifier stores each labeled training sample. That is, it populates the r-dimensional feature space with labeled points. This is the total extent of training. For on-line operation, a feature vector representing an unknown sample is associated with the nearest stored training sample and is assigned its label, where nearness is measured in the r-dimensional feature space. This assignment strategy is called the *nearest-neighbor decision rule*. To complete the classifier design, the notion of *nearest* must be made more explicit in the form of a distance metric or less formal as a distance measure. Several examples of distance measures follow, where $\mathbf{X}$ denotes the feature vector extracted from the as yet unlabeled sample, and $\mathbf{Z}$ denotes a labeled sample:

$d(\mathbf{X}, \mathbf{Z}) = \|\mathbf{X} - \mathbf{Z}\|_s^{1/s}$, where for the commonly used Euclidean distance $s = 2$.
$d(\mathbf{X}, \mathbf{Z}) = 1 - M(\mathbf{X}, \mathbf{Z})$, where $0 \leq M(\mathbf{X}, \mathbf{Z}) \leq 1$, and M denotes match.
$d(\mathbf{X}, \mathbf{Z}) =$ tangent distance along a manifold.[13]

The 1NN classifier can be expressed as a connectionist diagram, as shown in Fig. 3.10.

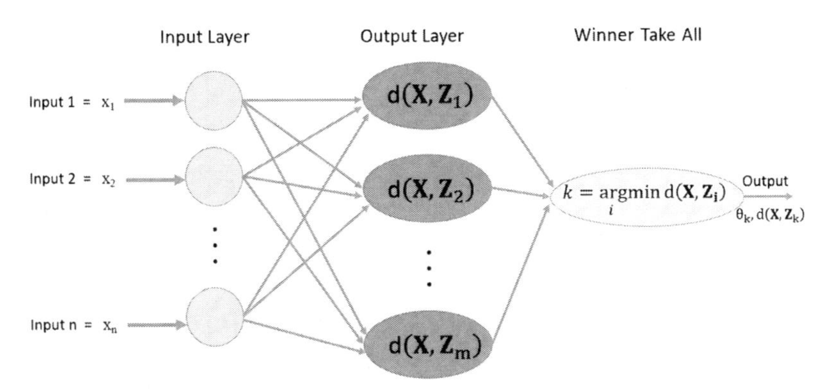

Figure 3.10 Connectionist diagram of single-nearest-neighbor classifier. **X** is the feature vector extracted from the unknown object, and $\mathbf{Z}_i$ are the stored labeled feature vectors.

The 1NN classifier classifies the training set perfectly. However, doing so does not necessarily produce a contest-winning classifier. It also matters how well the classifier generalizes to data that don't quite match training samples.

If training samples are learned templates (1D, 2D, or 3D) instead of vectors, then the distance function can be written as

$$d(\mathbf{X}, \mathbf{Z}) = \min_T [\mathbf{X}, T(\mathbf{Z})], \tag{3.9}$$

which denotes the minimum distance between **X** and a transform of **Z**, where T is a set of transformations, such as rotation, translation, and scale. Limits must be placed on the transformations, such as scale, so that, for example, a toy truck is not mistaken for a real truck. Note that template **Z** serves as a cookie cutter to cut the target embedded within **X** from its background. If instead of stored templates, a single CAD/CAM model of each target type is stored and templates are generated and transformed on-line as needed, the resulting ATR falls into the category of "model-based ATR."

The 1NN classifier is computationally intensive if each unknown has to be compared to each stored sample. There are various ways to speed up the 1NN classier. Instead of storing each training sample, the input feature vectors for each target subclass can be clustered, and then only cluster centroids (prototypes) stored. Another way to reduce the search for the nearest neighbor is to prune stored feature vectors that won't affect the search. A third approach is to store the data in a structured form, such as with a k-dimensional tree.

A k-nearest neighbors (kNN) classifier bases the classification decision on the labels of the k-nearest neighbors to the unknown. The decision can be based on a majority vote. Alternatively, the vote can be weighted by the nearness of the k neighbors to the unknown, so that nearer neighbors have a stronger vote than more distant neighbors.

Simple nearest-neighbor classifiers are becoming reasonable approaches for ATR, now that memory size, processing capacity, and programming costs are increasing. A nearest-neighbor classifier is easy to program, debug, and analyze. It fulfills the wish of "one-shot learning."

3.6.2.2 Naïve Bayes classifier

A true Bayesian classifier requires learning and storage of billions of parameters. A less-demanding approach is the naïve Bayes classifier (NBC), also known as idiot's Bayes. This classifier naïvely assumes that the input features are independent of each other. For example, a vehicle might be classified as a school bus if it is long, yellow, or has lots of windows. The NBC allows each of these features to contribute to the probability that the vehicle is a school bus, regardless of the presence or absence of any of the other features. This can be written as

prior probability

likelihood

posterior probability

$$p(C|, x_1, x_2, \dots, x_n) = \frac{p(C)p(x_1, x_1, \dots, x_n|C)}{p(x_1, x_1, \dots, x_n)}, \tag{3.10}$$

evidence

where $C = \{c_1, c_2, \dots, c_m\}$ is the set of target classes, and $\mathbf{X} = [x_1, x_2, \dots, x_n]^t$ is an input feature vector. The equation can also be written as

$$p(c_k|\mathbf{X}) = p(Class = c_k|\mathbf{X}) = \frac{p(c_k)p(\mathbf{X}|c_k)}{p(\mathbf{X})}$$

for each of the k possible classes.

It is the posterior probability that we are looking for. Since the evidence is the same for all target classes, it is the numerator of the right side of the equation that we shall focus on:

$$p(C)p(x_1, x_1, \dots, x_n|C) = \frac{1}{Q}p(C, x_1, x_1, \dots, x_n) = \frac{1}{Q}p(C)\prod_{i=1}^{n} p(x_i|C), \tag{3.11}$$

where $Q = p(\mathbf{X})$ is a scaling factor. The input feature vector $\mathbf{X}$ is classified as target type c_k by the following decision rule:

$$\underset{k}{\operatorname{argmax}}\, p\,(Class = c_k)\prod_{i=1}^{n} p(x_i|Class = c_k). \tag{3.12}$$

If the prior probabilities of target classes are unknown and assumed equal, then the term $p(Class = c_i)$ can be left out of the equation.

Suppose that we assume that features have Gaussian distribution. Let μ_k and σ_k^2 denote the estimated mean and variance for a feature x for a class c_k or, more realistically, a subclass c_k. Then,

$$p(x = h|c_k) = \frac{1}{\sqrt{2\pi\sigma_k^2}} \exp[-(h - \mu_k)^2/2\sigma_k^2]. \tag{3.13}$$

The NBC can be drawn in connectionist form. Each element of the input feature vector **X** is fed into a separate node of an input layer. Each input node feeds each output node. Each hidden node computes the probability of a target subclass. The final layer picks the strongest of these probabilities and outputs the corresponding target subclass and its probability estimate. Note that the topology of Fig. 3.11 is the same as that of Fig. 3.10.

3.6.2.3 Perceptron

The classical perceptron is a supervised method for learning a binary classifier. This model was designed into hardware in the 1950s under Office of Naval Research funds for target recognition.[14] It is an artificial neural "network" with just one neuron. During training, the error is backpropagated to the neuron to adjust its weights. It works quite well if the two classes are linearly separable. Several improved training techniques exist. The pocket training algorithm keeps the best solution seen thus far, rather than relying solely on the last training iteration.[15] Perceptron training can also be formulated to find the largest separating margin between the two classes. This *perceptron of optimal stability*, together with the kernel trick, serve as the basis of SVMs.

The perceptron makes its decision using a function of the biased dot product of a learned weight vector **W** with the feature vector **X**. If the function

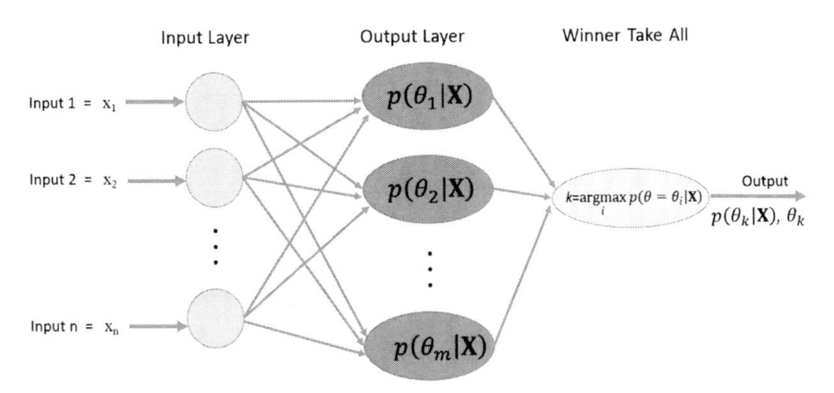

Figure 3.11 Connectionist diagram of the naïve Bayes classifier. **X** is the feature vector for the unknown input, and θ_i are subclasses. The final node just picks out the most likely subclass and can be left out of the design.

is the step function, the perceptron node outputs a binary conclusion: $\text{sgn}(\mathbf{W} \cdot \mathbf{X} - b)$. This traditional type of perceptron is called a discrete perceptron. With a smoother, e.g., sigmoidal function S, the node outputs a graded response: $S(\mathbf{W} \cdot \mathbf{X} - b)$. This type of perceptron is called a continuous perceptron. The bias term b is often left out of the equation by just considering it to be another weight and adding an additional element to the input feature vector having a value of 1.

Suppose that a single-node continuous perceptron is trained on perfectly registered and scaled images (or gradient images) of a target at a particular aspect angle, i.e., subclass θ_i, versus everything else. Once trained, its weight vector $\mathbf{W}$ will be a spatial domain template for the target. The single-node perceptron will compute $S(\mathbf{W} \cdot \mathbf{X}) = M(\mathbf{W} \cdot \mathbf{X})$, where $0 \leq M(\mathbf{W} \cdot \mathbf{X}) \leq 1$. A bank of m such nodes, one for each subclass, will form a bank of templates. With this approach, each of the weight vectors is an image that can be visualized and made sense of.

A perceptron with one input layer and one output layer is historically referred to as a single-layer perceptron network (Fig. 3.12). This confusing term stems from the fact that the input layer performs no computations and there is only one layer of links between the input and output layers. The output nodes can be trained independently of each other to form an ensemble of binary classifiers. Each classifier is trained on one particular class versus all other classes. This is called a one-versus-all (OvA) strategy. For a continuous perceptron OvA network, the strongest node output conveys the inferred target class. The output nodes of a discrete perceptron network use step functions to produce only zeros or ones. The N output nodes can be trained to represent 2^N possible categories. For example, with three output nodes, 011 would correspond to the third class.

Classifiers such as the single-layer perceptron that don't make assumptions about underlying distributions are called discriminant-based classifiers. These are popular among engineers who don't want to dig too deeply into the

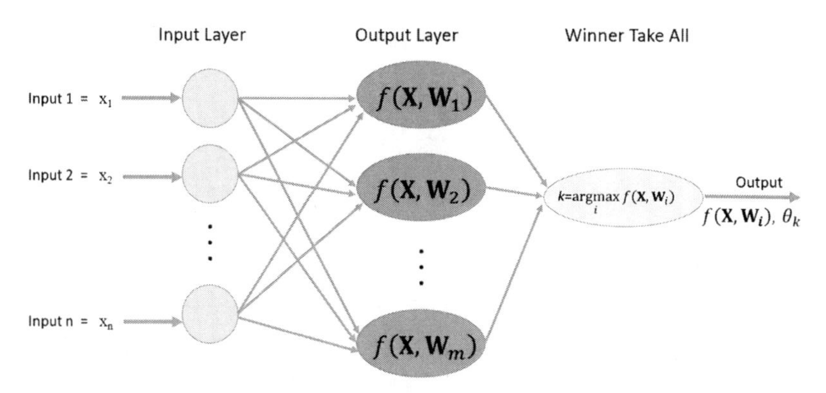

Figure 3.12 Connectionist diagram of single-layer perceptron.

statistics of the data. As will be covered later, the multilayer perceptron has supplanted the single-layer perceptron for most applications.

3.6.2.4 Learning vector quantization family of algorithms

Learning vector quantization (LVQ) is one of the many clustering-type neural networks invented by Teuvo Kohonen and others.[16] LVQ utilizes radial basis functions to delineate the feature space. The assumptions that underlie classification with basic LVQ are:

- A distance metric or, equivalently, a set of features can be selected so that sets of points clustered under the metric are in the same output subclass.
- The clusters are spherical in the selected feature space.

In LVQ the complete training algorithm consists of two phases. In the first phase, each target sublass θ_i is processed separately. A fixed number of clusters n_i are postulated for each target sublass θ_i. A set of n_i *codebook vectors* or cluster centroids is constructed and initialized to lie in the vicinity of the training vectors for target sublass θ_i. Let $\mathbf{m}_j$ be the jth codebook vector $(1 \leq j \leq n_i)$. Each training vector $\mathbf{X}$ of target subclass θ_i is selected in random order and compared with each of the n_i codebook vectors. The codebook vector $\mathbf{m}_w$ that is closest (in the distance metric) to the training vector is deemed the *winner*. The winning codebook vector is then adjusted to decrease the distance $\|\mathbf{m}_w - \mathbf{X}\|$. All of the other codebook vectors, $\mathbf{m}_j$, $1 \leq j \leq n_i, j \neq w$, are also adjusted to decrease the distances $\|\mathbf{m}j - \mathbf{X}\|$ but to considerably lesser degrees. The process is repeated with a decaying learning rate (the degree to which each training vector affects its codebook vector) until the sets of training vectors that are bound to each codebook vector stabilize. In order to discard outlying clusters (associated with anomalous data) and to develop the fine structure of larger clusters, all clusters are periodically examined. Codebook vectors that have insufficiently large following are removed. This approach flattens the distribution of the number of training vectors that are bound to each codebook vector and improves performance.

The second phase of training begins after sets of codebook vectors are found for each target subclass. The sets of codebook vectors are labeled with their respective ground truths at the broader class level and combined into a single set for supervised tuning. The tuning approach is known as *adaptive nearest neighbor*. Each training vector $\mathbf{X}$ is compared with all codebook vectors. The closest "in-class" $\mathbf{m}_{val}$ and the closest "out-of-class" $\mathbf{m}_{inv}$ codebook vectors are identified. If the ground-truth label of $\mathbf{m}_{val}$ is correct, i.e., it matches the ground-truth of $\mathbf{X}$, then no further processing is performed with vector $\mathbf{X}$, and the next training vector is examined. If the ground-truth labels do not match, then codebook vectors $\mathbf{m}_{val}$ and $\mathbf{m}_{inv}$ are adjusted such that they are closer to, and more distant from, $\mathbf{X}$, respectively. This process is iteratively repeated with diminishing adjustment. Once trained in this manner, on-line classification usually follows a nearest-neighbor paradigm.

3.6.2.5 Feedforward multilayer perceptron trained with backpropagation of error

Multilayer perceptron (MLP) networks, iteratively trained with error back-propagation (BP), are the most popular type of neural network for target recognition. As the term implies, during training error gradients propagate backward to output nodes and then to hidden nodes. Network weights are iteratively updated to minimize the output error over the training data. The training samples are shuffled in presentation order for each training epoch. Training continues until some stopping criterion is reached. The standard implementation of the MLP network shown in Fig. 3.13 uses a layer of input units, a single layer of hidden units, and a layer of output units. As usual, the input units are non-computational. The standard MLP uses a sigmoidal function in the hidden and output units of the form $f(x) = S(x) = 1/[1 + \exp(-\alpha x)]$, but many alternative nonlinear functions are suggested in the neural network literature. All connections are asymmetric, with input units feeding hidden units, and hidden units feeding output units. A simple variation also connects input units directly to output units. In extensive testing, we have found no advantage to this variation. Another variation is to inject progressively decreasing noise into feature vectors during each round of training. We have found no advantage to the injection of noise during training.

3.6.2.6 Mean-field theory networks

The class of mean-field theory (MFT) networks is large and diverse. The particular MFT networks that we tested consist of input, output, and hidden nodes, as shown in Fig. 3.14. Although the network state equations for the MFT network are nearly identical to those for the MLP network, the absence of feedback connections in the MLP test state allows identification of a

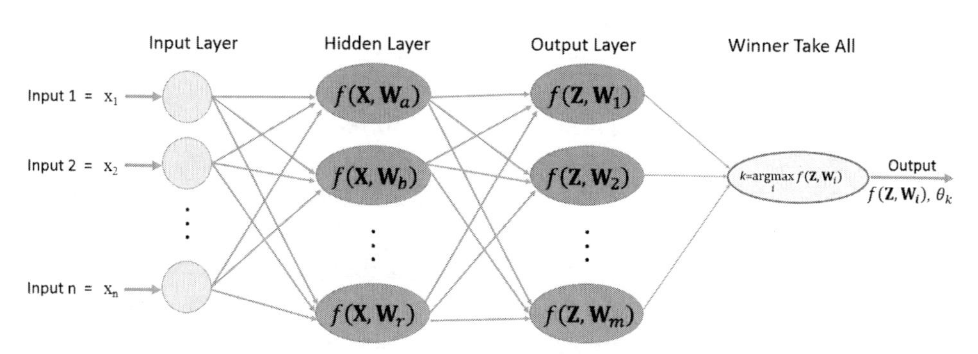

Figure 3.13 Connectionist diagram of multilayer perceptron.

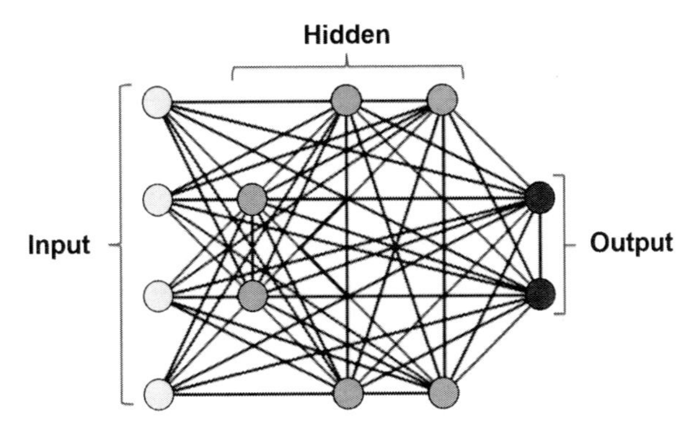

Figure 3.14 Architecture of mean-field theory neural network.

sequence of state dependencies and allows solving the system of state equations deterministically and without iteration. This is not the case for the MFT network.

The MFT input nodes serve solely to buffer input to the network. As usual, they perform no computations. The combined set of output and hidden nodes is referred to as the set of *computed nodes*. The computed nodes typically use a standard nonlinear sigmoidal activation function.

In order to guarantee network stability, i.e., convergence, we stipulate the symmetry of all connections w_{ij} between hidden and output nodes; that is, we have $w_{ij} = w_{ji}$. Connection symmetry is the hallmark of attractor networks. Since each computed node is connected to every other node, the set of equations characterizing MFT form a nonlinear system that ought to be solved simultaneously. There are several approaches, such as the m-processor parallel relaxation algorithm.

In the relaxation algorithm, nodes are partitioned into m groups and initialized. In parallel, every node from each group is selected, and its state is computed without regard to the fact that the known state values of all unselected nodes could have significant error, and the known state values of all selected nodes will be outdated as soon as the parallel computation is completed. The situation is analogous to a group of dogs simultaneously trying to catch each other's tail. Each dog moves in the direction of its target, which is also moving. After every node is updated, the newly updated states of all nodes are distributed and become available to all nodes on all m processors. The algorithm is repeated until either a convergence or non-convergence criterion is met. The exact number of iterations required for convergence is dependent on the input data. An extensive collection of techniques is used to minimize oscillation and yet retain high convergence speed. The state values of the output nodes are examined at the conclusion of relaxation. In our tests, the fully connected MFT network performed no better than the simpler feed-forward MLP network. Since the MLP network has lower latency, it is preferred for use in an ATR.

3.6.2.7 Model-based classifiers

The human brain has the ability to rotate representations of 2D and 3D objects.[17] This is called mental rotation. Even blind people have this ability. The response time is proportional to the angle that the mental object is rotated. This suggests that the mental model of the object is incrementally rotated in our brain, just as it would be in the physical world, to determine whether or not it corresponds to another image or stored model.

Following the mental rotation paradigm, one particular type of classifier has become known as the model-based approach. This classifier is exemplified by the DARPA/AFRL MSTAR algorithm. MSTAR identifies targets in SAR imagery. Radar-scattering models of the targets are derived from XPATCH® signature prediction software. XPATCH is provided with 3D target geometry in the form of a CAD/CAM model. The output is a 3D scattering representation of the target type characterized by the model. Scattering representations are used in the real-time MSTAR system. The scattering representations are rotated and projected onto the SAR image plane. Projections are used in an on-line hypothesize-and-test procedure that matches predicted target signatures against features from the image data. Figure 3.15 shows what targets look like in three different resolutions of SAR imagery.

3.6.2.8 Map-seeking circuits

David Arathorn's map-seeking circuit (MSC) is a means for efficiently matching a 3D model **M** to a 2D image **I**.[18] The MCS could be used in a model-based classifier. Following the notation of Murphy et al.,

$$c(T) = T(\mathbf{M}) \cdot \mathbf{I}, \tag{3.14}$$

where T is a particular set of transformations applied to model **M**.[19,20] $T(\mathbf{M})$ is 2D, as is the image **I**. Correspondence c is the dot product of manipulated **M** with target image **I**. The MSC manipulates the 3D model **M** with L transformations:

$$T = T_{i_L}^{(L)} \cdot \ldots \cdot T_{i_2}^{(2)} \cdot T_{i_L}^{(1)}. \tag{3.15}$$

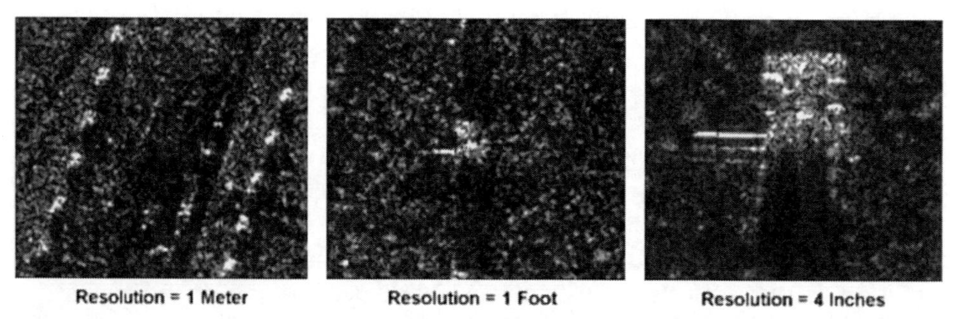

Figure 3.15 Target at three different resolutions. The key points on SAR images are the strong scattering centers. (General Atomics' Lynx® image from Sandia.gov.)

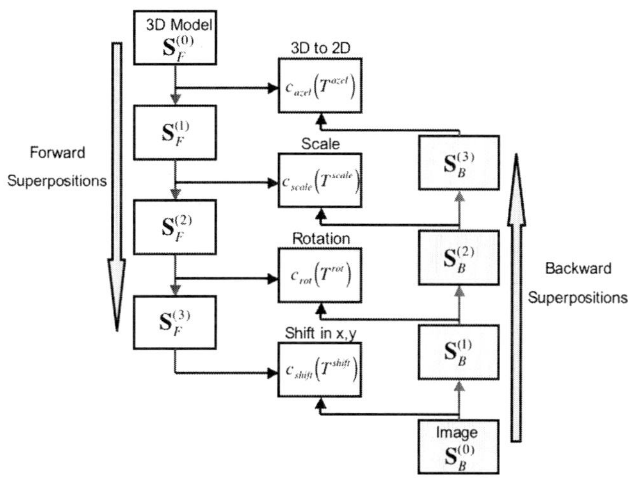

Figure 3.16 Four-layer map-seeking circuit. (Adapted from D. Arathorn.[18])

MSC is implemented as an L-level network (Fig. 3.16). The superscripts in Eq. (3.15) refer to the layer number of the network. Each layer performs a different type of transform. For example, one transformation may be 3D to 2D, another rotation, another translation, and another scale change. [Correlation or other forms of matching can be used in place of the dot product in Eq. (3.15) to eliminate the need for some of the transformations, as suggested by Overman and Hart.][21] The subscript i_K in Eq. (3.15) identifies a specific parameter value for a transformation $T^{(K)}$, e.g., the rotation of 359 deg rather than d deg, where $d = 0, \ldots, 358$.

The MSC relies on the ordering property of superposition, which states that for sparse vectors $\mathbf{v}$, the following holds:

$$\begin{aligned} &\text{For superposition } \mathbf{S} = \sum_{i=1}^{n} \mathbf{v}_i, \\ &\text{if } k \in [1, \ldots, n], \text{ and } j \notin [1, \ldots, n], \\ &\text{then } P\{(\mathbf{S} \cdot \mathbf{v}_k) > (\mathbf{S} \cdot \mathbf{v}_j)\}. \end{aligned} \tag{3.16}$$

This is just the common sense observation that the dot product of a sum of sparse vectors $\mathbf{S}$ with a vector $\mathbf{v}_k \in \mathbf{S}$ is on average higher than the dot product of $\mathbf{S}$ with a vector $\mathbf{v}_k \notin \mathbf{S}$. These vectors result from parameterized transformations of some type, not necessarily rotation, translation and scale.

The forward path through the network projects the 3D model $\mathbf{M}$ onto the image. The backward path projects the image onto the model. Back-and-forth iterations continue until just one nonzero instance of a particular kind of transformation survives at each layer of the network. The final set of values represents the computed set of transformations that provide a best match of the 3D model to the image.

3.6.2.9 Ensemble classifiers

Ensemble classifiers are constructed from a number of simpler classifiers. Ensemble classifiers are normally more robust than the best single classifier and often achieve better performance. The following types of ensemble classifiers are popular:

- A *binary classifier ensemble* decomposes the N-class problem into N separate classification decisions. Each classifier distinguishes between a single class and all other classes. With a winner-take-all strategy, the output class score is that of the binary classifier with the strongest decision.
- A democratic *committee machine* gives equal weight to the votes of the committee members, i.e., collection of classifiers. Performance is more stable than that of individual committee members.
- *Stacking* methods improve the performance by using a new classifier to correct the errors of a previous classifier.
- *Bagging* combines classifiers through weighted voting. Better performing classifiers are given higher weights.
- *Boosting* is a general method for combining weak classifiers and weak features to produce good classification results. With the popular AdaBoost algorithm, a large set of features are extracted. AdaBoost selects which features to use and how to combine them.[22] AdaBoost calls up additional copies of a classifier in sequence. The training set of each additional classifier is based on the performance of previously trained classifiers. Subsequent classifiers focus on the training samples misclassified by the previous classifiers.
- Any multistage classifier that decomposes a complex decision into a series of simpler decisions is called a *decision tree classifier*. Each non-leaf node represents a test, each branch node signifies the outcome of the test, and each leaf (terminal) node designates a class or subclass label. A *binary tree classifier* makes a binary decision at each node of the tree (Fig. 3.17).
- Random Forest™ classifiers were introduced by Breiman and Culter.[23,24] A Random Forest outputs the mode of the classes of individual decision trees. It is a weighted neighborhood scheme, akin to a kNN algorithm. In extensive testing, Random Forests performed better than stacking, bagging, and boosting techniques.[25]

3.6.3 Contest-winning and newly popular classifiers

There has been a recent explosion of interest in deep-learning (DL) networks (Fig. 3.18). DL does not refer to a specific neural network type. A DL network is just one with many hidden layers. A DL network can be hierarchical or compounded from different network types. The layers in a

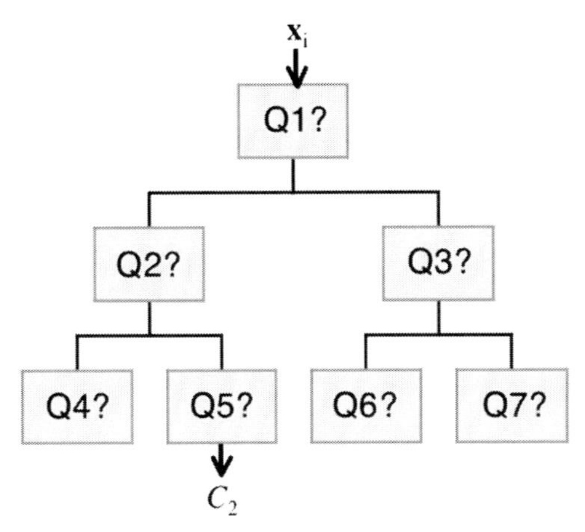

Figure 3.17 Example of a binary decision tree classifier. The answer to a question Q at each stage determines the next query.

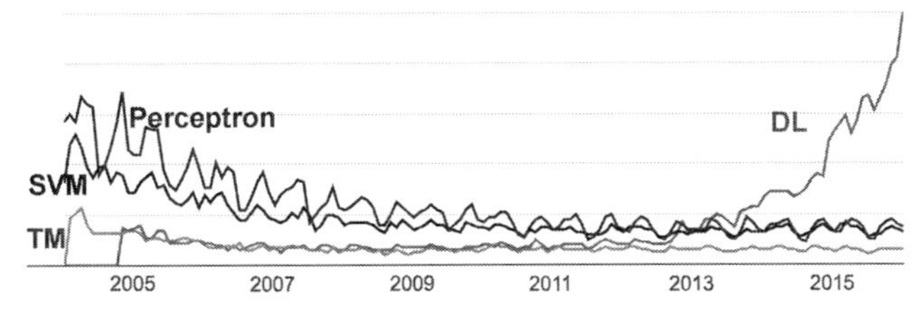

Figure 3.18 Classifier trends: Interest in several types of classifiers based on their on-line search history. Deep learning (DL) is trending up, while most other classifier types are trending down. TM denotes template matcher. (From a search using Google Trends.)

multilayer feed-forward DL network can be trained one at a time or simultaneously. Another type of DL network is a recurrent neural network (RNN). An RNN is deep over time. One philosophical difference between feed-forward DL networks and standard feed-forward neural networks is that, in its purest form, the raw input data is fed into the input layer of the DL network. The first layer or first few layers combine the raw input data into low-level features. Then the next layer combines these raw features into higher-level features, and so on. This approach makes sense if a tremendous amount of training data is available and the project engineers are befuddled as to which features to use. Academic and commercial DL networks are generally trained using graphics processing units (GPUs). Thermal cooling and latency limits in military systems suggest the use of lowest-power heterogeneous multicore processors, FPGAs or ASICs. Internet search companies are just now starting to look into FPGA solutions. Intel is working on FPGA/processor chip hybrids. It is also possible to train the network with GPUs and then download the set of weights to lower-power, real-time hardware.

3.6.3.1 Hierarchical temporal memory

One popular spatiotemporal classifier is D. George and J. Hawkins' hierarchical temporal memory (HTM), which is covered by a number of patents.[26,27] Davide Maltoni has developed a version of HTM that uses Gabor features as input as well as partially overlapping blocks at Level 1 and potentially also at higher levels (Fig. 3.19).[28] Maltoni's approach uses saccading over a test sample, which essentially randomly shifts each input sample, obtaining a separate output decision each time, and then chooses the strongest among these as the reported output. Saccading can also be used during training.

The basic operation of each node of Maltoni's HTM is given in Fig. 3.20. The approach is Bayesian belief propagation.[29]

3.6.3.2 Long short-term memory recurrent neural network

A recurrent neural network (RNN) operates over time. The most successful network of this type is called long short-term memory (LSTM).[30] LSTM learns rapidly changing temporal or spatiotemporal patterns via short-term memory and slowly changing patterns, going back many time steps via long-term memory. Its basic unit is the LSTM block.

Three gates govern a block's data flow. An input gate determines how much new content will be memorized. A forget gate determines how much old content will be forgotten. When the forget gate is nearly closed, the block's memory content will be retained over many time steps. When the forget gate is fully open, the memory will be reset. Thus, the block's memory can be expunged or updated. An output gate determines how much information escapes the block. LSTM networks most commonly have one hidden layer of LSTM blocks.

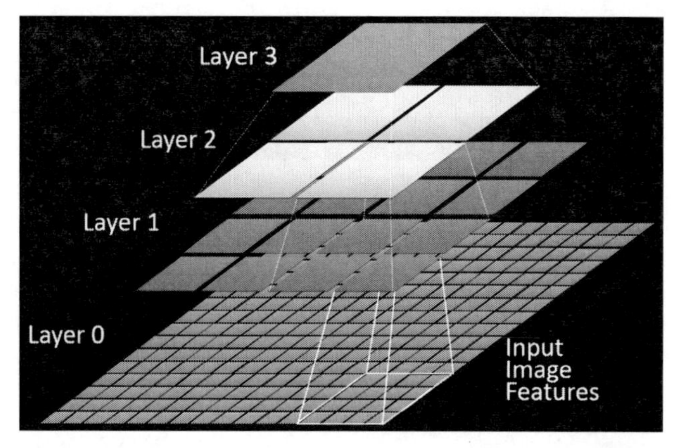

Figure 3.19 Four-level HTM. The input image is 16 × 16 pixels in this example. Level 0 has 16 × 16 input nodes, each associated with a single pixel or a single local image feature. Level 1 has 4 × 4 nodes, each of which has 4 × 4 child nodes. Level 2 has 2 × 2 nodes, each of which has 2 × 2 child nodes. Level 3 is the output layer and has a single output node, which has 2 × 2 child nodes.

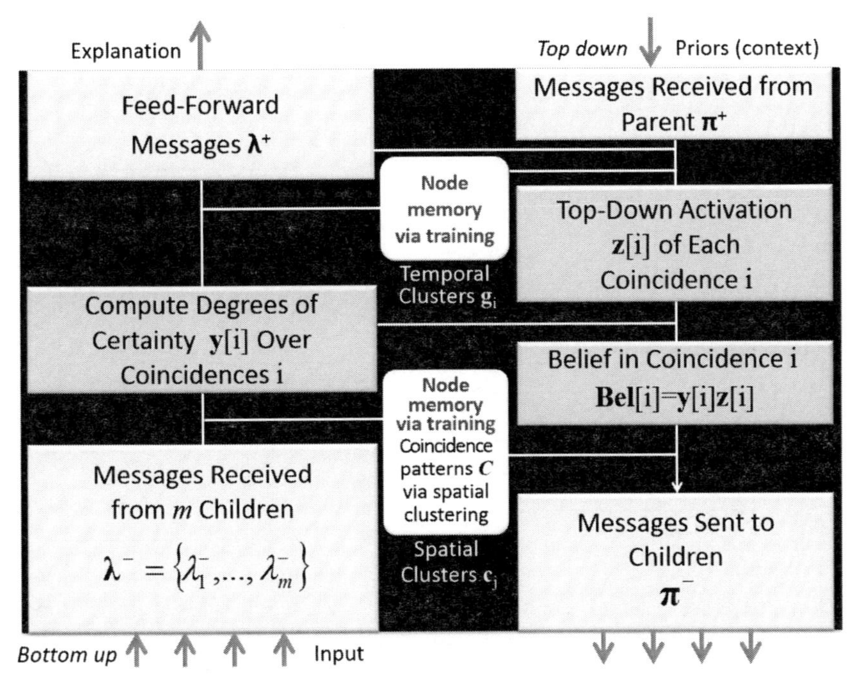

Figure 3.20 One node of Maltoni's HTM.

Network training is generally done by backpropagation of error through time. As with most neural networks, there are as many variations as there are graduate students needing thesis topics. Examples of these include variations in: training methods, internal connectivity structure within an LSTM block, activation functions, organization of blocks into a network; and combination with non-LSTM-type nodes, amalgam of LSTM and non-LSTM type networks, merger of networks operating at different time scales, bidirectional network, etc.

LSTM allows data to be stored across arbitrary time lags. Traditional LSTM has difficulty with input data streams that are not segmented into self-contained temporal subpatterns. Feedback loops called peephole connections let components inspect current internal states, allowing the duration of intervals between short-term events to be learned.[31] However, some researchers claim good performance without peephole connections.

Figure 3.21 shows a generalized version of a LSTM block, which we will call LSTM-dag, represented as a directed acyclic graph (dag) of its strongly connected components.[32,33] Unlike basic LSTM, LSTM-dag blocks can be arranged into a wide variety of architectures. This topic is covered more thoroughly in Chapter 5.

3.6.3.3 Convolutional neural network

The most popular deep-learning paradigm is the convolutional neural network (CNN or ConvNet). However, very deep CNNs, with 50-1000 layers

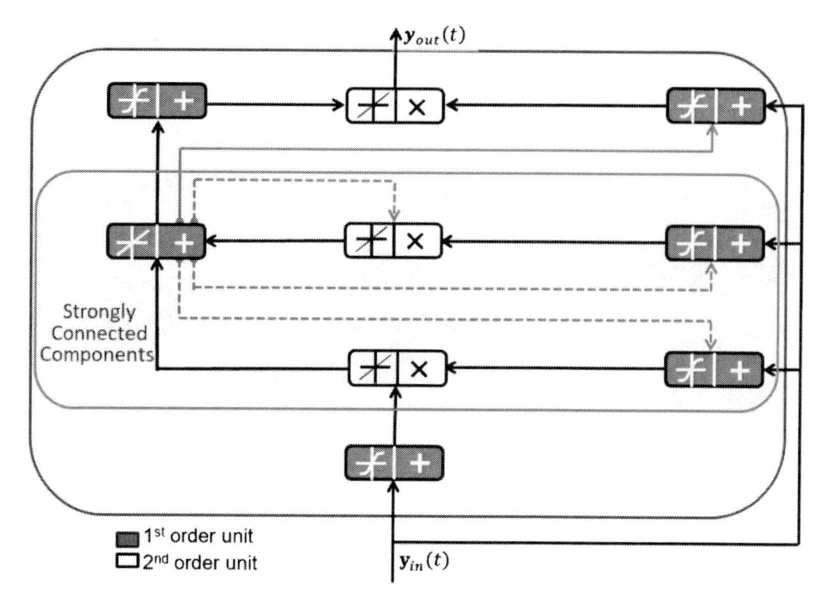

Figure 3.21 LSTM-dag block. Each edge connects two computational units. Peephole connections are left out of the equations for simplicity.

and millions or tens of millions of internal weights that must be learned, are better suited to the commercial world where training data is readily available than to military systems where data on foreign military vehicles is expensive to collect.

CNN is surrounded by much hype. It is a reasonable but not optimal classifier. It doesn't use all of the available information. For forward-looking imagery, it assumes that the same convolutions are needed for the target and its background. It assumes that the same convolutions are needed for a SAR target and its shadow. For a visible-band sensor, it doesn't make use of shadow location, which is determinable. Basic CNNs do not include mechanisms for dealing with *a priori* information, separation of target from background, known target-location-error, scale-error statistics, spatiotemporal patterns of indeterminable length, complex (magnitude, phase) data, and exploitation of metadata. Thus, CNN is a good tool for the ATR designer's toolbox but is not an all-encompassing solution as sometimes claimed.

A CNN uses a bank of convolutional filters for its front-end and a standard MLP as its back-end. Thus, it is really two types of networks stuck together. The convolutional front-end is interpreted as a trainable feature extraction machine, feeding a feature vector into the MLP. Thus, the paradigm is a familiar one: feature extraction followed by classification. In some implementations, the convolutional front-end of the classifier is trained without supervision. In other implementations, the network is trained as a unit, using supervision. It is common to transfer a CNN front-end from one application to another application.

A convolutional node's 1D, 2D, or 3D weight vector is constrained during training to be shift invariant. As it slides over the input data, point by point, a node performs the standard function of its input data and stored weight vector: $S(\mathbf{W} \cdot \mathbf{X})$. For 2D data, a weight vector might, for example, represent a 5×5 array of filter values. As the node's filter is applied to an input image, it outputs an image of equal size (ignoring border effects). That output image is called a feature map. A feature map is usually downsampled, taking the maximum value over an $n \times n$ neighborhood. Thus, the output 2D map is smaller than the input 2D map.We will denote this operation by: $S(\mathbf{W} \cdot \mathbf{X})\downarrow$. In CNN jargon, this kind of downsampling is referred to as max pooling.

For target classification, the input to a CNN is a scaled, centered, contrast-normalized ROI. A deep CNN can have multiple convolutional layers and multiple MLP layers. More commonly, there are just one or two convolutional layers. Often, each node in a convolutional layer is connected to only one or two nodes in the next layer. In the example shown in Fig. 3.22, each node in the first convolutional layer is connected to one node in the second convolutional layer. The outputs of the m nodes in the last convolutional layer are either downsampled to one value each to form an m element feature vector or are combined to form a larger feature vector. The entire network can be trained iteratively with gradient descent to minimize the discrepancy between the desired output and the actual output. As straightforward training often runs into problems, commercial companies are

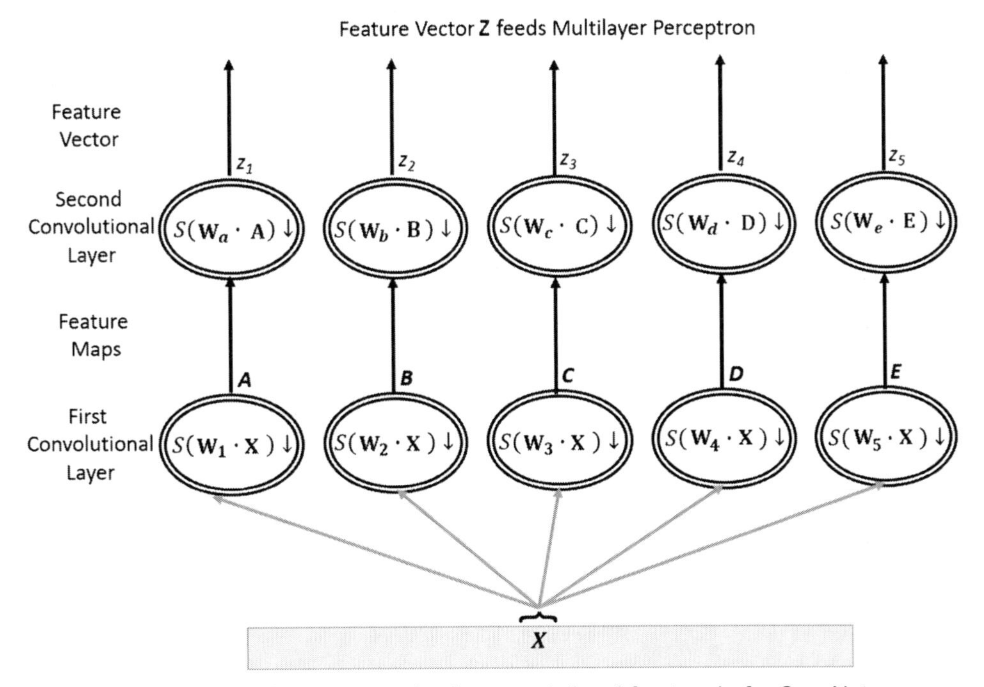

Figure 3.22 Simple example of a convolutional front-end of a ConvNet.

making huge investments into more efficient training methods. Popular improvements include "dropouts" and residual connections. The dropout method prevents overfitting by randomly removing connections during training to thin the network. Bypass, shortcut or residual connections provide pathways for sending data between non adjacent layers of the network. Commercial solutions are applied to big-data problems using tens or hundreds of millions of training samples.

An approach that we successfully flight tested in the 1990s was to individually train each convolutional node as a subclass template matcher (TM). The max pooling was done over the size of the input ROI image. The convolutional layer then outputs a vector of match values feeding the MLP network. The military processor was designed to implement Fourier transforms efficiently, so the convolutions were done using fast Fourier transforms (FFTs).

One reason for applying CNN to ATR problems is low cost. Many CNN chips are under development and many software packages are available for free.

3.6.3.4 Sentient ATR

The sentient ATR (Fig. 3.23) is a hypothetical machine that exhibits behavior at least as skillful and flexible as a trained human operator. It attends the

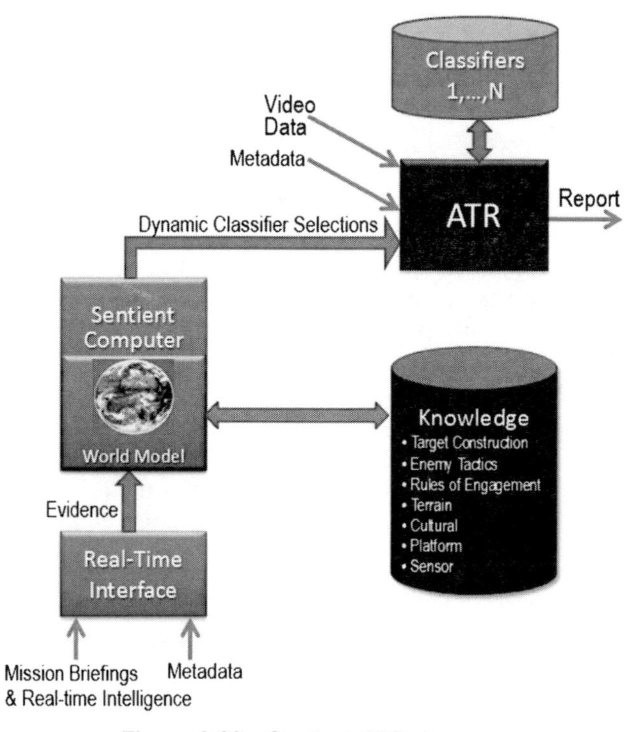

Figure 3.23 Sentient ATR diagram.

mission briefings, is in communication with the commander on the ground, and receives all forms of intelligence updates. It has a knowledge base at least as good as that of a pilot or image analyst. It uses all of this information to form and continuously update a world model. It chooses one or more classifiers from those it has stored. Classifier choice is based on all factors that it knows about, such as sensor mode, mission target list, depression angle, weather conditions, and recently received intelligence.

3.7 Discussion

Target classification consists of a training stage to *produce* the classifier model and a prediction stage that *uses* the classifier model (Fig. 3.24). This is best accomplished with an understanding of the overall problem, including sensor, platform, source of range/scale data, target set, and ConOps. Contractual details specify performance requirements. To start building the ATR, a sufficiently comprehensive set of training data (including time-synchronized metadata), test data, and sequestered blind test data need to be assembled. This is a budgetary issue. Synthetic training data can be used as a last resort, for example, to supply some data for "denied targets" or to augment real data, or as necessitated by financial constraints. A set of features needs to be found that concisely explains the differences between object classes. The ATR engineer's skill is revealed in the choice of features. Use of a classifier with built-in feature extraction is sometimes acceptable as long as the engineering team has a good grasp of how the classifier is making its decisions. If the engineers don't understand what the classifier is learning, a visible-band DL classifier might, for example, make heavy use of the color of the flowering bushes that happened to be behind the targets in the training samples. Or, it might make use of the orange flags placed in the ground to indicate where targets are to be parked.

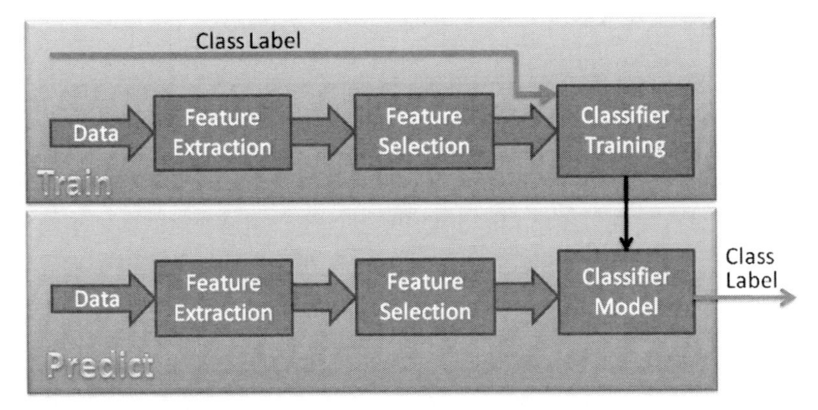

Figure 3.24 The typical target classifier is trained offline. Once trained, it can be used in a real-time ATR.

The next step is choosing a classifier (model) for distinguishing among target classes and usually also clutter. Models based on signal processing, popular in the defense electronics literature, are typically unsuitable for high-resolution targets, complex backgrounds, widely varying conditions, and exploitation of semantic concepts. Models based on the human brain are immature and simplistic, since not enough is known yet about how the brain operates. Real neurons are extremely numerous, extraordinarily complex, varied in design, and highly interconnected. However, human perception models have long been a source of good ideas for machine perception. Biological perception provides a roadmap for ATR advancement.

A classifier is developed for the express purpose of making a prediction of what will be encountered in the future. As the Danish physicist Neils Bohr noted: "Prediction is very difficult, especially about the future." The operational military situation is likely to be unpredictable. Uncertainty results from not knowing what the enemy might do or what the weather will be. There are risks that the ATR designers are not aware of—the so called "unknown unknowns." All models are imperfect and assumption laden. Models are gross oversimplifications of reality. But, models are necessary for ATR design.

The main challenge in classifier selection is identifying the classifiers that offer the best predictive performance and least chance of catastrophic failure. It is not obvious how this can be done. Complex models provide a better fit to the training data than simple models. But, complex models may result in overfit; that is, they may confuse noise for signal. Complexity is the enemy of clarity. Occam's razor suggests that everything else being equal, simple models should be favored over complex models. The No Free Lunch theorem advises that the best classifier for one type of data will not necessarily be best for another type of data.

Classifiers based on Bayesian inference need informative priors. They need to know the probability of encountering target classes and conditions. This is a belief or prediction about the future battlespace—without a crystal ball. Thus, Bayesian inference is all about the analysis of beliefs. Because information about the future is unavailable, Bayesian approaches have a degree of subjectivity. Alternatively, frequentist methods can be used. These methods meticulously rely on the information in the training data. The idea is to collect a *sufficiently comprehensive* training set so that classification error goes to zero. The frequentist philosophy is that the future will repeat the past. This concept is more convincing for sensors that produce fairly consistent data, such as radar, LADAR, and sonar and less so for EO/IR and acoustic sensors.

Most frequentist classifiers have a Bayesian counterpart. A Bayesian classifier loses much of its "Bayesian-ness" if it is assumed that all target classes are equally likely. A neural network can be based on a Bayesian or frequentist philosophy. Some neural networks are just connectionist models of standard statistical classifiers. Hybrid approaches are common. Other approaches to classification exist, such as: problem-specific methods, logistic regression, transductive inference, advanced correlation filters, expert systems, genetic algorithms, and spiking neural networks. Model selection requires deep thinking about the problem and evaluation of alternatives. This chapter covers only a sampling of approaches.

The performance of a target classifier is limited by SNR and true spatial resolution (ground-resolvable distance). Classification is not possible with too much noise, too few pixels on target, and too little observable detail. System designers, whose job is to minimize size, weight, power, and cost, often offer gross overestimations of the performance that can be achieved with particular sensors. The world's best classifier is of no use if the target is inadequately represented in the data.

Good engineering practice requires a well-designed test plan and key performance parameters agreed to by all stakeholders. Risk reduction starts with laboratory blind tests conducted by independent organizations. It continues with field trials.

Choosing a classifier comes down to several important issues:

- The size of the training set: The rule is, the smaller the training set, the simpler the classifier.
- Insensitivity to differences between training and testing data: The ultimate test set is enemy targets imaged live during combat. These may look different from laboratory training data due to modifications of targets (e.g., external fuel tank, metal skirts), variations in positions of articulated parts (turret rotated, gun raised, hatch open), objects on targets (cargo on open flatbed of truck, fuel drums, branches on vehicles, soldiers sitting atop vehicles, bundles tied to vehicles), weather conditions (sun heating one corner of vehicle, splattered mud), natural obscuration (tall grass, tree in front of vehicle), terrain variations (tilt of ground, rough vegetated terrain, large rocks, background features blending into target), unexpected down-look angle, uncalibrated, aging, or modified sensor, objects next to the target, degraded visual environment, long range, etc. Inability to predict or model the exact appearance of a target in combat suggests the use of a simpler, less highly tuned, classifier.
- Hardware limitations: The time between initial design and service release is extensive for military ground vehicles, ships, and aircraft. Processors used in military systems are often years behind those available in civilian life. Even if a more recent processor is used, the clock rate is generally reduced to lower heat dissipation. Advances in technology will eventually diminish this problem. Platforms will one day be more autonomous, and ATRs more "brain-like."

- Software limitations: It is extremely expensive to write, regression test, and document software for a military system. This includes heterogeneous multicore processors, FPGAs, and programmable application-specific integrated circuits (ASICs). The contract specifies which software standards must be followed. After a system is delivered, it may be uncertain who will correct future software bugs and who will pay for this work. This suggests the use of a simpler, more transparent classifier.
- Data rate: Performance in frames-per-second or ground area covered per second is specified by contract and must be met. This suggests not using the most elaborate algorithms envisioned.
- Metadata: An approach that can use all available metadata has advantages over approaches that rely solely on single-sensor data. Examples of metadata include digital terrain elevation data, weather, inertial navigation system (INS) and GPS information, time-of-day and season, laser range, road networks, target handoff from other sensors, and military intelligence. Approaches that can, for example, control sensor mode, call up data from another sensor (perhaps on another platform), or control ownship to get a better look at a target have obvious advantages.

What has worked in the past? Let us consider popular techniques by decade:

- 1970s: Statistical pattern recognition
- 1980s: Template matching, advanced correlation filters (including optical correlators)
- 1990s: Combinations of template matchers and neural networks (including human vision models and model-based approaches)
- 2000s: Support vector machine
- 2010s: Deep learning

Whether these techniques were popular because they performed well or performed well because they were popular (i.e., were well-funded) is unclear. Research funds go to techniques buoyed by the most hype. Classification schemes that achieved stable performance in the past include a committee of neural networks, each fed different feature data, combined with a template matcher (Fig. 3.25). A shallow convolutional neural network worked well as a one-class classifier. A deep-learning CNN appears well-suited as a rudimentary multiclass classifier, but one must consider the cost of collecting the huge data set required to properly train it.

In the decade starting in 2010, commercial companies have spent orders-of-magnitude more on research into object classification than defense contractors have spent. Search engine and social media companies are using CNNs.[34,35] LSTM recurrent neural networks have won many international

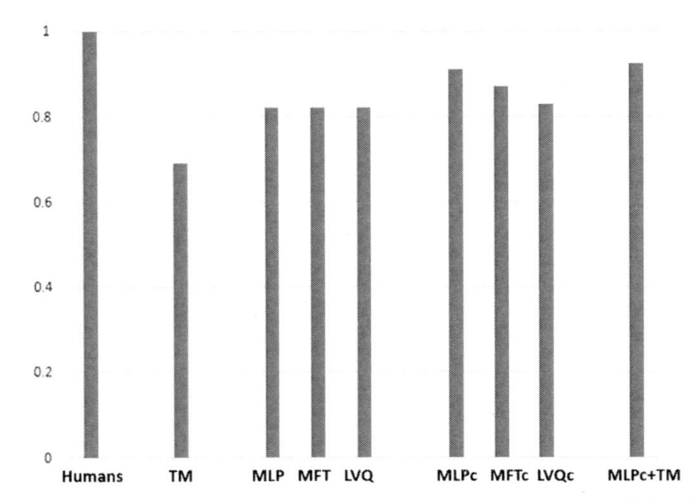

Figure 3.25 Results of classifier performance tests in the mid-1990s. Test details: performance results normalized by average performance of four trained human observers, LWIR data, 328 test images, 3 target classes, ±20% scale error, low grazing angle, wavelet features, and very large training set. TM denotes edge-vector-based template matcher. Subscript c denotes weighted vote of committee of 6 classifiers, each tuned to a different spatial frequency band.

competitions on spatiotemporal data.[35] Self-driving cars and auto safety will require algorithms, sensors, processors, and electronic packaging similar to those used by the defense industry. The automotive industry accounts for 3.5% of U.S. gross domestic product (GDP), about the same as total defense expenditures. The Internet economy is about 5% of GDP. The ATR designer needs to monitor commercial developments.

The usual ATR approach is to train the classifier off-line, and then test the classifier on-line in real-time, in platform or ground-station, (Fig. 3.24). Hundreds of different classification algorithms and schemes have been reported to perform well. A new project usually only has the time and money to competitively test a few candidates. One question is: How strong is the evidence supporting a particular classifier paradigm?

Weak evidence: The vast majority of the classifiers reported to perform well in the engineering literature make their claim based on self-test. The rationale for the publication of a technical paper is that an advancement has been made in the state-of-the-art. The developers invariably compare their well-tuned favorite new paradigm to less-tuned alternatives. There is often an unclear or dubious distinction between training data and testing data. Often, the favorite algorithm is tweaked each time it is applied to the test data, with only the best results reported. Rarely is the training data as different from the test data as it will be during an actual military operation. Nearly all of these experiments ignore the types of metadata that are available on a military platform.

Some of the startup companies reporting good classification results are seeking venture funds or are positioning themselves to be bought out by a larger company. In evaluating claimed results, one must take into account the degrees of freedom of the data. Is the data affected by shadows, paint, season, time-of-day, terrain background, nearness to other objects, target operating history, atmospheric conditions, etc.? In general, data from active sensors have fewer degrees-of-freedom than data from passive sensors.

As an example, suppose that training data is collected with a visible-band hyperspectral sensor one day and the next day test data is collected with the same sensor and platform from the same altitude and depression angle at the same time of day against the exact same vehicles near the same spot. The resulting training and testing data will be unrealistically similar. MWIR data is more affected by shadows, dust, smoke, and fire than LWIR. IR data in general is affected by a vehicle's operating history. SAR data is not affected by solar shadows, fires, engine temperature, and time-of-day. The radar is the illuminator. But SAR and HRR radar data are affected by vehicle motion. Camouflage netting will cloak better against certain sensors than others. The premise that the battlespace will be composed of a closed set of well-characterized military vehicles might be wrong-headed to begin with. Insurgents might use an open set of civilian vehicles, occasionally capturing a military vehicle, or might be on foot, blending into the local population. An unclear mix of civilian and military vehicles suggests looser classification categories of small vehicles, medium-sized vehicles, large vehicles, and dismounts with large weapons.

Stronger evidence: Algorithms winning numerous civilian tests such as CNN and LSTM are worthy of consideration.[34,35] The parallel Random Forest classifier achieved best performance in an extensive test of 179 classifiers from 17 families.[23] Ensemble classifiers that can readily adapt to different target sets without excessive retraining might be appropriate for certain ConOps. Laboratory blind tests and field tests by a well-qualified military T&E organization following a well-designed test plan provide strong evidence.

Competitive tests provide better evidence than tests of a single contractor's approach. But it must be realized that such tests will never be completely fair, since some contractors will also be the sensor, platform, or processor supplier; others may be working closely with the customer for years, repeatedly testing at the customer's test sites and helping to formulate the test plan. The classifier winning a competition might not necessarily be the best paradigm. It may just be the one that benefitted from the biggest investment in time and money. It may be the one that tested the most times on the same data, with parameters tweaked each time. The winning algorithm might be the luckiest, in that its training data by chance matched the character of the test data. Only a peek into the code will determine if the winning approach is true to the named paradigm.

The military target classification problem differs from the academic problem in that the enemy is actively trying to defeat detection and recognition. Battlespace targets and conditions have a high degree of volatility and unpredictability. So, even what appears as strong evidence might not be as strong as seemed. This doesn't mean that target classification is a hopeless problem. It just means that sound engineering practices are required.

In conclusion, once again, ATR is a system design problem, not an algorithm design problem.

References

1. D. H. Wolpert and W. G. Macready, "No free lunch theorems for optimization," *IEEE Transactions on Evolutionary Computing* **1**(1), 67–82 (1997).
2. S. Watanabe, *Knowing and Guessing: A Quantitative Study of Inference and Information,* New York, John Wiley & Sons, pp. 376–377 (1969).
3. Y. C. Ho and D. L. Pepyne, "Simple explanation of the no free lunch theorem and its implications," *Journal of Optimization Theory and Applications* **115**(3), 549–570 (2002).
4. S. Das, "Filters, wrappers and a boosting-based hybrid for feature selection," *Proc. 18th International Conference on Machine Learning*, pp. 74–81 (2001).
5. Y. Bengio, "Learning deep architectures for AI," *Foundations and Trends in Machine Learning* **2**(1), 1–127 (2009).
6. M. K. Hu, "Visual pattern recognition by moment invariants," *IRE Transactions on Information Theory* **8**, pp. 179–187 (1962).
7. B. Schachter, "A survey and evaluation of FLIR target detection/ segmentation algorithms," *Proc. DARPA Image Understanding Workshop* (1982).
8. N. Dal and B. Trigs, "Histogram of oriented gradients for human detection," *IEEE Conference on Computer Vision and Pattern Recognition* **1**, 886–893 (2005).
9. B. D. Lucas and T. Kanade, "An iterative image registration technique with an application to stereo vision," *Seventh International Joint Conference on Artificial Intelligence* **81**, 674–679 (1981).
10. I. Laptev, M. Marszalek, C. Schmid, and B. Rozenfeld, "Learning realistic human action from movies," *IEEE Conference on Computer Vision and Pattern Recognition*, 1–8 (2008).
11. C. Cortes and V. Vapnik, "Support vector networks," *Machine Learning* **20**, 273–297 (1995).

12. R. C. Souza, "Kernel functions in machine learning applications," *March* 17, 2010; Web: http://crsouza.blogspot.com/2010/03/kernel-functions-for-machine-learning.html.
13. S. Rifai, Y. N. Dauphin, P. Vincent, Y. Bengio, and X. Muller, "The manifold tangent classifier," in *Advances in Neural Information Processing Systems* (NIPS 2011 Proceedings), pp. 2294–2302 (2011).
14. F. Rosenblatt, "The Perceptron—A Perceiving and Recognizing Automaton," Report 85-460-1, Cornell Aeronautical Laboratory (1957).
15. S. I. Gallant, "Perceptron-based learning algorithms," *IEEE Trans. Neural Networks* **1**(2), 179–191 (1990).
16. T. Kohonen, *Self-Organizing Maps*, Springer, Berlin (1997).
17. R. N. Shepard and J. Metzler, "Mental rotation of three-dimensional objects," in *Cognitve Psychology: Key Readings*, D. A. Balota and E. J. Marsh, Eds., Psychology Press, New York, 701–703 (2004).
18. D. Arathorn, *Map Seeking Circuits in Visual Cognition, A Computational Mechanism for Biological Machine Vision*, Stanford University Press, Stanford (2002).
19. P. K. Murphy, P. A. Rodriguez, and S. R. Martin, "Detection and recognition of 3D targets in panchromatic gray scale imagery using a biologically inspired algorithm," *IEEE Applied Imagery Pattern Recognition Workshop*, 1–6 (Oct. 2009).
20. C. K. Peterson, P. Murphy, and P. Rodriguez, "Target classification in synthetic aperture radar using map-seeking circuitry technology," *Proc. SPIE* **8051**, 805113 (2011) [doi: 10.1117/12.884015].
21. T. L. Overman and M. Hart, "Sensor agnostic object recognition using a map seeking circuit," *Proc. SPIE* **8391**, 83910N (2012) [doi: 10.1117/12.917640].
22. Y. Freund and R. E. Schapire, "A decision-theoretic generalization of on-line learning and application to boosting," *Journal of Computer and System Sciences* **55**(1), 119–139 (1997).
23. L. Breiman, "Random Forests," *Machine Learning* **45**(1), 5–32 (2001).
24. A. Liaw, *Documentation for R. Package Random Forest*, software package and documentation available on-line, regularly revised.
25. M. Fernandez-Delgado, E. Cernadas, S. Barro, and D. Amorim, "Do we need hundreds of classifiers to solve real world classification problems?" *Journal of Machine Learning* **15**, 3133–3181 (2014).
26. D. George, "How the Brain Might Work: A Hierarchical and Temporal Model for Learning and Recognition," Ph.D. thesis, Stanford University, Stanford, California (June 2008).
27. D. George and J. Hawkins, "A hierarchical Bayesian model of invariant pattern recognition in the visual cortex," *Proc. Int. Joint. Conf. on Neural Networks*, 1812–1817 (2005).

28. D. Maltoni, *Pattern Recognition by Hierarchical Temporal Memory*, Universita degli Studi di Bologna DEIS Biometric System Laboratory, Technical Report, Bologna, Italy (April 13, 2011).
29. J. Pearl, *Probabilistic Reasoning in Intelligent Systems: Networks of Plausible Inference*, Second edition, Morgan Kaufmann Publishers, San Francisco (1988).
30. S. Hochreiter and J. Schmidhuber, "Long short-term memory," *Neural Computation* **9**(8), 1735–1780 (1997).
31. F. Gers, "Long Short-Term Memory in Recurrent Neural Networks," Ph.D. thesis, University of Hannover, Hannover, Germany (2001).
32. D. Monner and J. A. Reggia, "A generalized LSTM-like training algorithm for second-order recurrent neural networks," *Neural Networks* **25**, 70–83 (2012).
33. K. Hwang and W. Sung, "Single stream parallelization of generalized LSTM-like RNNs on a GPU," *arXiv* preprint arXiv:1503.02852 (2015).
34. "Results of Large Scale Visual Recognition Challenge 2014 (ILSVRC2014)," http://image-net.org/challenges/LSVRC/2014/results. [Updated regularly.]
35. J. Schmidhuber, "Deep learning in neural networks: An overview," *arXiv* preprint arXiv:1404.7828 (2014).

Chapter 4
Unification of Automatic Target Tracking and Automatic Target Recognition

4.1 Introduction

The owl tracks the mouse, not the blowing leaf. The outfielder tracks the fly ball, not the bird flying by. The jet fighter tracks the strategic missile launcher, not the school bus. Tracking is inexorably tied to object recognition. It is important to track the pickup truck headed straight toward the tactical operation center, not a similar truck headed into the farm field. So, it is not only the identity of the object that is important, but also the activity that the object is engaged in.

This chapter is not the usual tale of tracking point-like targets. The subject being addressed is whether the automatic target tracker (ATT), the automatic target recognizer (ATR), and the activity recognizer (AR) should be treated as independent cooperating modules or should be fused together so tightly and so well that their distinctiveness becomes lost in the merger. The latter approach has historically not been the case outside of biology and a few academic papers. There are a lot of open questions that need to be tackled in the years to come. Is it the low-level statistics that are important or the high-level semantics? Or, to put it another way: Does every picture tell a story? Is tracking the end goal or is it an intermediate task leading to motor control, as in all biological systems? Should single-actor activity recognition be treated as no more than a natural temporal generalization of target recognition? Can complex multi-actor scenarios be discerned using queries to a track file database? Should the ATR and ATT be designed by independent groups, which is often the case, or are they best not considered as separate entities in the broader system design? We will reflect on these issues.

The concept of automatic target tracking arose with the invention of radar systems. Radar returns are processed to produce a collection of candidate targets. These raw detections result from target-like objects, background noise, and spurious returns. In radar terminology, sensors provide *measurements* to the tracker. The term *observations* sometimes refers to basic measurements plus more complex target data. Different kinds of radar systems/modes produce different kinds of data. The measurements are reported in successive recording intervals or scans. The update rate is not as fast as for an EO/IR sensor. The radar tracker cannot rely on visual intelligence (pictorial data) as does a video-based tracker. Traditionally, radar-based tracking research and development focused on *where* rather than *what*.

A radar target tracker updates a track by forming a weighted average of the current detected position and the last predicted position, both of which have unknown errors. The issue is how to deal with the errors. There are essentially two kinds of errors: measurement errors and uncertainty within the target motion model that is being used. A radar target state tracker develops a filtered (smoothed) estimate of the current and predicted target state, where state can include not only position, but also velocity, acceleration, etc. Thus, the tracker estimates the kinematics of a potential target based on consecutive radar observations. In tracking jargon, the object being tracked is commonly referred to as a *target*, rather than a raw detection, even when it is not known whether the object tracked is on the list of military targets being sought. A detector with good performance (low target location error, high P_d, low P_{fa}) helps a tracker to perform well. If the detection threshold is extremely low, targets may be declared only after they are tracked. This is referred to as *track before detect*. Thus, *target detection* (or detection track) can be said to occur in a radar tracker only after a track has been convincingly established.

The underlying assumption in radar trackers is that the situation can be modeled. The problem is treated as one of statistics and signal processing. Estimates are updated with subsequent observations. The underlying assumption in video-based trackers is that the real world is extremely complex, infinitely variable, and difficult to model. The situation is affected by time of day, season, sun angle, camouflage, cloud shadows, dust trails, target operating history, enemy tactics, 3D scene structure, range error, atmospheric turbulence, and attenuation, just to name of few operating conditions and complexities. Algorithm choice is best based on testing over large volumes of data.

To combine an ATR and an ATT in a serious way, we need to have both a recognition problem and a tracking problem. For EO/IR imagery there has to be sufficient pixels across the critical dimension of the target to recognize it. To also have a tracking problem, the target has to move in a nontrivial way from one frame time to the next. With a FLIR camera operating at 120 frames per second, a human or ground vehicle will overlap its own image from one frame to the next. However, by tightly combining ATR and ATT, we can violate the usual rules-of-thumb of what it takes to recognize a target and what it takes to track a target.

An EO/IR target detector locates objects of possible interest on successive frames of imagery. These potential targets are referred to by the ATR community as raw detections, pre-screener outputs, or "blobs"—until a classification decision is made via further processing. An EO/IR target detector reports the positions of its detections in image coordinates, which may eventually be converted to world coordinates. Unlike radar-based tracking, traditionally, ATR research and development focused on *what* rather than *where*.

Target tracking algorithms have one overriding objective: to correctly *associate* a new detection with an existing track, while not being distracted by clutter or noise (Fig. 4.1). This is also referred to as observation-to-track assignment. Association keeps a track alive. Traditional trackers only track moving point-like objects. Trackers tied to EO/IR ATRs commonly track both stationary and moving objects, from stationary or moving platforms, where the object is more substantially represented than just a point. Future unmanned platforms with extremely high-data-rate sensors and relatively low-bandwidth data links will need to transmit actionable intelligence and video snippets rather than a fire-hose-like stream of raw data. Future, more intelligent, ATRs will need to meticulously deal with the *what*, the *where*, the *when*, and the *why* problem.

Just because objects are being tracked doesn't necessarily mean that they are targets of military interest. They can be four-legged animals, birds, swaying trees, dust devils, smoke, contrails, cloud shadows, clouds themselves, farm tractors, or tumbleweed. They can be kids carrying lacrosse sticks rather than insurgents carrying rifles. Nonetheless, object motion is an important clue that a detected object might be of military interest. Trajectory

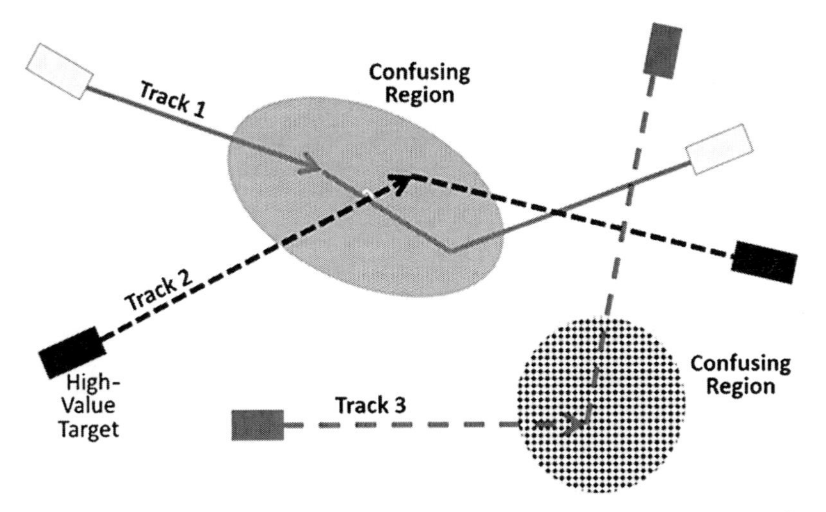

Figure 4.1 During confusing periods of surveillance, the tracker attempts to determine which detection to assign to which track.

is sometimes the only information required, for example, in the case of inward bound missiles.

Detection and tracking are both plagued by the same problems: noise, clutter, occlusion, bad data samples, unknown target densities, variable target and background appearance, decoys, countermeasures, and inability to understand the gist of the scene. Tracking has the additional problems of move-stop-move motions, targets moving in and out of the FOV, and targets crossing each other's paths. An EO/IR sensor may have variable search patterns in cases where a target is viewed for a short time; the sensor is then slewed in another direction, only to later revisit the target area.

A tracker may be a separate entity operating independently of an ATR as is usually the case. Or, as we suggest, a tracker could be integral to the ATR, fully fused with ATR code. Combining tracking with ATR functions can improve detection, recognition, pose estimation, and track quality, while reducing false alarms. A tracker can also be tightly integrated with sensor and platform controls, such as in the case of a closed-loop tracker. Here, the tracker predicts target location so that the control system can keep the tracked target at the center of the FOV. A missile seeker uses closed-loop tracking. The human visual system keeps a tracked target in the foveal area of the retina.

Target tracking requires multiple frames of data in EO/IR imagery or multiple observations (detection data, measurements of some type) with other kinds of sensors. There are many different kinds of tracking problems involving different types of targets and different types of sensors. Air vehicles move freely, while ground vehicles must maneuver around obstacles and are limited by terrain, sometimes following road networks with intersections, merging lanes, bridges, jay walkers, stop signs, and traffic lights. The problem of tracking incoming munitions is different from that of tracking submarine periscopes or dismounted soldiers. The non-cooperative tracking problem differs from the GPS-enabled Blue Force (friendly forces) tracking problem. There are various kinds of comparable nonmilitary problems, such as tracking whales, asteroids, pedestrians, eyes and faces (Fig. 4.2). Civilian air traffic control is largely a tracking problem. Despite the differences, there are some common elements in algorithms and mathematics used to solve the various kinds of tracking problems.

4.2 Categories of Tracking Problems

Tracking problems can be categorized in various ways according to different criteria. These criteria are explained in the next four subsections.

4.2.1 Number of targets

A single-target tracker is focused on tracking the object of highest interest. The objective is to continually associate the single track with the correct

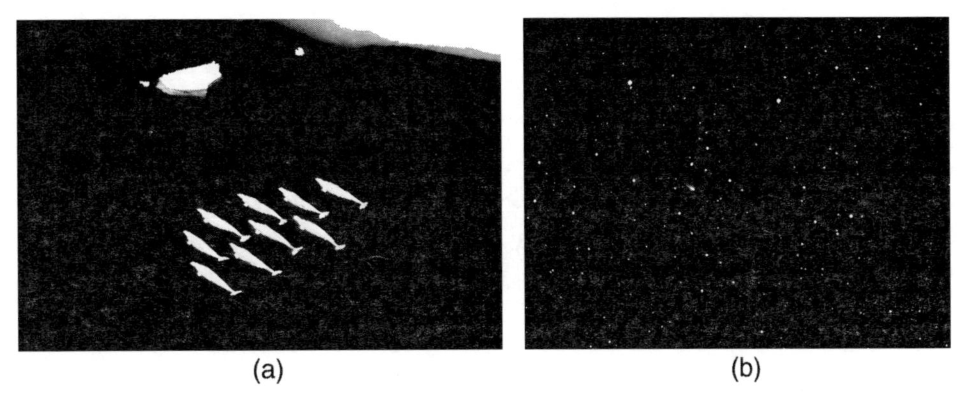

(a) (b)

Figure 4.2 Examples of civilian tracking problems: (a) tracking whales (NOAA image) and (b) the comet ISON at center of the image. (NASA image.)

elements of the new data as they arrive. Detection can be limited to just a window (gate) about the predicted location of the single target at the next time slice. Some trackers further simplify the problem by limiting the association to just the one detection nearest to the predicted target location. Multiple-target tracking methods track a few to a small number of objects. A multiple-hypothesis tracker (MHT) in its pure form considers every possible hypothesis; that is, it considers the possibility that each of N tracked targets might associate with each of M detections extracted from the new data. An independent nearest-neighbor (INN) tracker simplifies the association problem by considering only the nearest neighbor to each tracked target's predicted next position. Certain radar systems and a tracker associated with down-looking EO/IR wide-area motion imagery (WAMI) track all of the targets in a city-sized area. Association is then a challenging problem if the update rate is low. With a city-sized region, buildings, road networks, and traffic patterns provide useful clues and constraints.

4.2.2 Size of targets

Long-range tracking is done with certain types of radar systems/modes and infrared search and track (IRST) systems. Long-range tracking results in little information on target. Point-like and very small (i.e., low-resolution) targets are not amenable to processing with an advanced ATR due to scarcity of spatial feature data. Medium-sized (i.e., extended) targets are best characterized by very simple features such as length and 1D range profile. Target recognition is possible, but there is not much for a leading-edge ATR to work with. Multisensor approaches are common. A ground moving-target indicator (GMTI) radar could, for example, detect a moving target with good geolocation and then cross-cue an EO/IR sensor operating in narrow-FOV mode. With large (i.e., high-resolution) targets, detection, classification, recognition, identification, and "fingerprinting" are possibilities. That is the focus of this chapter. Trackers associated with ATRs generally utilize hundreds of data samples or pixels on target. These high-resolution targets provide sufficient information to apply sophisticated pattern recognition and image understanding techniques. However, segmentation is generally problematic. It is also not obvious whether to use the detection point or another point on the target as a stable key point in a tracking algorithm. Objects can be rigid, articulated, or deformable. They change in size, shape, and appearance while being tracked. These types of problems are handled well by the human vision system. Human vision often tracks a high-resolution object by rapidly determining its most salient point or selecting a suitable key point. Human vision keeps track of background motion, even while head, eyes, and body are moving.

For commercial security systems and academic research, event detection generally refers to actions that people are engaged in, such as fighting, digging, firing a weapon, cutting a chain link fence, or handoff of a package. This type of event detection requires detection of limbs, pose, and held object. For air-to-ground operation at long range, people cover too few pixels to expose such details. Short-term event detection and longer-term activity recognition then rely on target tracks. Examples of track-based activity recognition are: (1) a vehicle evading a check point, (2) people rendezvousing in the desert, and (3) one vehicle following behind another.

4.2.3 Sensor type

Different types of sensors provide different types of information to a tracker. Active sensors are said to provide *measurements* to the tracker. All tracked objects can be referred to as *targets*, even when no further processing is done to confirm that they are what is actually being sought. An ATR-type detection algorithm may not be used *per se*. Instead, a signal processor thresholds raw data to yield potential targets. False or noisy signals are an issue. Active sensors usually provide range data directly.

A (moving or stationary) target detection algorithm processes data from passive sensors to provide candidates to the tracker. The detection algorithm can be the front-end of an ATR. The clutter-reduced output of the back-end of an ATR can also be reported to a tracker. Passive sensors don't directly provide range information; this is a key problem to solve when handing off target location in geographic coordinates.

With multitarget, multisensor trackers, data must be correlated; i.e., each part of a target seen by one sensor data must be matched to a piece of the data seen by another sensor type. Even with good target location error, this is more difficult if sensors are of different types or are on different moving platforms. Different sensor types don't necessarily "see" the same parts of a target, the same features, or the same types of clutter, and they don't provide data in the same coordinate system. A target may be occluded when viewed from one platform but not the other.

4.2.4 Target type

Ground targets do not move smoothly and are surrounded by background clutter. With a forward-looking sensor, targets may be momentarily occluded by other targets, dips in terrain, or manmade or natural clutter. For other scenarios, such as with a clear sky background, natural clutter isn't an issue. Aircraft trackers are sometimes referred to as maneuvering target trackers to indicate that the tracker is not based solely on a constant velocity model. Some trackers are designed to handle both maneuvering and non-maneuvering portions of a trajectory.

Humans are difficult to track because they group, ungroup, and change in apparent shape as they move over rough terrain. Dismounted combatants (dismounts) often carry large weapons, backpacks, tools, supplies, and construction material. They drop off, pick up and hand off equipment. This changes their profile and center point. In thermal infrared imagery, parts of a person, such as a face and bare legs might show up well, while areas covered by an insulated jacket might disappear because they match the background temperature.

4.3 Tracking Problems

Several categories of tracking problems are reviewed.

4.3.1 Point target tracking

Suppose that processed sensor data yields measurements for one or several targets of interest. A point target has position x, velocity $\dot{x}$, and acceleration $\ddot{x}$ at each moment in time, where the variables are each vectors. These express the state of the target. The accumulated data forms a time series. Target motion is described by a vector of state variables $\mathbf{x}(k) = [x(k)\dot{x}(k)\ddot{x}(k)]^T$, or just $\mathbf{x}(k) = [x(k)\dot{x}(k)]^T$, where k denotes discrete time. $x(k)$ can, for example, be positioned in a 2D or 3D Cartesian coordinate system. The state vector can be augmented with additional elements, such as a target identification label.

The Kalman filter (KF) (named after one if its inventors Rudy Kálmán) often serves as the basis of an algorithm to track a single target. The KF estimates a target's current state vector and its state error covariance matrix. Then, when the next measurements are received, these estimates are updated. The KF cycles back and forth between the prediction and correction phases. It is referred to as a filter because it filters (smooths) out the noise in the time domain. The KF can be viewed as a type of Bayes filter in which the variables are linear and normally distributed. The process is Markovian in that each estimate is based on the previous estimate in time, with an infinite but decaying memory.

The basic Kalman equations are given in Table 4.1.[1] In practice, some terms are left out or set to constants, simplifying the model.

Many different types of kinematic trackers have been developed, most of which have a familial resemblance to the KF. These trackers differ in their underlying statistical assumptions, estimation criteria, processing and association schemes, number of targets tracked, and whether the number of targets present is assumed known or unknown. Tracking approaches are commonly combined, blended, modified, and improved on in various ways to tackle specific military problems. A more subtle point is that kinematic tracking models are generally chosen by engineering judgment, possibly with limited testing, and then are rigidly coded in software. They are generally not developed using the more ATR-like approaches of (1) slow learning over massive amounts of data, from combinations of relations that co-occur or (2) competitive testing a large number of alternatives using massive amounts of data.

The extended Kalman filter (EKF) is a modified nonlinear version of the standard single-target tracking KF (p. 387 of Ref. 1). Its state transition and measurement models are not linear as indicated below:

$$\begin{aligned} \text{State transition equation: } \mathbf{x}(k+1) &= f_1[\mathbf{x}(k)] + \mathbf{v}(k), \\ \text{Measurement equation: } \mathbf{y}(k) &= f_2[\mathbf{x}(k)] + \mathbf{w}(k), \end{aligned} \tag{4.1}$$

where f_1 relates the states at times $k+1$ and k, f_2 relates the state to the measurement, and $\mathbf{v}$ and $\mathbf{w}$ are the process and observation noises, respectively, both assumed to be zero mean multivariate Gaussian.

Particle filters (PFs) are a generalization of traditional Kalman filters. PFs use Monte Carlo sampling methods to represent probability densities. A PF generates a set of samples (particles) to represent the posterior probability density function of the target state, $p[\mathbf{X}(k) \mid \mathbf{Y}(k)]$, where $\mathbf{Y}(k)$ is the set of all measurements received up to time k. The unscented Kalman filter (UKF) is an improved version of the EKF. The UKF differs from Monte Carlo approaches, such as PFs, which require many more samples to propagate an accurate (possibly non-Gaussian) probability density of the target state. The UKF offers a compromise between the low computational effort of the KF and the potentially better performance of the PF.

The KF is used for tracking a single target. The probability data association filter (PDAF) is a statistical approach to the problem of track-to-measurement association. PDAF assumes that only one of the new detections is a target, and the rest are false alarms. All of the potential candidates for association-to-a-track are combined in a single, statistically most-probable update, taking into account the statistical distribution of the track errors and

Table 4.1 Kalman filter equations.[1]

Time Update (Prediction)	**Measurement Update (Correction)**
 State Prediction: $\hat{\mathbf{x}}(k+1\|k) = \mathbf{A}(k)\hat{\mathbf{x}}(k\|k) + \mathbf{G}(k)\mathbf{u}(k)$, where $\hat{\mathbf{x}}(k+1\|k)$ denotes the estimate (prediction) of the state vector **x** at a discrete time $k+1$ given noisy measurements $z(k)$, $z(k-1)$, ... $\mathbf{A}(k)$ is the state transition matrix. $\mathbf{G}(k)$ is a input transition matrix. $\mathbf{u}(k)$ is the known input (control) vector.	❸ Compute Measurement Prediction Covariance Matrix: $\mathbf{S}(k+1) = \mathbf{H}(k+1)\mathbf{P}(k+1\|k)\mathbf{H}(k+1)^T + \mathbf{R}(k+1)$, where **H** is the observation matrix, and **R** is the known measurement error covariance due to noise. Compute Kalman Gain: $\mathbf{K}(k+1) = \mathbf{P}(k+1\|k)\mathbf{H}(k+1)^T\mathbf{S}(k+1)^{-1}$, where **K** is the optimal weight matrix for combing the new data with the prior estimate to obtain a new estimate.
 Update State Error Covariance Matrix: $\mathbf{P}(k+1\|k) = \mathbf{A}(k)\mathbf{P}(k\|k)\mathbf{A}(k)^T + \mathbf{Q}(k)$, where $\mathbf{P}(k+1\|k)$ denotes the estimate of the error at discrete time $k+1$ given noisy measurements $z(k)$, $z(k-1)$, ... The goal is to minimize **P**, the covariance matrix of the state error. **P** is a measurement of state uncertainty. **Q** is process noise variance matrix (error due to process). **A**, **G**, and **H** are generally specified during model formation. Note: To start the process, initial values must be known for **x** and **P**.	 Measurement Prediction: $\mathbf{z}(k+1\|k) = \mathbf{H}(k)\hat{\mathbf{x}}(k+1\|k)$, where **z** is an observation or noisy measurement, at time $k+1$, of the state vector made with a sensor system; **H** is the observation matrix, and multiplying a state vector by **H** transforms it into a measurement vector. Measurement Residual: $\mathbf{v}(k+1) = \mathbf{z}(k+1) - \hat{\mathbf{z}}(k+1\|k)$. Update State Estimate with Measurement $z(k+1)$: $\hat{\mathbf{x}}(k+1\|k+1) = \hat{\mathbf{x}}(k+1\|k) + \mathbf{K}(k+1)\mathbf{v}(k+1)$  Update State Covariance Matrix: $\mathbf{P}(k+1\|k+1) = \mathbf{P}(k+1\|k) - \mathbf{K}(k+1)\mathbf{S}(k+1)\mathbf{K}(k+1)^T$, where $\mathbf{P}(k+1\|k+1)$ is the new estimate of the error.

clutter. The PDAF algorithm is based on the KF when the state and measurement equations are assumed to be linear. If the state or measurement equations follow a nonlinear model, then the PDAF is based on the EKF.

The interacting multiple model (IMM) algorithm uses multiple KFs to track multiple targets. Multiple targets can also be tracked with a multiple-hypothesis tracker (MHT) or a joint probabilistic data association filter (JPDAF). MHT and JPDAF maintain multiple hypotheses until enough evidence is accumulated to resolve the ambiguity in detection to track associations. The JPDAF is the multitarget extension of the PDAF. The JPDAF assumes that the number of targets present is known. It approximates the probability density function of each target state at each time step by a Gaussian.[2] The state estimator update is an expectation. Target location is estimated as the expected value of target location. The algorithm updates the filter for each track based on a joint probability of association between the latest set of detections and each track. With this algorithm, new tracks are not generated, and old ones are not terminated. Similarly, the MHT is a statistical framework that evaluates the likelihood of each hypothesis, representing a set of assignments of detections to tracks. The MHT maintains multiple correspondence hypotheses for each tracked object at each time step. To bound the computational complexity, some means of limiting the number of associations is required. The final track for a target is the most likely set of correspondences over the time period of its observations. The MHT is designed to introduce new tracks and terminate unworthy tracks.

A target tracker consists of a number of modules (Fig. 4.3), each with a given task. Not all target trackers have all the following capabilities.

1. **Track Management**: The set of tracks is maintained over time. This is done with several submodules.
   a. Track initiation: New tracks are initiated by detections that do not associate with current tracks or known clutter objects.[3] Thus, every new detection can spawn a track. It may be a true track from a good target detection or a false track from clutter or noise. Initiation can occur when a target enters the FOV, starts moving, or emerges from obscuration. Initiated tracks are characterized as tentative until certain criteria are met.

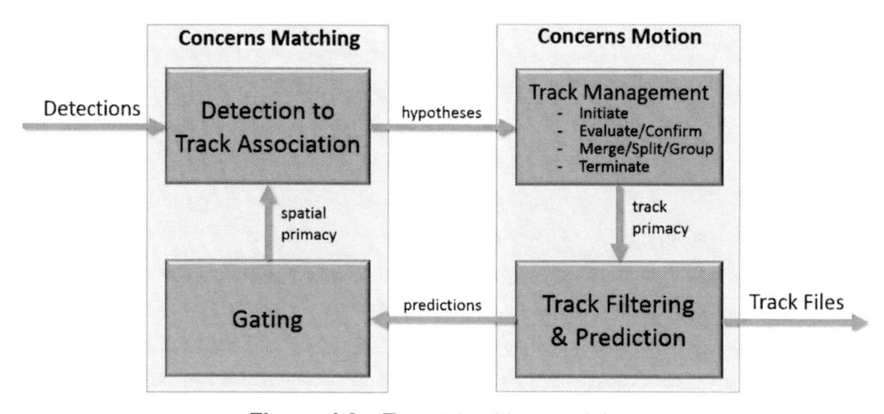

Figure 4.3 Target tracking modules.

b. Track splitting: Splitting can result from true physical phenomena, such as a launched missile separating into parts. It can also result from two different targets that are traveling so close together, as to be detected at first as a single target, later diverging in path.
c. Track merging: Two parts of a vehicle, such as the front and back of a tractor-trailer, might at first be considered as separate targets, later to be judged as part of the same target. Tracks that leave the FOV of one sensor and enter that of another sensor can be joined. Tracklet merging is the chaining together of track fragments.
d. Deletion (termination): Weak tracks or tracks that haven't been recently updated are pruned away. Tracks can be dropped or suspended if the target leaves the sensed area or no longer meets the evaluation criteria. For example, if only tanks are being tracked, a target that accelerates rapidly to 70 mph is not likely to be a tank and can be dropped from consideration.
e. Track grouping: Tracks with collective movement are grouped, such as in a convoy.
f. Track confirmation: Tracks are categorized as good, bad, or unsure according to various criteria. Track confirmation is achieved when track quality measures rise above a predefined threshold. A track is returned to an unconfirmed status when its quality falls below a lower bound. Track confirmation is defined in terms of system requirements. The number of tracks that the system can maintain is limited by computational and memory resources.

2. **Gating**: A track gate is an elliptical, hyper-ellipsoidal, or rectangular window about the predicted next position of a tracked target. The number of detections processed for data association is often limited by gating. An alternative to using kinematic gates is nearest-neighbor assignment of tracks to detections. A greedy nearest-neighbor algorithm analyzes all detection to track pairings allowed by the track gate and minimizes some overall cost metric involving distance and other criteria.
3. **Measurement (detection) to track association**: A set of measurement vectors is determined for a new frame or scan. Association determines which tracked targets generated which of these measurements, and which measurements cannot be attributed to any track. Association incorporates non-kinematic information if available.
4. **Track filtering and prediction**: This is the process of estimating the current target position and velocity (and perhaps also acceleration), and predicting the expected target position on subsequent radar scans or image frames. This process is required because measurements are imperfect and periodic. A number of methods are available, for example, KF.

   a. Filtering is the smoothed estimation of the present state vector based on past measurements.
   b. Prediction is the estimation of the state vector at a future time, usually for the next time step.
5. **Track file generation**: Each tracked object is represented by a track record, generally in accordance with NATO standard STANAG 4676. The track record contains current, historical, and other pertinent information. The collection of all track records is the *track file*. Targets are reported in track files. Track files can be communicated outside the tracking system, for example, for visual display or for handoff to another platform or system. Track files can be incorporated into or combined with ATR data in the form of target reports.

With such a sound engineering design as described above, it is reasonable to ask: What can go wrong? There are two common errors: lost track and track switching. When tracking multiple similar objects, a tracker can easily switch to the wrong object. This can be because the target changes appearance as it is being tracked. This happens when the target suddenly changes direction, articulates, or reconfigures. Targets can pass in front of each other, be occluded by shadows and other scene objects, or temporarily exit the FOV. It is common to lose track after a relatively short period of time. An attempt is then made to regain track. This can be some time later. It is difficult to determine if the re-associated object is really the same object as previously tracked. With a HRR, air-to-ground radar tracker, track can be lost when a ground vehicle stops. At that point, a SAR mode can be used for target identification. If the target slows down to go around a bend, an inverse-SAR mode can be used for classification. Merging the tracker with the ATR has the potential for improving both recognition and tracking in all situations.

4.3.2 Video tracking

Target tracking is a motion (or kinematic) problem as well as an association (or matching) problem. When so little information is available about a target that it can only be treated as a point, a scarcity of clues exists to support association. That is, *point target tracking* assumes that target tracking information is primarily that of unexceptional featureless dots moving through space. In the single-target tracking problem, data association usually involves finding the nearest neighbor (or strongest near neighbor) to predicted target position on the new frame or scan. The association problem becomes complicated with many tracks and many real, false, and missed target detections. The tracker then has the opportunity to make many incorrect detection-to-track assignments at each time step. The number of possible detection-to-track associations grows geometrically over time. Predicting each target's new position to be within a kinematic gate trims down the association problem.

If only one new detection falls within the gate, measurement-to-track association is straightforward. Complicatedness arises when more than one or no detections fall within the gate and also when a detection falls within multiple gates.

An ATR has little to offer a point-like target tracker; it has too little data to work with. But, this is not the case in video tracking, where the target covers a relatively large number of pixels. Features can be extracted from hypothesized target regions. However, there are still a number of problems that must be dealt with. There is no known optimal set of features with which to model the problem. The problem involves statistics and semantics. The target key point, e.g., center, needs to be found to support motion estimation. Obtaining a key point is problematic in infrared imagery, where, for example, a hot engine might dominate appearance at one moment, then a hot tail pipe at the next moment, as the vehicle rounds a bend. There are numerous other problems with track association with EO/IR imagery, such as hot indentations in the ground from a moving vehicle's rubber wheels or steel modular treads, dust trails kicked up by a ground vehicle moving through the desert, diesel exhaust, shadows from objects and clouds, occlusion, ill-defined notion of target center, battlefield fires and smoke, etc.

4.3.2.1 Correlation tracking (video data)

The simplest video tracker places a gate or box around a detected target. Gate position is not predicted, so kinematic tracking is not taking place. The target is expected to move less than the gate size from frame to frame. The gate is repositioned on the next detection falling within the gate. This approach is limited to tracking one or several dispersed targets against a benign background.

A somewhat more complicated approach places a small window around the detected target. The window is matched against an area about the predicted target location on the next frame, continuing frame to frame. Matching involves shifting the window within a neighborhood defined by the track gate. Target tracking can benefit from similar algorithms developed for commercial video processing and what is referred to as *video analytics*. This type of operation takes place in motion video compression. Matching simply involves subtracting the pixel values within the window at the current location from those in the window at the predicted location on the next frame. This is repeated at several shifts about the predicted location. The best match corresponds to the minimum absolute difference. This is referred to as block matching and prediction.

The next step up in correlation tracking complexity is to segment the object from its background and correlate the segmented region with an area about the predicted target location. A further step up is to model the target as well as the background to which the target is headed with a guard region to separate target from background. At this point, the tracker is using some of the tools of an ATR. The ultimate approach is to throw the whole ATR into the target association problem. Then features are extracted. The target is identified. Its motion helps determine front from back. Its pose is estimated. Its ID sets bounds on its predicted motion. Target ID can be smoothed over the individual IDs as the target is tracked, using a belief propagation approach, increasing the synergy between the recognition and tracking components of the ATR. This topic will be elaborated on later.

4.3.2.2 Feature-vector-aided tracking (video data)

Using features from the area about each detection point along with kinematic data is known as *feature-aided tracking*. The features help disambiguate the target from other targets and the background scene. Simplistic feature-aided tracking uses rudimentary features such as target length or 1D target signature. An ATR could supply more sophisticated features. Before an ATR makes its recognition decision, a feature vector or feature image is usually extracted about the detected target. Feature types are determined by competitive testing during a design phase, selected using pattern recognition techniques from a large number of pre-specified candidate features, or automatically determined with an auto-encoder. A working ATR has a very good feature vector to represent each detected target. The feature vector representing the current appearance of the tracked target can be matched against the feature vector from the newly detected object. Feature extraction and matching can be repeated, each time shifted about the predicted target location, with or without target segmentation. Whatever features are useful for target recognition and clutter rejection are most likely viable for the target association module of a tracker.

Typical features are:

- color, intensity, texture
- strong edge vectors or corner points
- shape, length, width
- moments, wavelets, Fourier, HOG or HOF features (see Chapter 3).

When the target is a deformable or articulated object, such as a person or an animal, other types of features can be added. These features include:

- gait
- estimated pose of limbs.

When picking blueberries, features are initiated top-down: blue, round, small. When a security guard spots a shoplifter, he may analyze the situation and pick one salient feature bottom up; for example, to track the guy with the red cap, who he saw putting the watch in his pocket. The supposition is that there is not likely to be another person in the area wearing a red cap. Mahadevan and Vasconcelos use the latter approach.[4] Their tracker iterates between bottom-up learning of maximally discriminant features in one frame and top-down use of these features in the next frame.

4.3.2.3 Mean-shift-based moving object tracker (video tracking)

The mean-shift tracker is a popular algorithm for motion video imagery. It works best with simple scenes, stabilized video, single slow-moving targets, little occlusion, and high frame rate. The object must be distinct from its background with little change in appearance or scale from frame to frame. In mean-shift tracker terminology, the target at the initial location (defined as location $\mathbf{x} = 0$), is called the *model*. The potential target centered at or near the predicted location $\mathbf{y}$ is called the *candidate*. In our description, the model will come from frame k and the candidate from frame $k + 1$. The mean-shift tracker tries to find the candidate that best matches the model. Models and candidates can be segmented blobs or rectangular regions-of-interest (ROIs).

We will describe the strategy of Comaniciu, Ramesh, and Meer.[5] The probability density functions (pdfs) of the model and candidate are approximated by m bin histograms, denoted by $\mathbf{q}$ and $\mathbf{p}$, respectively. These normalized histograms can be thought of as vectors with m elements:

$$\text{Target model:} \quad \mathbf{q} = \mathbf{q}(\mathbf{x} = 0) = \{q_u\}, \qquad u = 1, \ldots, m \quad \sum_{u=1}^{m} q_u = 1. \tag{4.2}$$

$$\text{Target candidate:} \quad \mathbf{p}(\mathbf{y}) = \{p_u(\mathbf{y})\}, \qquad u = 1, \ldots, m \quad \sum_{u=1}^{m} p_u = 1. \tag{4.3}$$

The objective is to find the position $\mathbf{y}'$ in which the Bhattacharyya coefficient between the model and candidate reaches a maximum value:

$$\mathbf{y}' = \max_{\mathbf{y}} \rho[\mathbf{p}(\mathbf{y}), \mathbf{q}] \approx \max_{\mathbf{y}} \sum_{u=1}^{m} \sqrt{p_u(\mathbf{y}) q_u}\,. \tag{4.4}$$

Like all fundamental tracking algorithms, there are numerous variations, improvements, and combinations developed each year.[6]

4.4 Extensions of Target Tracking

Moving entities as well as stationary entities can be tracked. Target tracks by themselves are uninteresting. They offer little in the way of situational understanding. Further analysis is required.

4.4.1 Activity recognition (AR)

Object tracking is often just a starting point for the analysis of the military situation. The critical questions are: *What* activity are the tracked objects engaged in? *Why* are they doing this activity? *Where* and *when* is this taking place?

An *activity* is defined by the National Geospatial-Intelligence Agency (NGA) as a sequence of *events* (or *actions*) with spatial and temporal relationship to each other. The events form a spatiotemporal pattern, such as one car following another car or a truck approaching a forbidden zone. Most AR algorithms use video data from forward- or downward-looking EO/IR sensors. However, AR can also refer to activities determined from radar track files, for example, three trucks rendezvousing near a sensitive location.

An event is spatially and temporally localized. An *activity* is a concatenation of short-term events (or action primitives) having complex relationships to each other. AR determines if a particular activity has occurred, is occurring, or is predicted to occur. Event detection integrates over seconds, while AR integrates over longer time scales and involves reasoning. A *scenario* is a multifaceted set of activities covering even larger scales of time and space.

One school of thought is: "Brain circuits compute grammars. Grammars encode time and structure relations hierarchically." – Richard Granger, 2015.

A grammar is a set of production rules for strings of symbols in a formal language. Production rules specify how sentences (activities) can be constructed from words (action primitives). Automata are often used as the recognizers of formal languages. Recurrent neural networks can be trained to act like automata. Recognition is done by decomposing the observed activity into a sequence of symbols. Each symbol represents an action primitive. The string of symbols is fed into an automaton or recurrent neural network to infer the activity. Inferences about a single-actor activity are embedded into the actor's track file, where an actor can be a person or a vehicle. Multi-actor activities, or scenarios, commonly result from queries into the database of track files, using a rule-based expert system to examine patterns and relationships. Consider some examples of events and activities shown in Table 4.2.

A track file by itself has limited intelligence value. Tracks are analyzed and often combined with other data to indicate an event. Scene context is important in determining if an event has occurred. To determine if a vehicle is going off-road, the road network must be known. Activities are combinations of events in complex spatial and temporal relationship. Consider an example in chronological order of actions. Even though the example is very simple, it requires a surprisingly large number of steps to describe:

1. Two vehicles are tracked.
2. The two tracked vehicles approach each other from different directions.

Table 4.2 Examples of events and activities. Those discernable solely by track data and location of fixed entities are denoted by an asterisk.

Events/Actions			
Humans		**Vehicles**	**Activity**
			Evade check point*
Run*	Throw	Stop*	Plant IED
Walk*	Dig	Start moving*	Circle block*
Carry	Drop	Accelerate*	Form convoy*
Handoff	Enter*	Decelerate*	Rendezvousing*
Receive	Leave*	Turn off road*	Launch missile*
Shoot	Hide*	Approach check point*	Unload vehicle
Rake	Pick up	Raise missile	Holdup
Open	Drape	Blow up	Swarm out of area*
Close	Lay down	U-turn*	Approach vehicle*
Pull gun	Crawl		Drape camo netting

3. The two vehicles go off road.
4. The two vehicles stop near each other.
5. One person exits each vehicle.
6. Each person walks to the back end of his vehicle.
7. Each person opens the trunk of his vehicle.
8. One person removes an object from his trunk.
9. That person hands off the object to the other person.
10. The other person receives the object.
11. That person places the object into his vehicle's trunk.
12. Both persons close the trunks of their vehicles.
13. Both persons walk to the front door of their vehicles.
14. Both persons enter their vehicles.
15. Both vehicles are tracked driving away in opposite directions.

Simple actions (such as digging) are usually learned from labeled training data. Once learned, they are fit to new data using such techniques as spatiotemporal template matching or hierarchical temporal memory (HTM) neural networks. Some events, such as a vehicle accelerating, can be discerned solely from kinematics. Events and activities have associated uncertainties in start and stop times, location, and whether or not they actually occurred.

An anomalous or dangerous activity is labeled as a *threat*. A threat may be declared because an activity is taking place near a sensitive facility or on a noteworthy date. If a threat activity exceeds a threshold probability, it triggers an *alert*. An alert is transmitted to a ground station from a UAV or unattended ground system. Further analysis and intelligence data are required to determine a larger meaning. This is now done by humans, rather than machine. Why are the people gathering into a large group? Why are they digging a hole? What are they placing into the hole? What alliance are they from? Are their intensions hostile? What response is called for?

Just as a tracker can support event/AR, event/AR can support the tracking function. The person sitting on a bench at a gas station and smoking a cigarette might not move for a while. Sitting on a motorcycle suggests other kinematics. Event/AR requires sufficient pixels on target. At short range, human pose and posture are intimately tied to both event/AR and tracking. At long range, event and AR are by necessity determined from track files and location in relation to important objects (e.g., facilities or roads).

Event and AR are starting to be accepted as ATR functions. This incentivizes tight integration of ATT, AR, and ATR. Taking this notion a little farther, and breaking from tradition: Are target detection/recognition and event detection/recognition really that different? Can they be accomplished by the same spatiotemporal processing, such as with third-generation spatiotemporal neural models?

4.4.2 Patterns-of-life and forensics

Radar and EO/IR sensors have been developed that cover very large ground areas containing thousands or tens of thousands of moving objects. If these sensors produce the data continuously for long periods, the capability is referred to as *persistent surveillance*. Consider sensors that produce wide-area motion imagery. City-sized fields-of-regard are downlinked to ground stations and mobile ground units. The ground units employ rugged military versions of desktop computers, laptops, tablets, or smart phones. The movement of people and vehicles is analyzed for days, weeks, or months. Activity varies by day of the week, weather, time of day, and holidays. The normal spatiotemporal pattern for the area is referred to as its *pattern of life*. Departures from the normal pattern indicate unusual or suspicious activity. Post-event backtracking (forensics) involves delving into collected data to determine the origin of tracks that led to an event (e.g., an IED explosion). Patterns-of-life and forensics again expand the notion of what constitutes an ATR and break down the barrier between ATT and ATR. Another way of looking at this is that the time scale over which ATRs are expected to operate have increased since early designs. In order of increasing time scale, we have:

1. Stationary target detection/recognition
2. Moving target detection/recognition
3. Event primitive (short term event, action primitive) detection/recognition
4. Event detection/recognition
5. Target tracking (yielding tracklets)
6. Linking/stitching of tracklets
7. Activity recognition
8. Forensics

9. Scenario recognition
10. Patterns-of-life.

4.5 Collaborative ATT and ATR (ATT↔ATR)

The collaborative ATT↔ATR associates kinematic features from the tracker with appearance features from the ATR to improve the performance of each (Fig. 4.4). Information is passed between the ATT and ATR at each time step. The remainder of this section considers the kinds of data produced by each that are useful to the other.

4.5.1 ATT data useful to ATR

The velocity vector of a ground vehicle is usually well-aligned with its major axis (this is often not so for a helicopter). Ground vehicle pose can be estimated from the velocity vector, sensor look angle, and digital terrain elevation data (DTED). Pose can sometimes be refined by correlation of vehicle motion with stored road network data. Road and ground slope data help to predict future vehicle pose, accelerations, and stops. However, road networks have limited use for battlefield analysis. A tank in battle is unlikely to be influenced by a stop sign and might not be traveling along a road.

Information that can reduce the decision space of the ATR is of great benefit to the ATR. For example, consider a template matcher. In the worst case, it has to match a huge number of templates against each detected blob. Depression angle and object pose reduce the number of templates that need to be applied to the detected object. Vehicle pose can also be used as input data for a neural-type classifier.

Target velocity, acceleration, and turn radius are useful data for an ATR. They eliminate, or reduce in probability, target types incompatible with tracker-derived kinematic data. Narrowing down the ATR's search space reduces its computational burden and chance for spurious matches.

Target kinematics provide the ATR with information about weapon readiness. A Scud launcher does not travel missile-up in launch position.

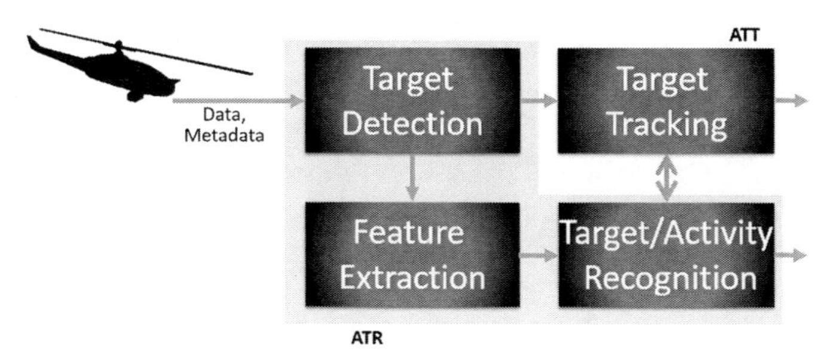

Figure 4.4 Block diagram of basic ATT↔ATR.

Conversely, a Scud launcher that emerges from hiding, quickly travels to a spot, and then stops may be preparing to launch. Helicopters and tanks often fire from stationary stances.

Target kinematics tell an infrared ATR that a ground vehicle is in motion. The vehicle's axles, wheels, engine, and exhaust pipe are very hot. This reveals a great deal about its appearance in the thermal band. Track data gives the ATR the opportunity to make a multilook classification decision for moving and stationary objects. In particular, classification performance builds up rapidly as the target is viewed at a diversity of aspect angles. Multiple looks at a target provide the possibility of multiframe super-resolution and 3D model construction. As in the parable of the blind men trying to recognize the elephant, each glimpse of a heavily occluded (e.g., under tree canopy) target will reveal just part of it. The separate pieces can be combined to form a more complete picture of the target. Target kinematics and desert location tell the ATR that a ground vehicle or low-flying helicopter will be kicking up dust from the desert floor. This changes the types of algorithms that are best-suited for detection, recognition, and video enhancement.

4.5.2 ATR data useful to ATT

Track association is an important module of the ATT. Track association is all about how N tracks can be correlated with M detections. Kinematic constraints narrow the number of possibilities that must be considered. However, if the revisit time interval is large, and target velocity is high, kinematic tracking is bound to fail. Additional information from the ATR can help resolve detection-to-track ambiguities. A tracker with a sensor such as a GMTI radar might lose a target if the target stops moving. By switching the radar to another mode such as 2D SAR, the ATR can detect and classify the stationary object and help preserve the track.

The ATR can provide the tracker with high-quality detections to initiate tracks or re-acquire targets momentarily occluded. It can also provide the tracker with a high-quality classification vector for each level of a decision tree. At a high frame rate, the classification vector is likely to remain fairly stable from look to look. Thus, the classification vector is very useful for associating a track with a detection.

The ATR can also provide the tracker with a single classification decision. This topic is heavily covered in the tracking literature, and the approach is called *classification-aided tracking*. Note that the classification decisions will not be robust if the current conditions don't match the training conditions of the classifier. If the classification probability is very low, the feature vector will provide more stable associations from look to look than from the estimated class, particularly if an optimal set of features is chosen and re-chosen on the fly. However, with certain kinds of sensors, such as HRR radar, a slight change in aspect angle will result in a drastic change in target signature. With these types of sensors, it is reasonable to rely on the classification decision. When re-acquiring a target that reappears from occlusion or camouflage netting, or that re-enters the FOV, association is better performed with a classification decision than a feature vector because the target aspect angle can change significantly since the last time the target was seen.

The ATT↔ATR is a step in the right direction. It indicates good engineering practice but doesn't represent a paradigm shift.

4.6 Unification of ATT and ATR (ATT∪ATR)

Now we are getting to the heart of the matter. Getting an ATT and ATR to work well together doesn't rise to the level of unification of ATT and ATR. With ATT↔ATR, the ATT and ATR are like voyagers passing in the night, whispering messages to each other. This has historically been the case, mainly because the expertise of some engineering groups resides in ATT and for others in ATR. But, can ATTs and ATRs be so tightly interwoven that the fabric of the fused ATT∪ATR cannot be unraveled?

Evolution is a long-term trial-and-error process with the errors now extinct and the best ideas conveniently left for us to study. First, let us consider how tracking is accomplished in biology, starting with visual pursuit as done by humans—with some caveats that we now note. Mimicking biological pursuit is an ambitious goal for reverse engineering, but it is one that can't now be fully achieved in military systems. For one thing, the mechanism is far from being wholly understood. Also, everything is tightly integrated in biological systems: multiple sensors, platform motion control, "gimbaled" head and eyes, learning, processing, and memory. The brain doesn't have separate memory and processing modules like computers. Military cameras don't work like eyes. They don't come in matched pairs. They aren't foveated. Current FLIR systems don't approach the number of photoreceptors of eyes; nonetheless, they have to recognize targets at much longer ranges than eyes. Military cameras cannot be pointed as quickly or precisely as human eyes. Yet, on the other hand, biological systems have survived and improved over a long time and are the only successful models for fully fused ATT∪ATR. There are many good ideas to learn from. Creatures with different sensors types and behaviors provide a plethora of interesting models.[7]

4.6.1 Visual pursuit

Humans excel at target tracking. The coach tells the young player to "follow the ball." This advice holds for a diversity of sports including tennis, ping pong, rugby, soccer, and baseball. A good batter tries to follow the ball from the pitcher's hand all the way home, only losing sight of it just before the bat is swung. A surprising new study suggests that head movement, rather than eye movement, plays the major role in tracking pitched balls.[8] Pitches that take an erratic path, defying visual tracking, are quite remarkable. Phil Niekro fooled many a batter with his knuckle ball.

The human retina contains a central fovea with a high density of photoreceptors. When an object moves, humans and many other animals move their head and eyes to track it. During *smooth pursuit eye movement*, human eyes slowly and precisely rotate to track the moving object. With the head stationary, smooth pursuit eye movement is a standard doctor's office test of wellbeing as well as sobriety. Birds such as eagles and owls can't rotate their eyes, so instead they swivel their heads. In either case, the high-acuity fovea is kept on the moving target. Fixation on stationary targets from a moving body is controlled by similar mechanisms. The computations contributing to smooth pursuit take place over the neural circuitry of the retina and a wide variety of cortical areas, eventually sending control signals to motor areas.[9] There is no distinct self-contained target tracking module within the human brain! Target tracking and recognition are not independent functions. Of the many areas involved in smooth pursuit, an intermediate-level cortical region known as MT+ is considered essential to motion processing. Other contributing functions are dispersed and include processing of non-retinal (e.g., vestibular) data and expectations. Smooth pursuit is largely an automatic behavior. However, humans have some ability to selectively track a particular moving target in the presence of more conspicuous movers.

The human visual system chooses a salient target for pursuit and jerks the eyes toward that target with a saccade, which is a rapid movement of the eyes. At the onset of pursuit, eye movement results from individual neurons responding to the motion of local image features. These local motions are soon grouped into the consolidated motion of the single moving target to be pursued. That is, the target is motion-segmented from the background. The gain of the target area is increased. This keeps the eyes pointed at the moving target despite the drag of the surrounding clutter moving in the opposite direction. If the target is momentarily occluded, the eyes continue their smooth motion (they coast). Tracking of a close-in target relies on a salient or selected key point. A smaller or distant target, such as a lightening bug, is tracked as a single point.

Two phases of pursuit are thus (1) initiation and (2) steady state. Initiation takes about 100 ms. In steady state pursuit, the eye velocity matches the imaged target velocity quite well. It is not known how the brain estimates target speed and direction from a population of noisy neurons. It is known that the motor system receives inputs from the perception system, and then follows speed and directional commands with near perfection, whether the received data are noisy or not.

During steady state pursuit, eye velocity in the head is combined with head velocity in the world. Thus, a world coordinate system is involved in computing signals that drive pursuit. So, somewhere in the brain, there is what engineers call a "track file" in world coordinates. It is not clear how the track file is encoded or passed to other brain areas.

One of the functions of smooth pursuit is to bring the tracked target into the fovea for detailed analysis. Spatial attention is centered on the tracked target rather than ahead of or behind it. But, because eye speed is higher during pursuit than during fixation, there is more eye jitter during pursuit. Little is known about the relative performance of visual target recognition during pursuit as compared to recognizing a stationary target during fixation. However, it is clear that smooth pursuit reduces motion blur.

4.6.2 A bat's echolocation of flying insects

Smooth pursuit is a reasonable model for EO/IR sensing, taking into account all of the caveats noted above. But what about active sensors such as radar and sonar? A good active model for the fusion of ATT and ATR is the bat. Over the last 50 million years of evolution, bats have developed a successful mechanism for detecting, selecting, recognizing, and catching maneuvering prey. There is a sizable body of literature on the subject, although considerably less than on human vision. Bottle-nose dolphins and whales provide additional models.

There are different types of bats with different lifestyles. Certain insect-eating echo-locating bats direct their beam at a target of interest with an accuracy of about three deg [Fig 4.5(a)]. Ghose and Moss draw an analogy between the orienting of their sonar beam and primates orienting gaze by saccades.[10] The amount of power reflected from the target gives the bat information on the target size. Modulations in the amplitude of the echo by insect wing beat provide additional information. Each species of insect has a characteristic micro-Doppler signature.[11] Both the temporal and spectral structure of the echoes are used as recognition features. In the terminal phase of tracking, the bat transmits with lower power and higher pulse repetition frequency to keep tracking the target until the final attack. Bats change their echolocation strategy based on conditions such as interfering noise. Certain bats can steer their sonar beam to either side of a target to pinpoint its location, trading off better tracking for reduced target detection.[12]

Certain species of bats feed on nectar and in the process pollinate plants [Fig. 4.5(b)]. The evolution of nectar-feeding bats and the plants needing them for pollination is an example of co-evolution, or to put it another way, cooperative target tracking, detection, and recognition. How do bats locate, classify, choose the best candidates, home in, and land on silent swaying

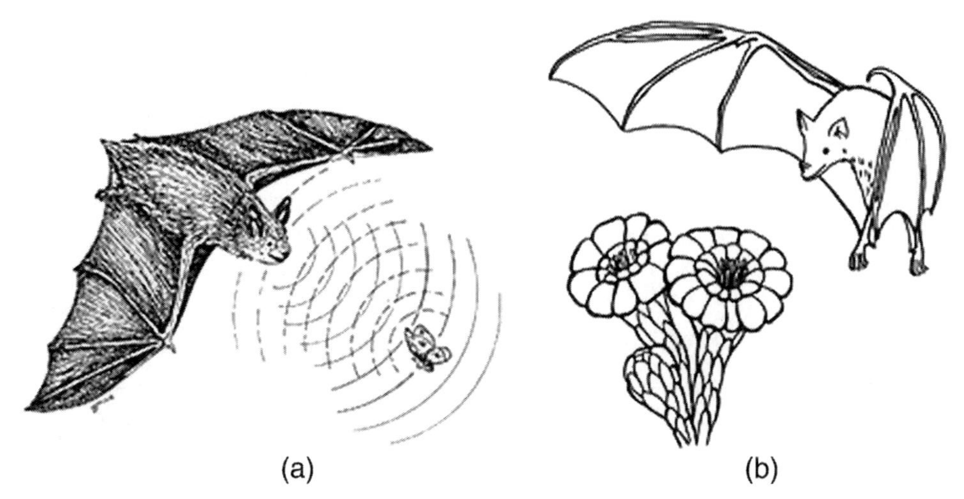

Figure 4.5 All hearing vertebrates perform auditory scene analysis. Echo-locating bats do this extraordinarily well. They probe the environment with vocal pulses and extract precise information from the sonar reflections. They develop a 3D representation of a complex scene and moving target within the scene. Some types of bats specialize in tracking, classifying, and capturing insects in flight. The insects make evasive maneuvers. The bat keeps track of its prey despite numerous echoes from other scene objects such as leaves, branches, and other bats. The bat predicts the insect's flight path, computes the best interception point, and adjusts its flight plan accordingly. Other kinds of bats are very important pollinators in desert and tropical environments. (a) Little brown bat. (Courtesy of http://dnr.wi.gov/eek/critter/mammal/wiscbat.htm.) (b) Lesser long-nosed bat. (Courtesy of http://www.pima.gov/cmo/sdcp/kids/color/llb.html.)

flowers in a highly cluttered environment? Sight, smell, echo-location, and excellent spatial memory are combined to find the most suitable flowers. Moss concludes that there is a three-way tradeoff: target detection performance, target location performance during tracking, and control of the FOV during scanning.[12]

4.6.3 Fused ATT∪ATR

The fully fused ATT∪ATR will use spatiotemporal algorithms. It will use sensors with the highest frame rate and highest resolution available; e.g., $\geq$2000 $\times$ 2000 pixel FLIR at $\geq$120 frames per second, or much higher-data-rate visible-band cameras. We will sketch out what a fully fused ATT∪ATR could look like, with a focus on multiband EO/IR sensors (Fig. 4.6).

4.6.3.1 Spatiotemporal target detection

Track initiation is the process of creating a new track from a new detection. New detections are also used to keep existing tracks alive. To detect targets with the fused ATT∪ATR requires a unified treatment of the different aspects of video imagery. These aspects include spatial (x, y), spatial

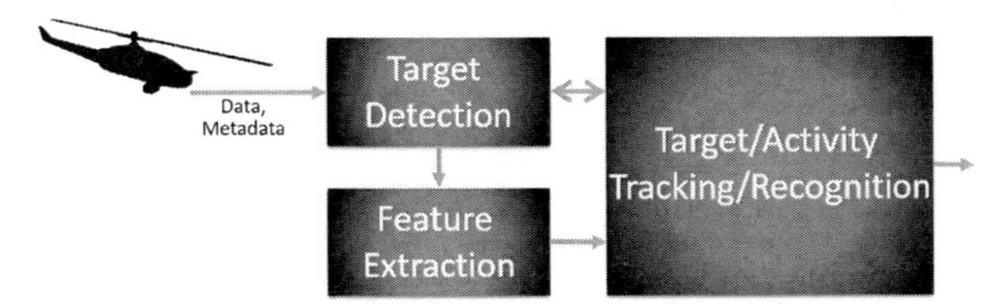

Figure 4.6 Block diagram of basic ATT∪ATR.

frequency (u, v), discrete temporal (k), and often color bands (b_1, b_2, b_3). Quaternions are a number system that extends the complex numbers. They form an algebra in which each object is represented by four scalar variables. A quaternion q is one way to achieve a uniform treatment of color and motion. A standard Fourier transform converts a grayscale image from the spatial to the spatial frequency domain. A generalization called the quaternion Fourier transform can be used for multiband video.[13,14] Let

$$q(k) = \beta(k) + \beta_1(k)\mu_1 + \beta_2(k)\mu_2 + \beta_3(k)\mu_3, \tag{4.5}$$

using visible color as an example:

$\beta_1 = RG$ is the red-minus-green color plane in the opponent color space,
$\beta_2 = BY$ is the blue-minus-yellow color plane in the opponent color space,
$\beta_3 = I$ is the intensity (grayscale) image,
$\beta = M = |I(k) - I(k-\Delta)|$ is a motion (change) image, and
$\mu_i^2 = -1,\ i = 1, 2, 3;\ \mu_1\mu_2 = \mu_3,\ \mu_2\mu_3 = \mu_1,\ \mu_3\mu_1 = \mu_2,\ \mu_1\mu_2\mu_3 = -1.$

The change image is most useful when the scene is stabilized and the target doesn't overlap from one time period to the next. More than two frames can be used to reduce ghosting:

$$\begin{aligned} M = {} & |I(k) - I(k-\Delta)| + |I(k) - I(k-2\Delta)| + |I(k) \\ & - I(k-3\Delta)| + |I(k) - I(k-4\Delta)| + |I(k-2\Delta) - I(k-4\Delta)| \\ & + |I(k-\Delta) - I(k-3\Delta)|. \end{aligned} \tag{4.6}$$

The modulus of a quaternion is $|q| = \sqrt{\beta^2 + \beta_1^2 + \beta_2^2 + \beta_3^2}$. A unit quaternion is a quaternion for which $|q| = 1$.

Equation (4.5) can be rewritten in symplectic form as

$$\begin{aligned} q(k) &= f_1(k) + f_2(k)\mu_2, \\ f_1(k) &= M(k) + RG(k)\mu_1, \\ f_2(k) &= BY(k) + I(k)\mu_1. \end{aligned} \tag{4.7}$$

The Fourier transform of a quaternion image $q(x, y, k)$ is given by

$$Q[u, v] = F_1[u, v] + F_2[u, v]\mu_2, \tag{4.8}$$

where

$F_i[u, v] = \frac{1}{\sqrt{MN}} \sum_{y=0}^{N-1} \sum_{x=0}^{M-1} \exp\left[-2\pi\mu_1\left(\frac{xu}{M} + \frac{yv}{N}\right)\right] f_i(x, y)$, and
$[u, v]$ are coordinates in the frequency domain,
(M, N) are number of pixels in height and width, and
k is left out of the equation for simplicity.

The form of the inverse quaternion Fourier transform, $\mathcal{F}^{-1}$ is an analog of that of the standard 2D Fourier transform:

$$f_i[x, y] = \mathcal{F}^{-1}\{F[u, v]\} = \frac{1}{\sqrt{MN}} \sum_{v=0}^{N-1} \sum_{u=0}^{M-1} \exp\left[2\pi\mu_1\left(\frac{xu}{M} + \frac{yv}{N}\right)\right] F_i(u, v). \tag{4.9}$$

Conveniently expressed in polar coordinates, the quaternion is the product of magnitude and phase terms. As in all Fourier forms, for a phase-only representation, the magnitude term has to be set to unity:

$$Q(k) = \|Q(k)\| \exp[\mu\phi(k)], \tag{4.10}$$

where $\phi(k)$ is the phase spectrum of $Q(k)$, and μ is a unit pure quanternion.

The reconstructed-phase-only version of $Q(k)$, using the inverse transform given in Eq. (4.10), is denoted by $q'(k)$. The saliency map $s(k)$ is the Gaussian-blurred magnitude of $q'(k)$:

$$s(k) = G_\sigma \|q'(k)\|, \tag{4.11}$$

where G_σ is a 2D Gaussian blur function of radius σ.

This approach can be modified to use both magnitude and phase terms in computing the saliency map s at varying scales λ:[15]

$$s(k) = G_\sigma \|\mathcal{F}^{-1}\{G_\lambda \|Q(k)\| \exp[\phi(k)]\}\|. \tag{4.12}$$

Detections are reported at local maxima of the saliency map. The blur widths and minimum distance between detections should be a function of the sizes of the targets sought, depression angle, and range. For forward-looking video and vehicular targets, the blur function should be elliptical: wider than high. The number of targets output per frame can be fixed or based on a threshold.

Several researchers have elaborated on this approach to spatiotemporal target detection in the usual ways: substituting another transform [e.g., discrete cosine transform (DCT), wavelet] for the Fourier transform, using a hierarchical representation, using various color or feature spaces, weighting quaternion components, using enhanced processing of saliency map, using top-down injection of higher-level knowledge, etc. The quaternion approach is analogous to various other spatiotemporal saliency paradigms, such as simply combining several types of saliency maps.

The quaternion was given as an example of how to produce a spatiotemporal saliency map for target detection. Recognizing a short-term event (one that is not discernable solely from track data) requires delving deeper into the time domain. Let $F[u, v, \tau]$ denote the Fourier domain representation of the 3D volume about a detection point. A spatiotemporal event template $T[u, v, \tau]$ is learned off-line in advance, based on a list of events of interest, using a very large training database for each type of event. Each template represents a particular object (e.g., a person) engaged in a particular action (e.g., boxing). Templates are stored for on-line use in Fourier form to save computations. The normalized cross spectrum is given by

$$C[u, v, \tau] = \frac{F[u, v, \tau]T^*[u, v, \tau]}{|F[u, v, \tau]T^*[u, v, \tau]|}, \tag{4.13}$$

where * denotes complex conjugate. The inverse Fourier transform of $C[u, v, \tau]$ provides the phase correlation matrix $PCM[u, v, \tau]$ with peak match point (x_0, y_0, t_0). It goes without saying that the technical literature is thick with improvements to this basic formulation, as well as dissimilar paradigms for achieving the same objective.

More-advanced alternatives are starting to emerge. Bio-inspired models of visual processing are in early stages of development. They propose a unified treatment of spatiotemporal object and event, detection, and recognition (e.g., see Refs. 16 and 17). Chapter 3 covers two biologically inspired spatiotemporal models: LSTM and HTM. Independent blind testing and field testing are required to pick out an appropriate paradigm for a particular application.

4.6.3.2 Forecast of features and classes

Current EO/IR sensors are capable of operating at a very high frame rate. As a target is tracked, the ATR produces ordered, equally spaced time series data. One objective of the fused ATT∪ATR is to produce a best estimate of current values based on past values. These best estimates are forecasts or forward-projections. Examples of values that can be forecast include classification vector and feature vector.

Consider a simple autoregressive model for a time series. The model is formulated as a least-squares regression problem. A predicted value of scalar x_k is based on the p previous values of the same time series. An autoregressive model of order p, $\mathcal{AR}(p)$, is defined as

$$x_k = \sum_{i=1}^{p} \varphi_i x_{k-1} + \varepsilon_k. \tag{4.14}$$

where $\varphi_1, \varphi_2, \ldots, \varphi_p$ are the parameters of the model, and ε_k is the noise term.

For a less-than-perfect tracker, recent values are more likely to be from the same tracked target compared to older values, so p should be finite, and $\varphi_1 > \varphi_2 > \cdots > \varphi_p$.

Equation (4.14) can be re-written in Yule–Walker form as

$$x_k = \sum_{i=1}^{p} \varphi_i x_{k-1} + \sigma_{\in}^2 \delta_{k,0}, \tag{4.15}$$

where

$k = 0, \ldots, p$ yields $p + 1$ equations, and

$\delta_{k,0}$ is the Kronecker delta function.

Consider Eq. (4.15) when $k = 0$:

$$x_k = \sum_{i=1}^{p} \varphi_i x_{-1} + \sigma_{\in}^2. \tag{4.16}$$

Now consider Eq. (4.15) in matrix form when $k > 0$:

$$\begin{bmatrix} x_1 \\ x_2 \\ x_3 \\ \vdots \\ x_p \end{bmatrix} = \begin{bmatrix} x_0 & x_{-1} & x_{-2} & & \\ x_1 & x_0 & x_{-1} & \cdots \\ x_2 & x_1 & x_0 & \\ & \vdots & & \ddots \\ x_{p-1} & x_{p-2} & x_{p-3} & \cdots \end{bmatrix} \begin{bmatrix} \varphi_1 \\ \varphi_2 \\ \varphi_3 \\ \vdots \\ \varphi_p \end{bmatrix}. \tag{4.17}$$

First, Eq. (4.17) is solved to provide the parameters of the model. Then Eq. (4.16) is solved to provide the noise term.

A value of p must be chosen to apply the model. This can be done by engineering judgment or experimentation. The $\mathcal{AR}(p)$ model could be applied to forecast each element of the classification vector. If the classification vector represents probability estimates, these estimates might need to be normalized to sum to 1.0. If the $\mathcal{AR}(p)$ process is applied to a feature vector, the parameters of the model can be averaged over all features. The $\mathcal{AR}$ process should not be applied to non-image features, such as time of day, day of year, depression angle, climatic zone, etc. Many alternative paradigms could be used for this same objective: simple temporal averaging or Winsorizing, theory of belief functions, Dempster–Shafer theory, vector autoregressive model $\mathcal{VAR}(p)$, or Kalman filtering.

4.6.3.3 Detection-to-track association

In this step of the processing, the ATT∪ATR determines which detections are associated with which tracks. This is done using more than just kinematics.

In some applications, a single detection is used to update each track. The first step is to bring each existing track to the current time step by predicting its position based on the most recent filter output. The target state estimate and the assumed target motion model are updated. The target motion model can be refined for a particular object being tracked once the ATT∪ATR deduces more information about its type. For example, gait can be used for a tracked dismount. Once the target state is estimated, detections are associated with tracks.

Future sensors will allow both high frame rate (e.g., 120 frames per second) and high pixel resolution [16M (for IR) or 100M (for visible) pixels/frame]. High frame rate makes tracking easier. High pixel resolution allows for a wide FOV. Detection should take place at a higher frame rate within track gates than for the rest of the scene. This is acceptable because new targets don't enter a wide-FOV scene very often. A scene gist model further constrains new detections and tracks. For example, if tanks are being tracked in forward-looking imagery, there is no sense in looking for or tracking objects much above the skyline.

The combined ATT∪ATR uses multiple forms of higher-level intelligence to aid detection-to-track association. 3D models of target and background (e.g., buildings) are built up as the sensor moves and the target is tracked. The tracked objects evolve slowly in form against a relatively more stable background. The object border shifts against the background as look angle varies, allowing temporal segmentation. 3D target appearance changes in predictable ways as range and pose slowly change and a known background is entered. Once a tracked target is motion segmented from its background, the ATT∪ATR uses the same forecasted feature vector and matching paradigm for both object association and recognition.

Suppose that the application requires more than just one detection-to-track association at each update step. The MHT is a reasonable paradigm, since it considers many detection-to-track hypotheses. As the process is repeated over new frames, the number of spawned tracks can grow exponentially. However, although the MHT calculates the probability of each potential track association, it is efficient for it to report only the most probable. Unlikely would-be track updates are killed off. Even after all of this, a number of detections can remain unassociated with existing tracks, and a number of tracks can remain without updates. Track maintenance deals with these problems.

4.6.3.4 Track maintenance

The track maintenance module causes new tracks to be initiated and some old tracks to be deleted. A decision is made about whether to end the life of a track. If a track is not associated with any detection for a certain number of steps, or *m* of the last *n* steps, the target may have become hidden, partially occluded, or have left the area; or the target may have just blended into the background texture; or it may have not been a true target in the first place, being the result of noise or clutter. Rules are designed to delete unconvincing tracks or put them into hibernation. The ATT∪ATR knows whether an identified object being tracked is on the list of targets currently of interest, based on mission planning data. If not, the track is deleted or placed into a nuisance object category. If the platform has two remaining missiles to fire, then the focus is limited to the two highest-value tracked targets.

4.6.3.5 Incorporation of higher-level knowledge

In biological systems, the equivalent of the ATT∪ATR utilizes information retrieved from memory, expectations, intelligent thought, processing priorities, goals, top-down direction, and adrenaline rushes to focus on real crises. Consider some more-general notions of how the ATT∪ATR could use information available to it. Such information includes mission-based stored data, newly arriving intelligence data, *a priori* target probabilities, known target motions, enemy tactics, and force structures. If vehicles are tracked going down a road in close proximity to each other, it is more likely that they are of the same type than the classification of each individually would indicate. This is only true for an actual battlefield, not a contrived test where there is likely to be one target of each type. If vehicles are tracked traveling off-road at about the same speed, heading in the same direction, it is more likely that they are of the same type than the classifier output would indicate. If the lead vehicle stops, slows down, rounds a bend, or crosses a stream at a particular spot, it can be anticipated that the following vehicles will make similar maneuvers. If small tracked objects are huddled together or moving in synchrony, and one is identified as a human carrying a large weapon, it can be assumed that the other ill-defined objects are also dismounts with weapons and probably not deer. Intelligence may indicate that enemy tanks are likely to be parked adjacent to schools, hospitals, and religious institutions, and missiles stored in the basements of homes. Old tank tracks (in this case, meaning indentions in the ground) headed toward a barn indicate that a tank is hiding in the barn. This type of military intelligence and reasoning has historically been easy for a pilot to utilize but difficult for algorithms to process.

Suppose that the objective is to track potential objects of interest in a wide FOV and then switch to a narrow FOV only for an object of probability greater than 0.7. If the objects being tracked all have a probability of about 0.5, then the FOV switch will never occur. However, if two tracked objects are momentarily in close proximity, then the probability that at least one is an object of interest is about 0.75, and the FOV switch should take place with the sensor pointed half-way between them.

4.6.3.6 Implementation

The fused ATT∪ATR is embedded into the sensor for minimum size, weight, power, cost, and latency. If multiple sensors contribute to tracking and recognition, their data is combined at a feature level rather than at a post-declaration level. Multisensor fusion should involve reasoning and feedback, rather than brainless statistical combination in a fusion box. One sensor tasks another sensor to obtain needed information, the way one's gaze jerks toward a loud sound.

4.7 Discussion

ATT systems have been in use since before World War II. ATR flight trials began in the early 1970s. Historically, integration of ATT and ATR has meant little more than separate units exchanging bits of information. The question is: Can ATT and ATR be so finely interwoven that they no longer appear as separate units? The standard collaborative ATT plus ATR was referred to as ATT↔ATR, while the suggested united ATT and ATR was referred to as ATT∪ATR.

The biological model of ATT∪ATR arises from dynamic patterns of activity distributed across many neural circuits and structures (including retina). The information that the brain receives from the eyes is "old news" at the time that it receives it. The eyes and brain forecast a tracked object's future position, rather than relying on received retinal position. That is, humans perceive objects where they "ought" to be.[18] Attention is tightly focused on the object of highest interest. If this process didn't work as well as it does, batting averages would be a lot lower, and outfielders would rarely catch a fly ball, making baseball tough to watch. Anticipation of the next moment—building up a consistent perception—is accomplished under difficult conditions: motion limitations (eyes, head, body, scene background, target) and processing limitations (neural noise, delays, eye jitter, distractions, abysmal clock speed of a neuron by computer standards). Not only does the human vision system surmount these problems, but it has innate mechanisms to exploit motion in support of target detection, target classification, event detection, and activity recognition.

When an object is tracked in biological vision, multiple coordinate systems are tied together, including eye-centered world coordinates. In fusion of tracking and recognition, visually tracked objects are quickly recognized and then ordinarily maintain a perceived stable identity. Recognition accuracy is traded off against continuity and stability. Although certain brain areas (such as MT+) do specialize, the neural version of ATT∪ATR is best described as a widely distributed assortment of cooperating multifunctional mechanisms.[19] The visual system is not composed of separate ATT and ATR in sporadic communication.

Biological vision doesn't normally operate on snapshots. Feature extraction, detection, and recognition are spatiotemporal. When vision is viewed as a spatiotemporal process, target recognition, tracking, and event recognition do not seem as distinct as they are in current ATT and ATR designs. They appear as similar mechanisms taking place at varying time scales. Using biological vision as a source of ideas, there are many as yet unexplored sources of information and ways of combining information for improving spatiotemporal processing. Different sensor and processor designs are used by different species, providing a rich variety of models for biomimicry.[7] It shouldn't be a matter of finding an ATR that works with a sensor, but rather having the ATT∪ATR and sensor designed by a single team to work well together.

The overall recommendation, which is just now advancing beyond the point of *thought experiment*, is to abandon the nearly 50-year-old paradigm of separate ATT, ATR, and AR black boxes. The concept of an ultrahigh-resolution, ultrawide-FOV, high-frame-rate, eye-like sensor with eye-like motion appears feasible. The retina and brain constitute a processing unit. Analogously, it is now possible to integrate substantial processing into a sensor, with no need for a separate processor box. Tightly interweaving cooperative subfunctions so that separate ATT, ATR, and AR modules are no longer discernable will be a major undertaking that can benefit from expected advances in brain modeling over the next decades.

References

1. Y. Bar-Shalom and L. Xiao-Rong, *Estimation and Tracking Principles*, Artech House, Boston (1993).
2. D. F. Crouse, Y. Bar-Shalom, P. Willett, and L. Svensson, "The JPDAF in practical systems: Computation and snake oil," *Proc. SPIE* **7698**, 769813 (2010) [doi: 10.1117/12.848895].
3. D. Musicki and S. Suvorova, "Target track initiation comparison and optimization," *Proc. Seventh International Conf. on Information Fusion*, Stockholm, Sweden, pp. 28–32 (2004).
4. V. Mahadevan and N. Vasconcelos, "Biologically inspired object tracking using center-surround saliency mechanisms," *IEEE Trans. on Pattern Analysis and Machine Intelligence* **35**(3), 541–554 (2013).
5. D. Comaniciu, V. Ramesh, and P. Meer, "Kernel-based object tracking," *IEEE Trans. on Pattern Analysis and Machine Intelligence* **25**(5), 564–577 (2003).
6. L. Wang, H. Yan, H. Y. Wu, and C. Pan, "Forward-backward mean-shift for visual tracking with local-background-weighted histogram," *IEEE Trans on Intelligent Transportation Systems* **14**(3), 1480–1489 (2013).
7. M. F. Land and D. E. Nilsson, *Animal Eyes*, Oxford University Press, Oxford (2012).

8. N. F. Fogt and A. B. Zimmerman, "A method to monitor eye and head tracking movements in college baseball players," *Optometry and Vision Science* **91**(2), 200–211 (2014).
9. S. Ohlendorf, "Cortical Control of Smooth Pursuit Eye Movements," Ph.D. dissertation, University of Basel, Switzerland (2007).
10. K. Ghose and C. F. Moss, "The sonar beam pattern of a flying bat as it tracks tethered insects," *Journal of the Acoustical Society of America* **114**, 1120–1131 (2003).
11. A. Balleri, "Biologically Inspired Radar and Sonar Target Classification," Ph.D. dissertation, University College London, UK (2010).
12. C. F. Moss and A. Surlykke, "Probing the natural scene by echolocation in bats," *Frontiers in Behavioral Neuroscience* **4**(33), 1–16 (2010).
13. C. Guo, Q. Ma, and L. Zhang, "Spatio-temporal saliency detection using phase spectrum of quaternion Fourier transform," *IEEE Conf. on Computer Vision and Pattern Recognition*, pp. 1–8 (2008).
14. C. Guo and L. Zhang, "A novel multiresolution spatiotemporal saliency detection model and its application in image and video compression," *IEEE Trans. on Image Processing* **19**(1), 185–198 (2010).
15. J. Li, M. D. Levine, X. An, X. Xu, and H. He, "Visual saliency based on scale-space analysis in the frequency domain," *IEEE Pattern Analysis and Machine Intelligence* **35**(4), 996–1010 (2013).
16. G. J. Rinkus, "A cortical sparse distributed coding model linking mini- and macrocolumn-scale functionality," *Frontiers in Neuroanatomy* **4**, 1–13 (2010).
17. S. Ji, W. Xu, M. Yang, and K. Yu, "3D convolution neural networks for human action recognition," *IEEE Trans. on Pattern Analysis and Machine Intelligence* **35**(1), 221–231 (2013).
18. M. Chargizi, *The Vision Revolution*, Benbella Books, Dallas, Texas (2009).
19. G. W. Maus, J. Fischer, and D. Whitney, "Motion dependent representation of space in area MT+," *Neuron* **78**(3), 554–562 (2013).

Chapter 5
Multisensor Fusion

5.1 Introduction

Suppose that you are visiting your Aunt Florence. You get hungry and meander into the kitchen. Sitting on the table is a plate of fish. It looks appealing, but something is a bit funky about it. So you poke it, smell it, and taste a tiny bit. It is tasty, but still doesn't seem quite right. Then comes the clincher. Your aunt yells from the next room: "Don't eat the fish!" The human brain is an excellent example of a multisensor fusion system. Fusion of data from your five senses kept you from eating the spoiled fish (Fig. 5.1).

But how did all five senses focus on the same object? This is called the binding problem.[1] All of the features and traits of the fish, in all of the sensor data, must have been segregated from all of the properties of other nearby objects and the background. Then the features must have been associated with the concept of "fish." Binding occurs in many different parts of the brain. There is no single algorithmic solution. Binding is a class of problems: binding over visual space, segregating one sound from others, cross-modal binding

Figure 5.1 All five senses are used to determine whether the fish is too far gone to eat.

associating the sound with the visual percept, and so on. There are at least seven different types of binding:

1. Location Binding: Objects are bound to their location, according to some 3D coordinate system.
2. Temporal Binding: Objects are bound, not only to a location, but also to an interval of time.
3. Parts Binding: The separate parts of an object must be segregated from their background and bound together, taking into account partial occlusion and voids.
4. Property Binding: The different properties of an object must be bound together to characterize the object. Properties can include color, texture, shape, and motion.
5. Conditional Binding: The interpretation of one property (e.g., size) may be conditioned on another property (e.g., range).
6. Hierarchical Binding: Perceptual categories are organized and bound together hierarchically. This applies to both hierarchically structured objects and events.[1] Tires are bound to a car. The batter is bound to a pitcher; each is bound to their respective team; both teams are bound to the same baseball game.
7. Conceptual Binding: The other forms of binding must be bound to or associated with a concept such as "target" or an event such as "bomb explosion."

How does this relate to ATR? Many different kinds of sensors can provide information to an ATR. Sensors are differentiated by their spectral band, operating range, coverage, emissions, latency; and size, weight, power, and cost (SWaP-C). One must also consider the impact on their imagery (or more generally, data) of weather, lighting conditions, and target motion. One sensor may be able to observe a target under certain conditions, e.g., fog, when another sensor can't see it at all. Multisensor data is combined (fused) to produce better ATR performance than can be achieved with a single sensor type. Multisensor fusion also provides robustness to sensor failure, degraded atmosphere, decoys, camouflage, and jamming.

The humblest form of fusion is the real-time combination of data from two sensors on the same platform. For ATR, as in biology, fusion presupposes a solution to the binding problem. Multiple sensory signals must be associated with or bound to the same object or event. The basic applications of fusion in ATR are improved target detection and recognition by a combination of cotemporal data. However, there are many other forms of fusion:

- Sequential Fusion: One sensor type may call up another sensor type or another mode of the same sensor to help make the decision. A person hears a loud sound and then turns his gaze toward the source of the sound.

- Fusion while Track: As a target is being tracked, each look provides a separate classification decision. The tracker binds one view of the target to the next. The sensor sequentially sees the target against different backgrounds from slightly different viewpoints. The separate classifier decisions can be combined over time.
- Handoff: A sensor on one platform provides target location and classification information to a sensor on another platform.
- Complimentary Sensors: One sensor type (e.g., laser rangefinder) may be used to provide a missing piece of information needed to process the data from another sensor (e.g., FLIR camera).
- Multilook: The same sensor views the target from multiple vantage points to construct a 3D model.
- Emission Control: An ATR uses a passive sensor until the last few moments when an active sensor is engaged.
- Anomaly Mitigation: One sensor is used to monitor the moments that another sensor has anomalous outputs. In some cases, data from one sensor is used to correct the anomalous or noisy outputs of the other sensor.
- Dynamic Sensor Selection: Adaptive algorithms select the most suitable sensor or sensors for the task at hand. The selection process is based on evaluation of the momentary data quality, efficacy, and energy consumption of each available sensor.
- Passive Radar: Any radar that detects, locates, and tracks targets without transmitting signals is called passive. A passive bistatic radar receives and processes reflections of waveforms originating from a noncooperative source of illumination, such as a commercial broadcast (Fig. 5.2). With two or more transmitters and/or receivers, the radar is called multistatic. ATR then depends on the fusion of signals from multiple spatially distributed transmitters and receivers. In addition, the system may detect, track, and recognize some targets by their radio frequency (RF) emissions.
- Target Location Error (TLE) minimization: One sensor provides good location in range, while another sensor provides good location in cross-range. Low TLE is achieved through their combination.

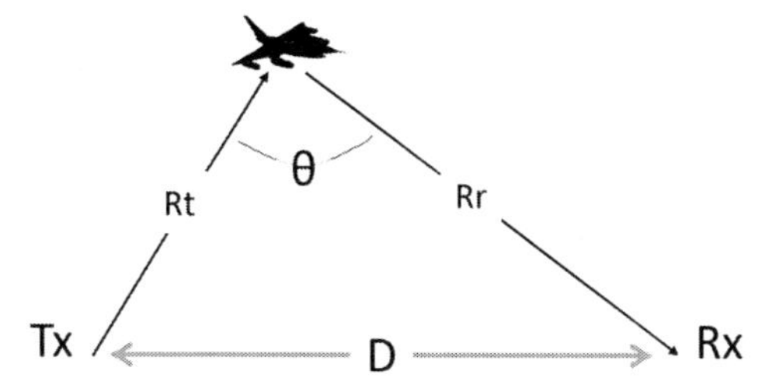

Figure 5.2 A bistatic radar includes an RF receiver Rx and transmitter Tx separated by a distance D. A passive bistatic radar includes a receiver, but not transmitter, instead using a noncooperative RF transmitter in the environment.

- Multifunction Radio Frequency (MFRF) Systems: Various functions normally performed by different equipment can be fused into a single system. MFRF systems bring together both radar and nonradar functions: Electromagnetic support measures (ESM), signals intelligence (SIGINT), passive electronic warfare (EW), and observations of communications intelligence (COMINT). For example, the DARPA Multifunction RF program sought to develop a common RF system using agile frequencies, waveforms, and apertures to optimally interweave different functions according to an aircraft's missions.[2]

5.2 Critical Fusion Issues Related to ATR

In regards to ATR, multisensor fusion can be more sophisticated than the often depicted brainless black box, crunching away on data from multiple sensors, spitting out consensus answers. Multisensor fusion pertains to many different types of problems, whose solutions require both engineering judgment and creativity. Multisensor fusion applies to different types of sensors, platforms, data, timelines, algorithmic models, and hardware architectures. Multisensor fusion can improve standard ATR functions such as detection, segmentation, recognition, and tracking, as well as complex processes such as scene modeling, GPS-denied navigation, target location estimation, missile warning, and landing in degraded visual environments. Here are some of the issues:

1. Use Case versus Problem Solving: Multisensor fusion is sometimes treated as a use case problem. A number of sensors are available on a platform. A number of algorithms, such as Kalman filters, scene registration, neural networks, expert systems, etc., are available in the software repository. The engineer is tasked with determining something useful that can be done with the available resources. This is shoddy engineering. A better approach is to talk with customers and end-users to determine the critical problems needing solution. Half of the work of developing a good solution to a problem is getting a precise definition of the problem. Once a clear-cut definition of the problem is agreed on by all stakeholders, the engineer performs a comprehensive survey of the resources that can be combined to solve the problem. The baseline discussion should be about what can be done with clever use of the tools on hand and a limited budget. If the required performance level cannot be achieved with existing resources (e.g., sensors, models, algorithms, databases), then the discussion should proceed to new ideas and use of new resources.
2. Centralized, Decentralized, or Distributed:
   a. Single-Platform Centralized Fusion [Fig. 5.3(a)]: The raw data (or possibly feature data) from multiple sensors are fed into a single fusion engine. The sensors and fusion engine are on the same platform. The fusion engine binds and combines the data to reach a fused conclusion.

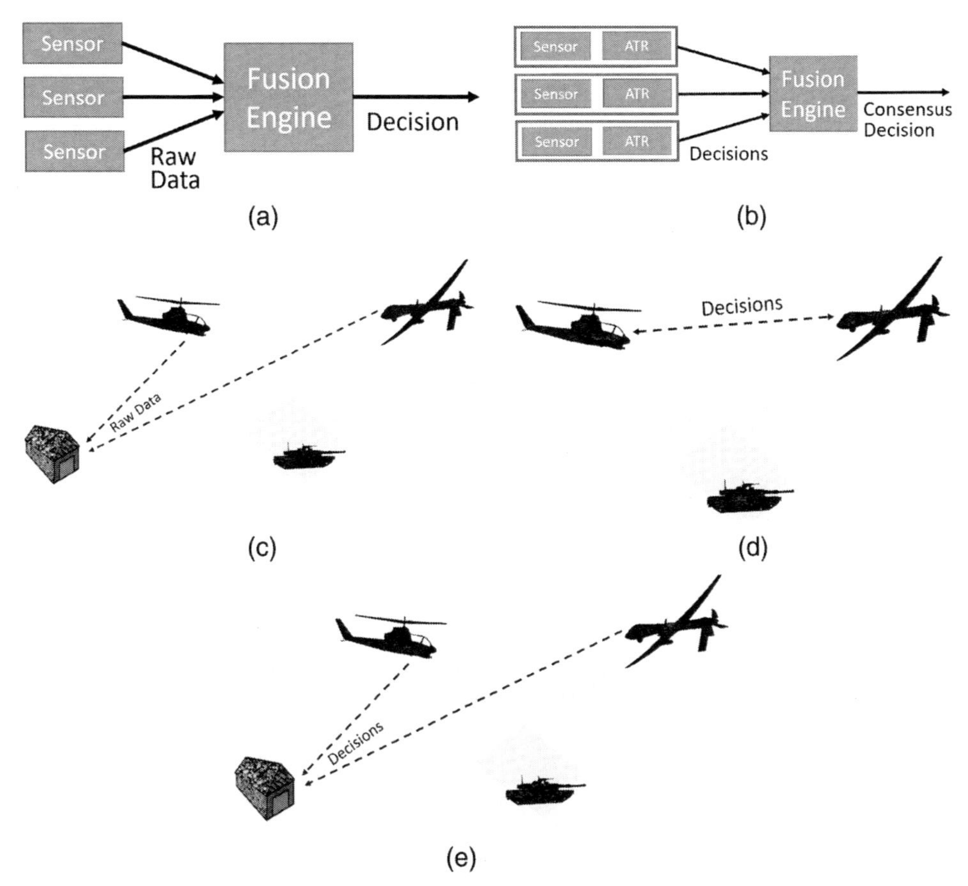

Figure 5.3 Several categories of multiclassifier fusion: (a) single-platform centralized fusion, (b) single-platform decentralized fusion, (c) multiplatform centralized fusion, (d) multiplatform distributed fusion, and (e) multiplatform decentralized fusion.

The cost of simultaneously collecting time-synchronized train and test data from multiple sensors is quite high and is the main issue determining whether this type of fusion is chosen.

b. Single-Platform Decentralized Fusion [Fig. 5.3(b)]: Each sensor is an intelligent agent. Each sensor has its own ATR (with integrated tracker) with a high degree of autonomy. The hard or soft decisions and track files output by the individual ATRs are combined by a simple fusion engine on the platform. Thus, most of the work is decentralized, but the final fusion step is centralized. The advantages of this approach are simplicity and reduction of the need to simultaneously collect training and testing data from multiple sensors.
c. Multiplatform Centralized Fusion [Fig. 5.3(c)]: The raw or feature data from multiple platforms are transmitted to a single fusion

center in a particular platform or ground-station. The fusion engine correlates and combines the data to reach a decision. This approach is rarely used due to high-bandwidth requirements.

d. Multiplatform Distributed Fusion [Fig. 5.3(d)]: Each platform has its own sensors, ATR, and fusion engine. All data are processed locally, with the hard or soft decisions transmitted to other platforms. Each receiving platform combines the received data with its local decisions to reach its own conclusions. Each platform has a high degree of autonomy. Conclusions reached on one platform may not be the same as those reached on other platforms. This approach is characterized by low-bandwidth requirements and low operational complexity.
e. Multiplatform Decentralized Fusion [Fig. 5.3(e)]: Each platform has its own sensors and ATR, and reaches its own conclusions. Hard or soft decisions are transmitted to a centralized fusion center, which simply combines the local decisions into a global one.

3. Registration and Association: If two cameras are used to identify an object, they should, if possible, have their optical axes aligned, the same field of view, the same pixel resolution, and be synchronized in time. This is not always the case, as in the picture-in-picture approach, where a high-resolution narrow-field-of-view sensor is looking at the target along with a low-resolution wider field-of-view sensor. This latter approach is similar to foveation used by the human eye and as such can be considered an effective use of limited resources.

   Any time that two or more sensors are used, one has to make sure that the data are converted to a common reference frame. Once this is accomplished, one still has the binding problem to deal with. If each sensor sees the target as a single point, the binding problem reduces to the traditional track association problem. With higher resolution, it must be determined which pieces of the observed target from one sensor correspond to which pieces from the other sensor. Association of target parts is difficult if, for example, an IR sensor is getting a strong signal from the hot exhaust of a tank, while a SAR sensor isn't seeing the exhaust, but is seeing the whole target, possibly with the pieces of the target scrambled in position by an effect called layover. Layover occurs when the air-to-ground radar beam is reflected from the top of a tall target before it is reflected from the bottom. The reflection from the top of the target will be received before that from the bottom. As a result, the top target component is displaced toward the radar, to lay over the image of the lower target part. Also, the shadows in SAR and EO/IR images don't align because in the former case the radar is the illuminator and in the latter case the sun (or moon) is the source of illumination.

4. Errors: The data from each sensor has different kinds of errors in target location, time stamp, and geometry. In this chapter, we are assuming that data from sensors being fused are roughly equally trustworthy. If this is not the case, the fusion solution must take into account the relative trustworthiness of sources. When registering images from two or more EO/IR sensors, one must also take into account optical distortion, parallax, atmospheric turbulence, interlacing if present, and bad or missing pixel data.
5. Latency and Time Synchrony: Different time scales, frame rates, and operating ranges complicate fusion. Some FLIR cameras can image at 120 frames/s or more, while a SAR system takes several seconds to form an image. Even with the best dual-band FLIR camera, the two bands of imagery will not be perfectly aligned in time due to the different integration times for each spectral band. Some dual-band cameras do worse, interleaving bands in time.
6. Statistical Independence: One of the basic tenets of pattern recognition is that improved performance can be achieved by combining statistically uncorrelated data. Data from a vibrometer, acoustic sensor, magnetic sensor, radar, and infrared sensor are fairly independent. At the other extreme, data from a LWIR and MWIR camera are highly correlated at night. Data independence is also a function of conditions. Blowing sand and dust will severely affect visible band data, less so for LWIR data, and hardly at all for passive millimeter-wave radar imagery.
7. Energy and Emission Minimization: The purpose of using multiple sensors may be to minimize energy and emissions. An acoustic sensor uses much less energy than an infrared camera and therefore may be used to wake up an IR camera in a smart landmine. For air-to-air operations, to minimize emissions, a radar sensor may be kept off until an IRST sensor detects a target.
8. Completeness of Data: Some sensors produce incomplete information. An additional sensor is required to fulfill mission requirements. Three examples follow: (1) A radar range profile is very useful for automatic target detection and recognition but less useful for human decision as required by the rules of engagement. Therefore, a narrow-field-of-view IR camera may be slewed toward the radar-detected target to provide literal imagery for the human observer. (b) An infrared camera does not inherently provide range to a distant target. A radar or laser rangefinder can provide the missing information. (c) With a third-generation dual-band (LWIR/MWIR) camera, each band will have some missing or bad pixels. Data from one band can help make a best estimate for missing pixel values in the other band.

5.3 Levels of Fusion

ATR engineers categorize multisensor fusion in terms of levels:

- Low Level: Data-Level Fusion
- Mid Level: Feature-Level Fusion
- High Level: Decision- or Score-Level Fusion

Each of these will be discussed in the following sections. But note that many other categorizations exist.[3,4] These categories typically rank fusion types by level number—often Levels 0 to 4. New fusion frameworks are introduced at the IEEE International Conferences on Multisensor Fusion, National or NATO symposia on fusion, and SPIE Defense + Commercial Sensing symposia. One widely used fusion framework is the Joint Directors of Laboratories (JDL) Data Fusion Working Group Model shown in Fig. 5.4.

5.3.1 Data-level fusion

Data-level fusion refers to the combination of raw or lightly processed data from two or more sensors or sensor components. Pixel-level fusion is a special case of data-level fusion. It often refers to images $\mathbf{I}_1, \ldots, \mathbf{I}_N$, each in a different spectral band, being combined to form a single multiband image $\mathbf{I}$, as shown in Fig. 5.5. Pixel-level fusion is best accomplished with simultaneously triggered sensors having the same field of view. Several common examples are reviewed.

Most standard color cameras use a Bayer filter pattern over an $H \times V$ element sensor chip as shown in Fig. 5.6, where H, V denote the respective number of pixels horizontally and vertically in the array. Sparsely sampled red, green, and blue images are formed each frame time. Resolution is reduced since only ¼ of the photosites capture red, ¼ blue, and ½ green. The three

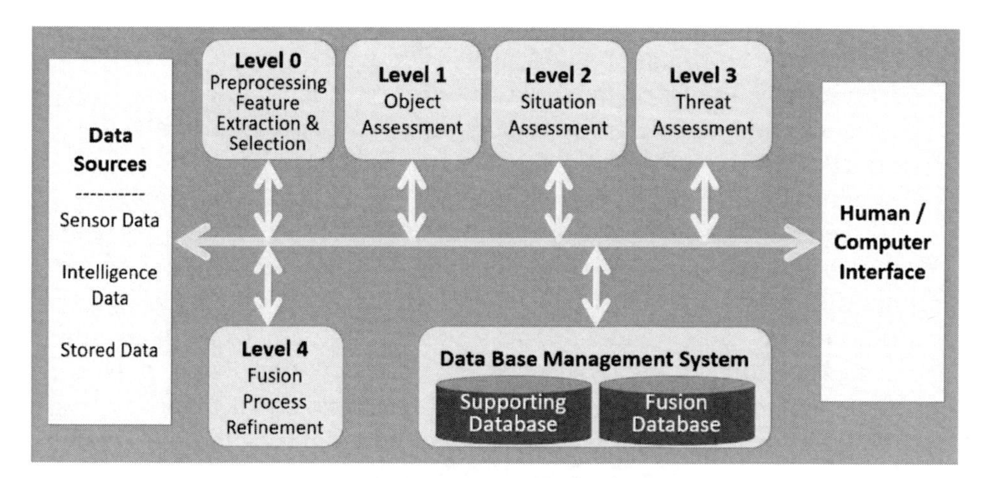

Figure 5.4 JDL data fusion model.

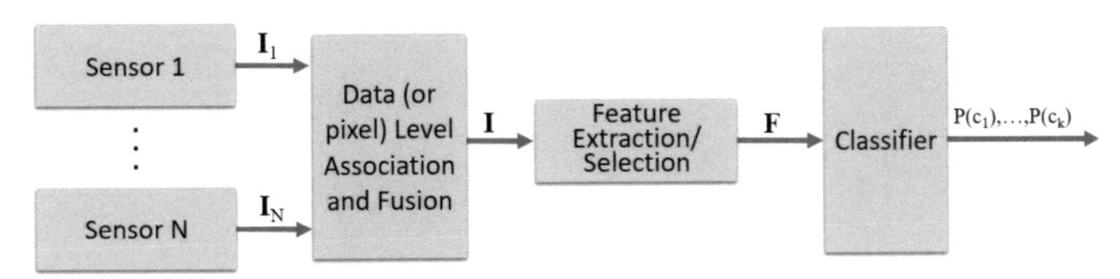

Figure 5.5 Pixel-level fusion most commonly refers to multiple time-synchronized imaging sensors feeding separate images $\mathbf{I}_1, \ldots \mathbf{I}_N$. to a fusion engine, resulting in a multiband image **I**.

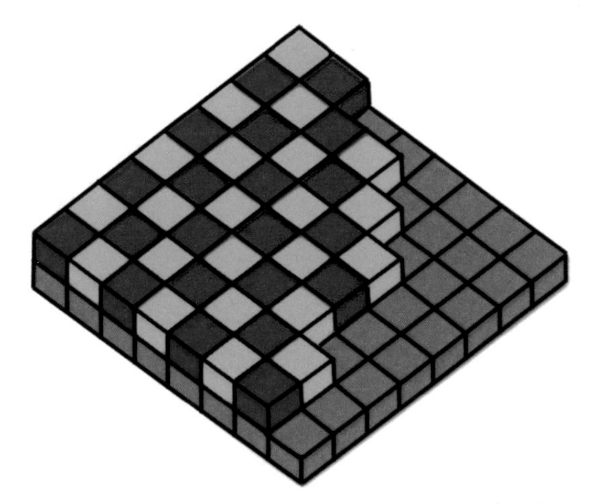

Figure 5.6 A Bayer filter array is a particular arrangement of red, green, and blue color filters over a single grid of photodetectors.

separate, sparsely sampled red, green, and blue images are interpolated to estimate the missing pixel values and then combined into a full-color image by a process known as demosaicking.[5] This produces an $H \times V$ pixel image, with a {red(x), green(x), blue(x)} vector at each pixel x. Note that the true resolution of the multiband image is less than $H \times V$.

The vast majority of color cameras use a single CCD or CMOS sensor chip overlaid with color filters in a Bayer or other pattern (Fig. 5.6). The advantages of the single sensor chip approach are small size and low cost. At higher size, weight, and cost, a three-chip camera can use prism optics to split the incoming light into three color channels, with one sensor chip for each color. The advantages of this approach are higher color resolution and better color accuracy. This approach is used in some high-end video cameras. It is also used in specialty cameras that produce imagery in bands other than {red, green, blue}.

A three-chip camera usually contains a trichroic prism assembly arranged as shown in Fig. 5.7. The broadband beam enters the first prism P1. The blue

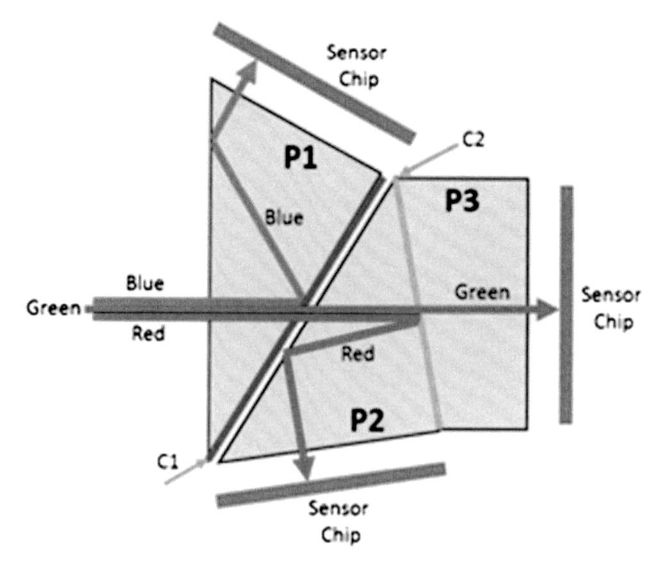

Figure 5.7 A three-CCD or three-CMOS camera uses an assembly of prisms to split the incoming light into three bands.

component of the beam hits coating C1, which reflects blue light, which then emerges from the side face of the prism. The coating C1 transmits the longer wavelengths. The light beam minus the blue components enters prism P2. The beam is then split by coating C2. The red light is reflected and passes out the side of prism P2, helped by a small air gap between prisms P1 and P2. This leaves the green component to be transmitted out of prism P3. The resulting three image components are captured by three separate sensor chips and are then generally combined into a vector image by an FPGA.

Another approach is to put several sensors into a single package. This approach is common in military systems since multiband cameras do not exist for certain desired combinations of bands. The cameras must be very close together and accurately aligned. For example, the Northrop Grumman high-resolution Lightweight Laser Designator Rangefinder (LLDR-2H) has the following components: celestial high-accuracy azimuth device, digital magnetic compass (DMC), embedded GPS/Selective Availability Anti-Spoofing Module (SAASM) receiver, day camera with high-resolution CCD, cooled MWIR camera, and eyesafe laser rangefinder (Fig. 5.8). Future LLDR systems will incorporate even more sensor types and use more advanced processing.

Bands other than {red, green, blue} have military application. The combination {ultraviolet, green, blue} is appropriate for undersea use since the red band is rapidly absorbed by seawater. Clay-like low-reflectivity paint is used on some military ground vehicles and helicopters. The combination {near infrared, red, green} is effective against this type of paint as well as dismounts with a vegetative backdrop (Fig. 5.9).

(a) (b)

Figure 5.8 (a) Northrop Grumman Lightweight Laser Designator Rangefinder with High Resolution (LLDR-2H) combines multiple sensors into a single package. (b) Northrop Grumman AN/PED-1 LLDR. [Photographs from: http://www.northropgrumman.com/Capabilities/ANPED1LLDR/Pages/default.aspx (accessed Nov. 25, 2017). Used by permission.]

Figure 5.9 A {near-infrared, red, green} image of dismount with vegetative background.

Perhaps the future of infrared ATR lies with third-generation (3rd Gen) high-resolution dual-band FLIR cameras, which operate in the LWIR and MWIR bands. The two bands are ideally captured simultaneously; however, some cameras capture them in an alternating sequence. The two bands can be combined using wavelet techniques to produce three-band imagery for improved ATR and visualization.[6]

Most animals have a pair of eyes, forming a stereoscopic vision system. Predatory animals tend to have front-facing eyes. Prey animals usually have eyes on the sides of their head. Binocular overlap refers to the part of the scene seen by both eyes. Humans have 120 deg of binocular overlap on a horizontal plane. Horses have only about 65 deg of binocular vision, with the remaining 285 deg being monocular vision. Thus, horses and other prey animals keep a lookout for predators coming from all directions, while predators use their greater stereo vision to detect, track, and rapidly home in on prey.

Binocular vision also enhances visual quality through binocular summation and provides backup for loss of one eye. The one-eyed racehorse named Patch ran in the 2017 Kentucky Derby. He started in 20th post position. All horses were to his blind side at the start the race. Patch, being quite the longshot, finished in 14th place. Patch did better at the Belmont Stakes, finishing in third place. The conclusion is that stereo vision is a definite enhancement, but mono vision is often adequate.

A stereo camera has two or more lenses with a sensor chip for each lens. Two nearby cameras on the same platform, looking in the same direction, also form a stereo pair. The cameras should be triggered simultaneously if the targets or platform are moving. Depth information can also be inferred from multiple 2D observations made by a single camera moving through the 3D environment, preferably with well-calibrated camera motion and a stationary scene. Binocular stereo and monocular motion stereo both use the principle of triangulation. The correspondence problem involves finding a precise match between the same scene elements in two images. The positions of the matched elements can then be triangulated in 3D space to provide range. The problem is two-fold: detecting corresponding points and accurately matching them. Doing so is very difficult, particularly for thermal infrared imagery, which is often rather fuzzy. Also, some 3D scene points are hidden from some camera views.

How useful is stereo vision to ATR? It is certainly a clue that most animals have stereo vision. But even for humans, only about 68% of the population has good to excellent stereo vision, while 32% have moderate to poor stereo vision. About 5% of the human population have amblyopia (lazy eye), causing poor or immeasurable stereo vision.[7] So, like the racehorse Patch, a sizeable portion of the human population manages to thrive without good stereo vision. Now let us consider only the part of the population performing military tasks. A 2014 Air Force study of studies concluded that "stereopsis plays some role in judging depth in the course of performing aviation tasks . . . "[8] "Tasks that are likely to depend on good depth perception" include "aerial refueling, clearing aircraft for landing, clearing aircraft during taxiing, and clearing aircraft during formation flight." . . . "While the need for stereo vision for tasks involving distance estimation seems intuitive, research examining the role of stereo vision and/or stereo displays often fails to show a clear relationship. Further, the results of different depth perception tests often differ substantially." It has traditionally been accepted that human eyes can perceive depth to about 6 m. However, one study hypothesizes that stereopsis provides an effective cue for distances up to 1 km. More conservatively, according to current USAF policy, stereopsis is assumed to be a helpful cue for distances within 200 m. This limits the usefulness of stereo vision to certain specific tasks such as landing, taxiing, and refueling, and leaves out the task of targeting at long range. Drivers of ground vehicles are dealing with shorter ranges.

The task-oriented cost versus benefit of stereo vision for undersea, air-to-ground, or ground-to-ground ATR has not yet been fully evaluated. The benefits of binocular stereo may be more in peripheral tasks such as image enhancement, automatic landing, collision avoidance, passive ranging at shorter ranges, or grasping with robot arms, rather than the key ATR task of long-range target identification. Beyond binocular stereo lies many-look imaging. Circling a stationary target, say from a drone, to obtain and transition multiple 2D images into a 3D target depiction with SAR or EO/IR sensors is an interesting research topic. DeGraaf shows that wide-angle SAR combined with volumetric backprojection imaging yields 3D radar imagery suitable for ATR.[9] He uses 3D superresolution techniques to improve sharpness and reduce speckle.

Electromagnetic waves can be classified according to their wavelengths λ or frequencies f; $\lambda = c/f$, where c is the speed of light. The electromagnetic spectrum ranges from RF waves to gamma rays. The atmosphere is fairly transparent to the particular bands used by military sensors: visible, SWIR, MWIR, LWIR, and RF. Visible light is only special in the sense that human eyes can detect it. Light has various properties, including:

- intensity (amplitude),
- frequency (spectrum),
- speed (300 million m/s in vacuum),
- polarization (orientation of light vibration),
- momentum (linear momentum in the direction of propagation, or angular momentum if the light is circularly polarized), and
- propagation direction.

An object is observed via emitted and/or reflected light. The illumination source can be a laser, sunlight, moonlight, starlight, street lamps, or a radar transmitter. ATRs often use intensity and/or frequency for target ID. But what about polarization?

All light is made of photons. A photon is a discrete quantum bundle of electromagnetic (or light) energy. A photon exhibits wave–particle duality, simultaneously behaving as a particle and a wave. The polarization of an electromagnetic wave refers to the direction of the electric field. Figure 5.10 provides a NASA pictorial on polarization. (Good animations of linear and circular polarization are given on various Wiki and YouTube web sites.)

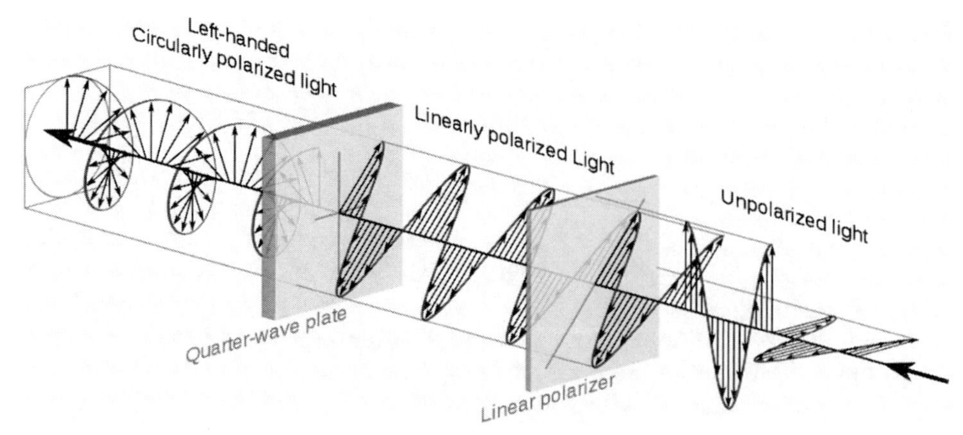

Figure 5.10 The waves of energy are called electromagnetic (EM) because they have oscillating electric and magnetic fields. The electric and magnetic field vectors point in directions perpendicular to each other. They are also perpendicular to the direction the wave is travelling. At any instant of time, the electric field vector of the wave describes a helix along the direction of propagation. A circularly polarized wave can be in one of two possible states: *right circular polarization*, in which the electric field vector rotates in a right-hand sense with respect to the direction of propagation, or *left circular polarization*, in which the vector rotates in a left-hand sense. (Image from NASA.gov—not subject to copyright.)

Sunlight, lamplight, and firelight are unpolarized. Unpolarized light can undergo polarization (1) by reflection off of nonmetallic surfaces, (2) by refraction, or (3) when the light is scattered while traveling through a medium. So when unpolarized light, for example, sunlight, strikes a smooth surface, such as a mirror or optics on a vehicle, or a puddle of water, it is reflected in polarized form, such that its vibration direction is parallel to the reflecting surface. Reflection off of rougher surfaces like soil and grass will be much less polarized. Polarization is also observed in the sky and undersea due to scattering by molecules and particles.

Many undersea creatures, birds, and insects sense polarization.[10] Polarization helps them navigate, find prey, and in some cases signal other members of their species. This is a hint that polarization information may be useful to an ATR. One way to form a polarization camera is with superpixels consisting of polarization filters in four directions, 45 deg apart (Fig. 5.11). The filters are placed over a monochrome focal plane array. The data in the four components of each superpixel are combined to yield such information as degree of linear polarization, angle of polarization, and degree of circular polarization. A less elegant approach to achieve the same result would be to use four cameras, each with a linear polarization filter rotated 45 deg apart. Military uses of polarization imaging include target detection, passive ranging, image enhancement in degraded environments, and, more specifically, detection or suppression of solar reflections off of mud puddles, lake or

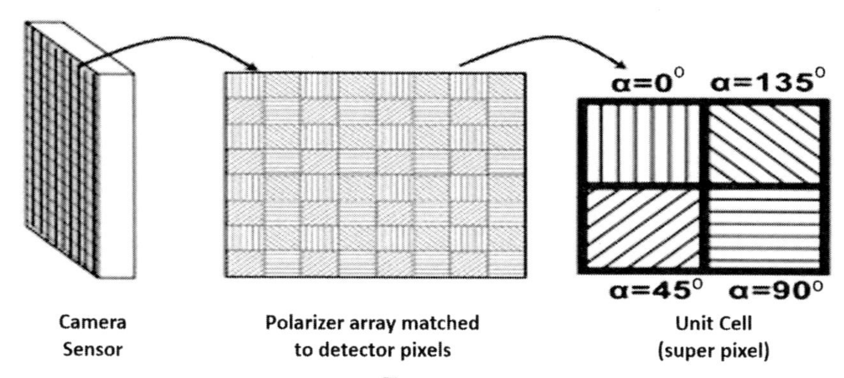

Figure 5.11 4D Technology PolarCam™ Micropolarizer Camera: superpixels, each with four discrete polarizations, tesselate the entire micropolarizer array. (Picture used by permission of 4D Technology Company.)

ocean surfaces, windshields, and optics.[11–13] The advantages of polarization imaging must be traded off against reduction in signal-to-noise ratio and/or resolution, and complicatedness of sensor design. The net benefit of EO/IR polarization to ATR remains unclear.

Change detection techniques are used to determine the differences between a pair of images that are often collected one day to the next. SAR-based change detection is used to determine the arrival or departure of vehicles, parked aircraft, surface mines, ships at port, building materials, or cargo. Change detection can also be used to detect scene changes due to bomb damage, earthquakes, hurricanes, downed aircraft, or more subtle changes due to footprints, tire tracks, or tank tread tracks. There are three basic types of SAR change detection:

1. a change in detected target locations (using ATR reports),
2. incoherent change detection (using pairs of registered SAR magnitude images), and
3. coherent change detection (using registered pairs of complex SAR images, where the term complex refers to magnitude and phase values at each pixel).

Technique 1 requires accurate target location. Techniques 2 and 3 require careful control of radar calibration and repeat pass geometry. Incoherent SAR change detection is essentially the same approach as used for optical imagery. Incoherent change detection is used when only (often compressed) SAR magnitude imagery is available. Coherent change detection is commonly formulated as a hypothesis test problem leading to a log likelihood solution. All three approaches are viable only if the effects of noise and minor scene changes can be mitigated. The same approaches can also be used for side-look sonar observing the sea bottom.

5.3.2 Feature-level fusion

Feature fusion is the process of combing two or more feature vectors into a single feature vector (Fig. 5.12). The objective is improved discriminating power as compared to each vector being used alone. For feature fusion to make sense, each of the feature vectors being combined should be of high quality, cotemporal, bounded to the same object or event, fairly nonredundant, and properly normalized. Feature fusion is the foundation of feature-level sensor fusion (Fig. 5.13).

Data-level sensor fusion is usually accomplished with commensurate sensors, like coboresighted MWIR and LWIR cameras. By contrast, feature-level sensor fusion works well with complementary sensors like: {LIDAR, hyperspectral}, {magnetometer, gyro, accelerometer}, {audio, video}, {acoustic, seismic}, or {ground-penetrating radar, electromagnetic induction sensor}.

Feature-level sensor fusion can be used in a bandwidth-reduction strategy. A communication channel with insufficient bandwidth to support transmitting the raw data may have sufficient bandwidth for features derived from the raw data. Bandwidth requirements are reduced, but not as much as when transmitting only class decisions made by an ATR associated with each sensor.

5.3.2.1 Sensor selection for feature-level sensor fusion

The more sensors that contribute to a classification decision the more information will be available to make the decision. The downside is the substantial cost of using multiple sensors.

The objective is to choose a set of sensors $S = \{S_1, S_2, \ldots, S_n\}$ from all potentially available sensors S'; $S \subseteq S'$ so that the chance of a correct

Figure 5.12 Feature fusion combines *N* feature vectors into a single feature vector **F**.

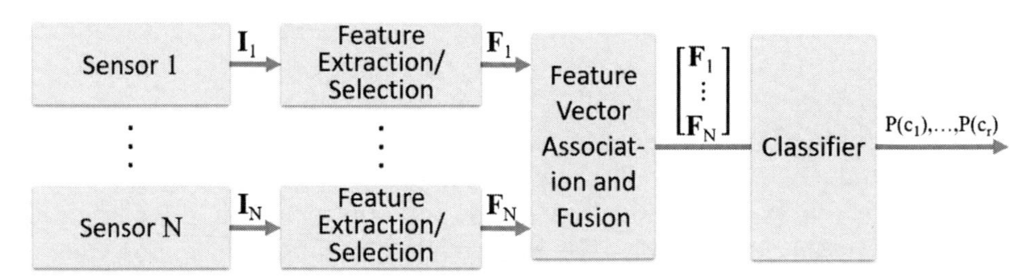

Figure 5.13 Feature-level sensor fusion first combines *N* feature vectors derived from data from *N* sensors into a single feature vector, which is then used to classify the target.

classification decision is maximized. The sensor selection problem is often formulated in terms of information theory.

The mutual (shared) information between variables $X = (x_1, x_2, \ldots, x_m)$ and $Y = (y_1, y_2, \ldots, y_o)$ is given by

$$I(X;Y) = \sum_{x_i \in X} \sum_{y_j \in Y} p(x_i, y_j) \log \frac{p(x_i, y_j)}{p(x_i)p(y_j)}, \tag{5.1}$$

where $p(x, y) = p(y, x) = p(x)p(y|x) = p(y)p(x|y)$ is the joint probability density function of X and Y, and $p(x)$ and $p(y)$ are the marginal probability density functions of X and Y, respectively.

The entropy of random variable X is given by

$$H(x) = \sum_{x} p(x) \log p(x). \tag{5.2}$$

$H(x)$ can be thought of as the expected information learned from one instance of the random variable X. The mutual information between variables X and Y can be expressed in terms of entropy:

$$I(X;Y) = H(X) - H(X|Y) = H(Y) - H(Y|X) = I(Y;X). \tag{5.3}$$

This is illustrated in the form of a Venn diagram (Fig. 5.14).

The mutual information between class vector $C = (c_1, \ldots, c_r)$ and feature vector $V = (v_1, \ldots, v_s)$ can also be expressed in terms of entropy:

$$I(C;V) = \sum_{c_i \in C} \sum_{v_j \in V} p(c_i, v_j) \log \frac{p(c_i, v_j)}{p(c_i)p(v_j)} = \sum_{c_i \in C} \sum_{v_j \in V} p(c_i, v_j) \log \frac{p(c_i|v_j)}{p(c_i)}. \tag{5.4}$$

Adding a sensor S_2 to a baseline sensor S_1 should improve the classification decision. But that is not the whole story. In order to decide whether it is worthwhile to add the second sensor, one must take into account (1) the redundancy in the features derived from the two sensors and (2) the total cost of adding the second sensor. Equation (5.4) measures the benefit of adding the

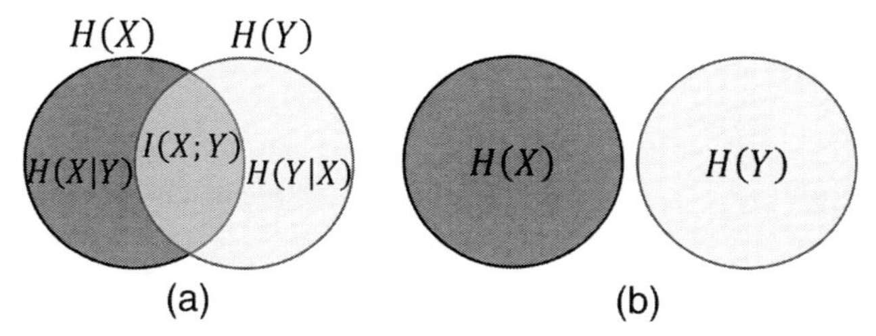

Figure 5.14 (a) The mutual information $I(X;Y)$ of random variables X and Y is a measure of the mutual dependence between the two variables. $H(X)$ and $H(Y)$ are the individual entropies of X and Y, respectively. (b) Two independent variables X and Y have no mutual information.

second sensor. Penalty Terms 1 and 2 (given below) measure the disadvantages of adding the second sensor.

The redundancy penalty is given by

$$\text{Penalty Term 1: } \frac{\beta_1}{|V_1||V_2|} I(V_1;V_2),$$

where β_1 is an appropriate constant, and $|V|$ denotes the number of elements in vector V.

$$\text{Penalty Term 2: } \beta_2 \times Cost(S_2),$$

where β_2 is an appropriate constant, and $Cost(S_2)$ includes all known costs: procurement, logistics, maintenance, risk, SWaP, data collection, test, and evaluation.

The analysis becomes more complex when choosing a best set of sensors $S = \{S_1, S_2, \ldots, S_n\}$. Equation (5.4) then becomes

$$I(C;V) = \sum_{c_i \in C} \sum_{v_j^{(1)}} \cdots \sum_{v_j^{(n)}} p(c_i, v_j^{(1)}, \ldots, v_j^{(n)}) \log \frac{p(c_i | v_j^{(1)}, \ldots, v_j^{(n)})}{p(c_i)}. \quad (5.5)$$

Various algorithms are available for efficient sensor selection, or similarly, efficient group feature selection.[14–17]

When two sensors supply statistically independent information, like range and cross-range location, the mutual information is zero [Fig. 5.14(b)]. For this case, each sensor can be evaluated separately.

Feature-Level Fusion Example: Networked unattended ground sensor (UGS) devices are used for perimeter protection around military camps and buildings. UGS nodes generally incorporate acoustic and seismic sensors. This sensor combination is useful for detecting and localizing approaching intruders on foot, as well as ground vehicles, helicopters, gunfire, and explosions. Other sensor types can augment the acoustic/seismic sensor array, such as EO/IR, geophone, accelerometer, magnetometer, and sensors on aerostats or tiny drones.

Typical acoustic features result from:

- ordinary human activity (putting on/off jacket or helmet, searching through backpack, eating, coughing, scratching, etc.),
- walking/running,
- weapons handling,
- talking in person or by radio,
- weapons fire or explosions, and
- ground vehicles, helicopters, and fixed-wing aircraft (each model of which may have a unique signature).

Typical seismic features result from:

- footsteps and
- wheeled vehicles, tracked vehicles, and aircraft.

Feature extraction and selection involve signal processing, spectrum analysis, and association of unreliable temporal data from sensors co-located on the same UGS device and the network of UGS devices. Typical features are Fourier or wavelet coefficients, statistics derived from Fourier or wavelet coefficients, power spectral density, and cadence. Fusion is a challenging multistage process, taking into account the threat level, background noise, context, weather, nearby roads and trails, positions of devices, and spatial and temporal reliability of information. Fusion may be centralized or take place in each separate device using only its local data. A detected threat triggers the transmission of an alert. For the UGS problem, feature-level fusion is preferred over decision-level fusion.[18–21]

5.4 Multiclassifier Fusion

Multiclassifier fusion operates on individual classifier outputs. It is also referred to as post-declaration fusion or decision-level fusion. The software or hardware doing the fusion is called a fusion engine, ensemble combiner, or committee machine. The classifiers being fused can be of the same type using different features, of the same type but trained differently over the same features or subsets of features, or of different types. As with individual classifiers, a multiclassifier decision falling below a threshold score can be rejected.

From a Bayesian viewpoint, the output of a proper classifier is a posterior probability,

$$P(y|x) = \frac{P(x|y) \times P(y)}{P(x)}, \tag{5.6}$$

which notionally represents

$$posterior = \frac{likelihood \times prior}{evidence}. \tag{5.7}$$

For ATR, we are trying to determine $P(c_i|\mathbf{x}) = P(\mathbf{x} \in c_i|\mathbf{x})$, which is the probability of input $\mathbf{x}$ belonging to class c_i, given the information in the input vector (or, more generally, data) $\mathbf{x}$. One may also have priors $P(c_i)$, which are the known, estimated, or assumed probabilities of the classes c_i; in this case,

$$\begin{aligned} &\textit{Probability of the } i^{th} \textit{ class conditioned} \\ &\textit{on the evidence } \mathbf{x} = P(c_i|\mathbf{x}) = \frac{P(\mathbf{x}|c_i) \times P(c_i)}{P(\mathbf{x})}. \end{aligned} \tag{5.8}$$

When the $P(c_i)$ are assumed to be equal, which is often the case, the equation loses some of its Bayesian flavor. With a still less Bayesian interpretation, a classifier is said to make a fuzzy class match, which can nonetheless be viewed as a rough approximation to a posterior probability $P(c_i|\mathbf{x})$. Proper normalization of classifier outputs is required. This can be achieved by dividing the outputs of any classifier by the sum of the output values, thus normalizing individual approximated posterior probabilities to the range [0, 1]. The sum of all possible (mutually exclusive) outputs should be 1. An alternative normalizer is the softmax function.

Any classifier is said to make soft decisions if its output is of the form

$$\begin{bmatrix} P(c_1|\mathbf{x}) \\ \vdots \\ P(c_r|\mathbf{x}) \end{bmatrix},$$

where $\mathbf{x}$ is the input vector, or, more generally, input data, and $\{c_1, \ldots, c_r\} \in C$ are the possible target classes, $0.0 \leq P(c_j|\mathbf{x}) \leq 1.0$, $\forall j = 1, \ldots, r$. The output (called a soft decision) is a vector of posterior probabilities conditioned on the input $\mathbf{x}$ to the classifier [Fig. 5.15(a)]. For a trainable classifier, these conditional probabilities are learned over a training data set, while the prior probabilities are handed to the classifier based on knowledge outside the training regime. (Alternatively, the classifier can be trained on a training set where the proportion of targets of each type represent the priors.) The validity of the function (e.g., neural network) mapping input vector to output posterior probabilities is sensitive to the train-versus-test environment, number of training samples, noise, and variable target appearance. For instance, with FLIR imagery, is the tank's engine on or off? If the training environment and target geometry don't adequately match the operational test environment and target geometry, then classifier performance will suffer. Or, to put it another way, the inferred conditional probabilities will have large errors. With a sufficiently large, representative, and comprehensive training database, the classifier will perform well. The trick (or cost) of ATR design involves coming up with a good training set. Big data is a term referring to a training database that is so

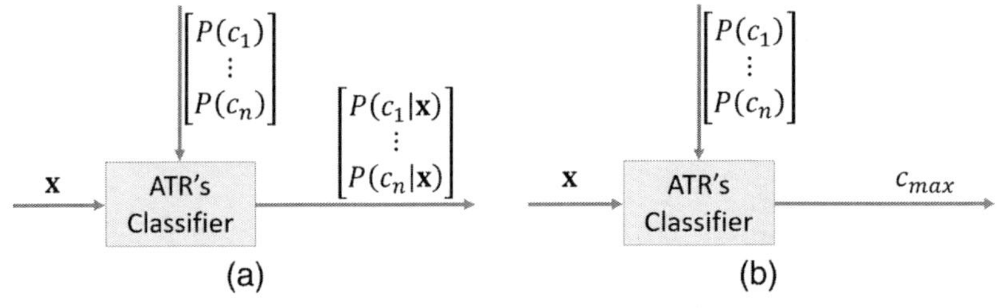

Figure 5.15 (a) ATR's classifier that makes soft decisions. (b) ATR's classifier that makes hard decisions.

large and complex that a traditional classifier is inadequate to deal with it, and a more complex classifier, such as a deep-learning neural network is required. Some commercial big-data problems involves training sets with hundreds of millions of samples. For military target recognition, obtaining a very large, perfectly ground-truthed training database of enemy targets (i.e., big data) is a difficult and sometimes prohibitively expensive undertaking.

A classifier is said to make hard decisions if it outputs a single class label c_{max} for each input **x**, such that

$$c_{max} = \underset{t=c_1,\ldots,c_r}{\operatorname{argmax}} P(t|\mathbf{x}), \tag{5.9}$$

where the argmax notation is defined as follows: $\operatorname{argmax}_{z \in D} f(z) = \{z | f(z) \geq f(y), \forall y \in D\}$.

Thus, max $f(z)$ is the maximum value of $f(z)$ as z varies through some domain, while argmax $f(z)$ is the value of z at which this maximum is obtained.

5.4.1 Fusion of classifiers making hard decisions

Suppose that there are N classifiers, as shown in Fig. 5.16. The classifiers can be of the same or different type. Each classifier is presented with a feature vector bound to the same object. Each classifier provides its hard decision, which is its target declaration. These declarations are combined by a post-declaration fusion engine. A consensus target declaration results. There are many different ways to fuse the possibly disparate decisions, as covered below.

5.4.1.1 Majority voting

Let the output of N classifiers making hard decisions be denoted by

$$\mathbf{d} = \begin{bmatrix} c_{max_1} \\ \vdots \\ c_{max_N} \end{bmatrix} = \begin{bmatrix} d_1 \\ \vdots \\ d_N \end{bmatrix}, \tag{5.10}$$

where the decision d_j of the jth classifier is denoted by $d_j \in \{c_1, \ldots, c_r\}$ for r classes.[22] An indicator function takes a value of 1 when an event happens, and

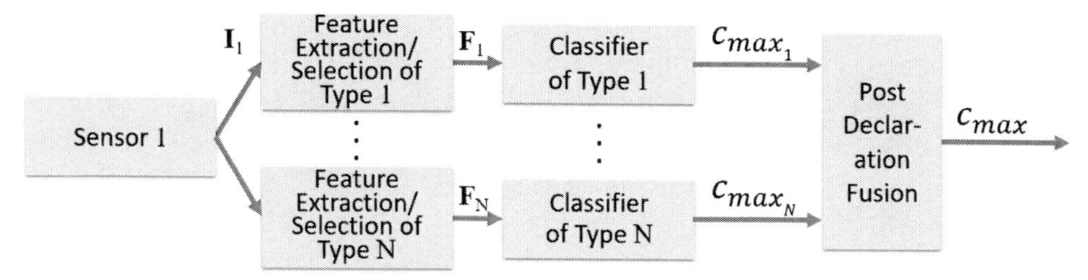

Figure 5.16 Hard class decisions from N classifiers can be combined into a single hard decision.

a value of 0 when an event doesn't happen. The binary characteristic function of the jth classifier is defined as

$$B_j(c_i) = \begin{cases} 1 & if \quad d_j = c_i \\ 0 & if \quad d_j \neq c_i \end{cases}. \tag{5.11}$$

This just says, for example, that if classifier $j = 3$ picks class $i = 2$, then class 2 gets one vote from classifier 3, and the other classes get 0 votes from classifier 3. To make a final decision, we just need to sum the votes from all N classifiers. The candidate winning the vote is given as

$$c_{max} = \underset{t=c_1,\ldots,c_r}{\text{argmax}} \sum_{j=1}^{N} B_j(t). \tag{5.12}$$

5.4.1.2 Combined class rankings

With a large number of target classes, say 100 or 1000, it may not be possible to make a clear-cut class declaration. Instead, let each of N classifiers output its k top class decisions for each input sample. Then the N lists can be merged and sorted, with the classes receiving the most votes moved to the top of the list. This narrows the number of possibilities that the final decision maker, being the human-in-the-loop, needs to consider.

5.4.1.3 Borda count method

The Borda count method seeks consensus among N classifiers, where each classifier ranks the classes using different criteria.[22] The Borda count $BC_j(c_i)$ is the sum of the number of classes ranked below class c_i by the jth classifier, $j = 1, \ldots, N$. The decision is given as

$$c_{final} = \underset{t=c_1,\ldots,c_r}{\text{argmax}} \sum_{j=1}^{N} BC_j(t). \tag{5.13}$$

Thus, the class c_i producing the maximum Borda count is the consensus decision.

5.4.1.4 Condorcet criterion

A Condorcet winner is the class that wins a two-class decision against each of the other classes. Thus, the Condorcet winning class is individually preferred to any of the other classes, which is a reasonable criterion for ATR. For r classes, this can be achieved by considering $r(r - 1)/2$ pairwise decisions. The existence of a Condorcet winning class is likely, but is not guaranteed. Some Condorcet methods work by having each classifier rank the classes top to bottom, in order of preference. Various algorithms can then be used to find a Condorcet winner if one exists. The Borda count, majority voting, plurality voting, and many other voting schemes do not satisfy the Condorcet criterion.

5.4.2 Fusion of classifiers making soft decisions

Now suppose that we use N classifiers, where the ith classifier receives feature vector $\mathbf{F}_i$ (Fig. 5.17). These feature vectors can be the same or different. Each classifier provides a vector of posterior class probabilities. Often, when the classifiers are of different types, they will be fed feature vectors of the same type, which will be our assumption.

5.4.2.1 Simple Bayes average

The simple Bayes average approach treats the outputs of each classifier as posterior probabilities, where $P(c_k|\mathbf{x}) = P(\mathbf{x} \in c_k|\mathbf{x})$ is the probability that an input sample $\mathbf{x}$ belongs to class c_k, $k = 1, \ldots, r$. The average posterior probability when using N classifiers is

$$P'(c_k|\mathbf{x}) = \frac{1}{N}\sum_{i=1}^{N} P_i(c_k|\mathbf{x}). \tag{5.14}$$

The sum rule selects the class with the highest average posterior probability:

$$c_{max} = \underset{t=c_1,\ldots,c_r}{\operatorname{argmax}} P'(t|\mathbf{x}). \tag{5.15}$$

This simple rule treats all class priors as equal and all classifiers as equally competent. For ATR, we once again note that simplicity is a virtue. This Bayes average approach is straightforward and performs quite well. One can replace the average operator in Eq. (5.14) with a maximum or median operator:

$$P'(c_k|\mathbf{x}) = \max_i P_i(c_k|\mathbf{x}), \tag{5.16a}$$

$$P'(c_k|\mathbf{x}) = \operatorname{med}_i P_i(c_k|\mathbf{x}). \tag{5.16b}$$

One may question the use of these simple rules rather than the optimal Bayes decision rule. Leaving out the class-independent normalization term, the optimal rule is

$$P'(c_k|\mathbf{x}) = \frac{1}{B}[P(c_k)]\prod_{i=1}^{N} P_i(c_k|\mathbf{x}), \tag{5.17}$$

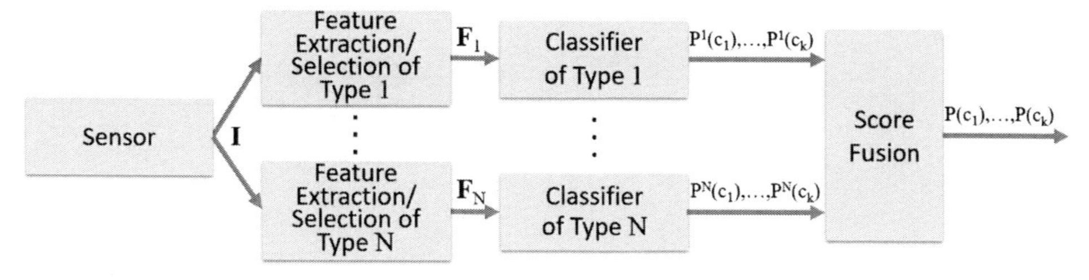

Figure 5.17 Fusion of soft decisions from N classifiers.

where B is a normalizing constant that is introduced so that the resulting probabilities sum up to one.

The reason that the optimal Bayes decision rule is seldom used for ATR is that if one classifier outputs a zero probability for a particular class c_k, then it doesn't matter how high a score the other classifiers give it. The product rule is said to suffer from the veto problem in that one classifier can veto the good work of all of the other classifiers.

5.4.2.2 Bayes belief integration

Now suppose that an open test set is available. The N classifiers can be run on the open test set. Each classifier's performance can be reported in the form of a confusion matrix, as shown in Table 5.1, where s_{ij}^k denotes the number of open test set samples from class c_i that were assigned to class c_j by the kth classifier, $k = 1, \ldots, r$. This is done in advance, off-line.

Now for on-line operation, let $\hat{c}^k$ denote the classification decision from the kth classifier. The belief in class c_i is obtained with a product rule,

$$Bel(c_i) = P(c_i)\frac{\prod_{k=1}^{r} P(\hat{c}^k|c_i)}{\prod_{k=1}^{r} P(\hat{c}^k)}, \tag{5.18}$$

where $P(\hat{c}^k|c_i) = P(\hat{c}^k|\mathbf{x} \in c_i)$ is the probability that the kth classifier output is $\hat{c}^k$ given that the unknown $\mathbf{x}$ was really in class c_i. Now using the predetermined confusion matrices s_{ij}^k,[23]

$$P(\hat{c}^k = c_l|c_i) = \frac{s_{il}^k}{\sum_{l=1}^{r} s_{il}^k}, \tag{5.19}$$

$$P(\hat{c}^k = c_l) = \frac{\sum_{i=1}^{r} s_{il}^k}{\sum_{i=1}^{r}\sum_{l=1}^{r} s_{il}^k}. \tag{5.20}$$

The class gathering the highest belief is the output of the fused classifiers:

$$c_{max} = \underset{i=1,\ldots,r}{\text{argmax}}\, Bel(c_i). \tag{5.21}$$

Equation (5.18) uses a product rule and as such is sensitive to errors in just one of the classifiers.

Table 5.1 Confusion matrix for k^{th} classifier.

		Reported by k^{th} Classifier		
Truth		c_1	$\ldots$	c_r
	c_1	s_{11}^k	$\ldots$	s_{ir}^k
	$\vdots$	$\vdots$	$\ddots$	$\vdots$
	c_r	s_{r1}^k	$\ldots$	s_{rr}^{jk}

5.4.2.3 Trainable classifier as a combiner

N classifiers output N vectors of conditional probabilities. These N probability vectors can be concatenated into a single vector, which can then be treated as a feature vector. Testing an ensemble of classifiers on S samples of test data will produce S such concatenated feature vectors. These S feature vectors can be used to train another classifier.

This approach is called a two-level classifier. At the lowest level, individual classifiers produce posterior probabilities that are treated as feature vectors. These feed the next level of the classifier. One could elaborate on this design to produce a hierarchy of classifiers. The bottom layer of classifiers would be trained first, then the next layer of classifiers would be trained on outputs from the bottom layer, and finally the upper layers would be trained. This approach is brittle to differences between train and test data.

5.4.2.4 Dempster–Shafer theory

Dempster–Shafer theory (DST) is a method for combining evidence from different sources.[24,25] The sources may provide different output labels, a situation not easily handled by other fusion techniques. DST outputs a degree of belief in a proposition. Recently, DST has been used as an umbrella term to cover a family of evidence combination techniques.[26,27] Some of the newer techniques combine strongly conflicting evidence better than the original approach.

We illustrate the basic approach with a simple example. Suppose that there exists a set of mutually exclusive hypotheses $\Theta = \{\theta_1, \ldots, \theta_r\}$. Under the closed-world assumption, one of the hypotheses must be true. From an ATR perspective, the θ_i represent target classes $c_i; \Theta = \{c_1, \ldots, c_r\}$.

For example, let $\Theta = \{H, F\}$, where H denotes helicopter, and F denotes fixed-wing aircraft. The power set $\mathbb{P}(\Theta)$ includes 2^r elements: all of the possible subsets of Θ plus the empty set φ. For our example, $\mathbb{P}(\Theta) = \{\varphi, \{H\}, \{F\}, \{H, F\}\}$. The number of possible subsets increases exponentially with the number of classes. So it is obvious that this approach is not viable for large r, e.g., for $r = 100$, since $2^{100} = 10^{30}$.

Suppose that the ATR associated with Sensor 1 says that the detected target is either H or F with an evidence measure of 0.7:

$$m_1(\{H, F\}) = 0.7.$$

Any evidence not assigned to a particular target or particular group of targets is assigned to Θ, so

$$m_1(\{\Theta\}) = 0.3$$

since the sum of the evidence measures has to be 1.0.

Suppose that the ATR associated with Sensor 2 says

$$m_2(\{H\}) = 0.9.$$

Then,

$$m_2(\{\Theta\}) = 0.1.$$

Let m_{12} denote the conjunctive consensus between the two sources of evidence. The DST rule combines evidence measures

$$m_{12} = m_1 \oplus m_2(X) = K \sum_{Y \cap Z = X} m_1(Y) m_2(Z), \tag{5.22}$$

where K represents the conflict between the two beliefs. $K = 1$ if there is no conflict, which is our assumption for this example. As an aside: Many other combination rules exist that sidestep consideration of conflict. For example, consider the alternative combination rule of Dubois and Prade:[28]

$$m_{12} = m_1 \oplus m_2(X) = \sum_{Y \cup Z = X} m_1(Y) m_2(Z).$$

Then,

$$m_{12}(\{H\}) = \text{fused evidence for } \{H\} = 0.63 + 0.27 = 0.9,$$

$$m_{12}(\{H, F\}) = \text{fused evidence for } \{H, F\} = 0.07,$$

$$m_{12}(\{\Theta\}) = \text{unassigned evidence} = 0.03.$$

So, the target declaration is helicopter. Table 5.2 provides the data for making the target declaration.

5.5 Multisensor Fusion Based on Multiclassifier Fusion

Techniques for combining N classifiers were provided in Section 5.4. It was assumed that several classifiers were presented with a feature data derived from a single sensor. Now suppose that we have N sensors. Sensor i feeds a feature vector $\mathbf{x}_i$ into the ith classifier. This converts multiclassifier fusion into multisensor fusion, as shown in Figs. 5.18 and 5.19.

For multiclassifier fusion, a diverse ensemble of classifier types is preferable to an ensemble of highly correlated classifiers. A diverse set of sensor types is preferable over those producing highly correlated output. For the simultaneous combination of classifier outputs, each representing a different sensor type

Table 5.2 Computations used in the simple example.

	$m_2(\{H\}) = 0.9$	$m_2(\{\Theta\}) = 0.1$
$m_1(\{H, F\}) = 0.7$	$\{H\} = 0.63$	$\{H, F\} = 0.07$
$m_1(\{\Theta\}) = 0.3$	$\{H\} = 0.27$	$\{\Theta\} = 0.03$

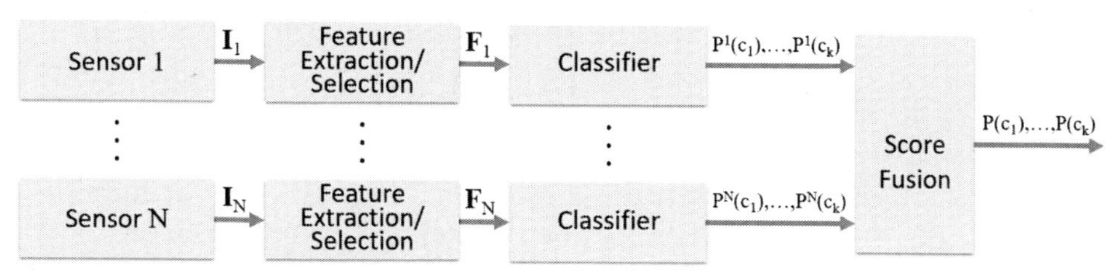

Figure 5.18 Fusion of the soft outputs of *N* classifiers, where each classifier is fed by a feature vector resulting from a different sensor.

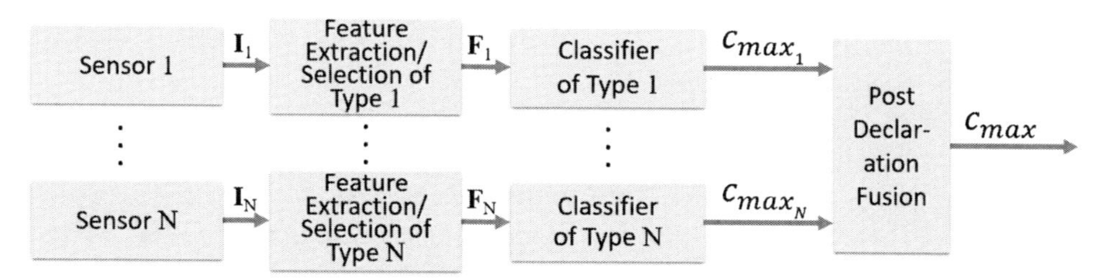

Figure 5.19 Fusion of the hard outputs of *N* classifiers, where each classifier is fed by a feature vector resulting from a different sensor.

(on platform) using different phenomenology, the various sensors being combined have to be able to image the same target, at the same range, at the same point in time. This is often difficult to achieve, as different sensor types have drastically different operating ranges, fields of view, and latencies. It is difficult to determine a target boundary in data from two different sensor types that see the world differently. It is hard to fuse data from a radar that normally operates at 50 km to target, with data from an uncooled FLIR sensor that normally operates at 0.5 km to target. It would be hard to simultaneously fuse decisions from an undersea color camera that normally operates at 15 feet from target, with a sonar that operates at 2 km to target. However, it may be possible to fuse the data over time as the platform moves closer to the target.

5.6 Test and Evaluation

Test and evaluation (T&E) of multisensor fusion systems is basically the same as that for single sensor systems, but with additional complexity. Testing should still include laboratory self-test, field test, and independent blind test by government agencies. Testing should take into account the effects of noise, countermeasures, decoys, and all forms of background and target variation. Test site items for one sensor type (like calibration panels, corner reflectors, orange flags, smoke generators, or smudge pots) should be reconsidered if they affect the data from another sensor type. We don't want the really clever multisensor ATR to learn to detect targets by the little orange flags used in a data collection to indicate where the vehicles should be parked.

First consider a test of a SAR-based ATR over foreign ground vehicles. Suppose that the engines are not working for two of the vehicles. They can be towed into place. No problem.

Now suppose that we add two sensors to the test: color visible and MWIR, each flying on a different helicopter. The two non-operational vehicles are no longer suitable, since their signatures cannot be varied with engine, heaters, headlights on/off. Substitutes will have to be found for these vehicles. Target signatures in the visible and thermal infrared bands vary over season, time of day, and climatic zone, and with rain, fog, and blowing sand. Testing now needs to take place at different locations and seasons.

Now suppose that two more sensors are added to the test: ground-based acoustic and seismic. The problem is that the acoustic and seismic signatures are ruined each time a helicopter flies by. There could be other tests or training exercises taking place in the area using additional helicopters.

Now suppose that we add three more sensor types to the fusion-based ATR tests: LIDAR, LWIR, and range-gated laser-illuminated SWIR. This brings the total number of sensor types at the test to eight. Use of lasers introduces eye safety issues. The test plan will need by be reviewed by an eye safety panel. Each person at the test site may have to get an eye exam before and after the test, and use laser safety glasses during the test. With eight sensor types, some of which are new designs, there is a high likelihood that one or more will fail during the test. If the contract requires an eight-sensor fusion test, then the test will have to be suspended until the broken sensor is repaired. That means that all of the test personnel (government testers, foreign vehicle drivers, ground crew, ground truthers, pilots, and contractors) will be sent home until the test can be rescheduled.

Multisensor ATR T&E will have more chance of success with a well thought out and detailed test plan, signed off by all stakeholders. The test plan needs to cover safety issues, environmental issues, and contingencies for bad weather, equipment failure, and illness of key personnel. Domain knowledge and good engineering practices are necessities.

5.7 Beyond Basic ATR Fusion

We will review several other kinds of fusion, beyond the usual archetype taking place within an ATR.

5.7.1 Track fusion

Track fusion most often refers to merging tracks from several sensors, each of which treats targets as points in 3D space (Fig. 5.20). This is typically the case for long-range air-to-air targeting. Track fusion is similar to fusion for target identification using higher resolution data, except that the focus of track fusion is target state (location and velocity). Even so, location and velocity are clues to target identity. For example, a helicopter can't fly as fast as a jet fighter or

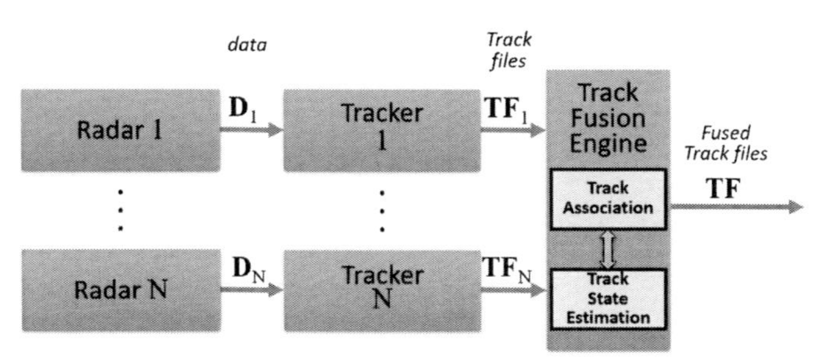

Figure 5.20 Track files resulting from *N* radar systems can be fused to provide more-accurate track files.

missile. Let's start with the simplest case, which can be viewed as a form of post-declaration fusion. Consider several radar systems observing an aircraft in flight. Each system initiates a target track after a detection on three or more radar scans. These detections are appraised to see how well they match the motions and trajectories of the targets being sought. Extraneous tracks are discarded. The tracker makes a smoothed (e.g., Kalman-filtered) estimate of each tracked target's position and velocity at each update cycle. The tracks from each radar system are kept in a track file database that is updated as each new detection is associated with an existing track. The separate track files feed a track fusion engine. The fusion engine has two main tasks:

1. Track-to-track association: Tracks from the different radar systems are associated to form consensus tracks, one for each target.
2. Target state estimation: Consensus state estimates are obtained by fusing the state estimates of the associated sensor tracks.

Several problems need to be solved for track-to-track association. There are likely to be range, azimuth, elevation, and time stamp errors, which may remain constant or vary over time. Track errors make track fusion more challenging. Track state estimation combines and smooths the correlated, but noisy, data to improve position and velocity estimation.[29–32]

Track fusion can involve different types of sensors, such as IRST, radar, and ESM (Fig. 5.21). Several problems must be worked out. The sensors won't have the same volume coverage and update rate, and, possibly, coordinate systems and accuracy (types of 3D location errors).

Simple Track Fusion Examples

A simple approach to multisensor tracking is to separately track the target with each sensor, associate the tracks, and then combine the state vectors.[30] Let

$$\mathbf{x}(t) = \begin{bmatrix} \mathbf{r}(t) \\ \mathbf{v}(t) \end{bmatrix} \tag{5.23}$$

represent the state estimate of position vector $\mathbf{r}$ and velocity vector $\mathbf{v}$ at time t.

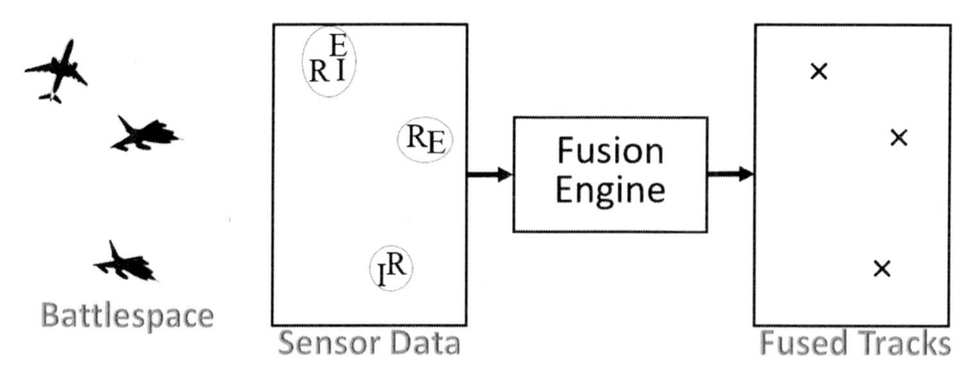

Figure 5.21 Multiple sensors can provide better target tracking than using a single sensor. One approach is to track the target with several sensor types and then fuse their tracks. Sensor types: IRST (I)—passive infrared, ESM (E)—passive electromagnetic radiation, and radar (R)—RF energy, transmitted, reflected & received.

Let $\mathbf{x}_1(t)$ and $\mathbf{x}_2(t)$ represent the state vector estimate resulting from two different sensors at time t. One state estimate may be noisier than the other. Let the measurement noise of sensor i be modeled as an independent Gaussian random variable with zero mean and variance σ_i^2, $i = 1, 2$. Thus,

$$\mathbf{x}_i(t) = \mathbf{x}_i'(t) + \mathbf{n}_i(t), i = 1, 2, \tag{5.24}$$

where $\mathbf{x}_i'(t)$ is the true state of the target at time t. For the simplest case, the noise $\mathbf{n}_i(t)$ is modeled by a scalar variance along each coordinate. The combined state vector is then

$$\mathbf{x}(t) = \sigma_2^2(\sigma_1^2 + \sigma_2^2)^{-1}\mathbf{x}_1(t) + \sigma_1^2(\sigma_1^2 + \sigma_2^2)^{-1}\mathbf{x}_2(t). \tag{5.25}$$

Thus, the state vector from the sensor with the lowest measurement error is weighted more heavily in the combined result than the state vector from the noisier sensor.

If the errors of the two sensors are uncorrelated but are best represented by covariance, the combined result becomes

$$\mathbf{x}(t) = \Sigma_2(\Sigma_1 + \Sigma_2)^{-1}\mathbf{x}_1(t) + \Sigma_1(\Sigma_1 + \Sigma_2)^{-1}\mathbf{x}_2(t), \tag{5.26}$$

where Σ_1 and Σ_2 are the error covariance matrices of the two sensors.

If the errors of the two sensors are assumed to be correlated, then the fused state estimate becomes

$$\mathbf{x}(t) = \mathbf{x}_1(t) + (\Sigma_1 - \Sigma_{12})\Sigma_2(\Sigma_1 + \Sigma_2 - \Sigma_{12} - \Sigma_{21})^{-1}[\mathbf{x}_2(t) - \mathbf{x}_1(t)], \tag{5.27}$$

where Σ_{12} and Σ_{21} are the error cross-covariance matrices.

The above examples assume Gaussian errors. Other approaches are needed when errors are not Gaussian or are varying over time.

So far, the track fusion engine has been described as a rather dumb black box. It is a passive recipient of data, making a decision based on whatever data it receives no matter how imperfect. More intelligence is needed to produce a better depiction of the battlespace. A so-called fifth-generation fusion engine actively tasks each sensor to get the required information to form a complete picture of the battlespace. Targets are engaged via sensor fusion even when no single sensor yields sufficient targeting information.

5.7.2 Multifunction RF systems

Multifunction radio frequency (MFRF) and cognitive radar systems are similar in concept, although, as implemented, each particular RF system has its own unique set of functions, hardware, software, and architecture. These new types of radar systems perform more complex fusion of functions and capabilities than older, more conventional radar systems.

5.7.2.1 Multifunction RF

A MFRF system is equivalent to a suite of radar and nonradar RF systems. Various functions normally performed by the different equipment are fused into a single system. The exact functions fused depend on the application and would be different for a radar employed on a fighter jet, Navy ship, or helicopter. MFRF systems can encompass such radar ATR functions as target detection, recognition, and tracking, and such nonradar functions as:

- electromagnetic support measures, also called electronic surveillance measures (ESM), to passively locate sources of radiated electromagnetic energy;
- signal intelligence (SIGINT) to analyze and identify intercepted RF radiation. SIGINT has two subcategories:
    - communications intelligence (COMINT) to gather electronic signals and
    - electronic intelligence (ELINT) for signals not directly used in communication;
- high-rate-data communication (Comms);
- electronic attack (EA), also called electronic countermeasures, such as jamming; and
- electronic self-protection (EP) to protect its own platform from EA.

Measurement and signature intelligence (MASINT) also operates over collected signals. However, the term MASINT usually refers to a broader discipline emphasizing analysis.

MFRF requires a higher performance scheduler than a conventional radar. Functions and resources must be managed to maximum efficiency. Some functions are interleaved. Other functions are performed concurrently.

For example, the DARPA MFRF program's goal was to develop a common RF system using agile frequencies, waveforms, and apertures to optimally interweave different functions according to an aircraft's missions.[33]

MFRF systems perform function-level fusion. An MFRF system can also be part of a highly structured multisensor fusion system. MFRF data can be fused with other types of sensor data, such as FLIR and LIDAR, terrain databases, and onboard platform navigation data, to enable a helicopter pilot to take off and land in degraded visual environments, as well as to improve ATR functionality.

An air target can be handed off to an MFRF system by another sensor in coordinates of:

- azimuth (by ESM)
  - Azimuth is the horizontal angle, where 0 deg azimuth is true north, and 180-deg azimuth is true south;
- azimuth and range (by 2D radar)
  - Range is the distance between the radar and the target measured along the line-of-sight;
- azimuth and elevation (by IRST)
  - The IRST system determines the compass bearing relative to true north of a point on the horizon directly beneath the observed target, then measures the angle between that point and the target, from the reference frame of the observer; or
- azimuth, elevation, and range (by 3D radar).

5.7.2.2 Cognitive radar

The key components of a conventional radar are its transmitter (Tx) and its receiver (Rx). Tx illuminates the scene with an emitted RF waveform. The emitted energy is reflected mainly by clutter, but occasionally also by targets. The received echo is then processed by a feed-forward chain of algorithms to detect, track, and ID targets.

A more advanced biologically inspired cognitive radar implements a perception–action cycle.[33,34] The perception–action cycle is more thoroughly discussed in the next chapter (e.g., see Fig. 6.9). What differentiates the cognitive radar from the perception–action cycle with passive sensors is that the radar's transmitted waveform can be dynamically adjusted to maximize P_{ID} and P_d versus P_{FA}. Thus, an ATR is needed in the system. The cognitive radar's reconfigurable, multifunctional hardware/software provides more on-line decision making and environmental adaptability than conventional radars. It requires considerably more processing power, memory, and intelligence (for ATR, spectrum awareness, and scheduling).

5.7.3 Autonomous land vehicles

Various types of autonomous systems are synergistic with ATR. These include autonomous {land [L], undersea [U], aircraft [A], surface [S]} vehicles (generically called AxV); vehicles with less autonomy are generically called UxV, where the U denotes unmanned. S, meaning systems, is sometimes substituted for the V in these acronyms. Thus, UAS stands for unmanned aircraft system. In the early 1980s under the original DARPA ALV program, the Northrop Grumman Auto-Q-II ATR was demonstrated finding the edges of roads in real-time. Progress made by many contractors in a string of DARPA programs indicated the feasibility of off-road and on-road driverless vehicles. High-technology companies (Google, Apple, Baidu, etc.), chip companies (Intel/MobileEye, Nvidia, etc.), auto suppliers (ZF, Delphi, etc.), and most car manufacturers have brought driverless cars to the forefront. Some car companies are (perhaps much too optimistically) claiming that driverless cars will be on the road by 2020. Even if that is not the case, driverless car requirements are pushing advances in AI, neural networks, and multisensor fusion. Sensor types, formerly finding mainly military application, are now being advanced and compactly packaged for driverless cars. These sensors include LIDAR, radar, FLIR, ultrasonic, INS/GPS, and distributed apertures systems (DAS) (Fig. 5.22).

A Defense Science Board report says that the operation of ALVs should not be viewed in isolation, but rather in terms of human–robot collaboration.[35]

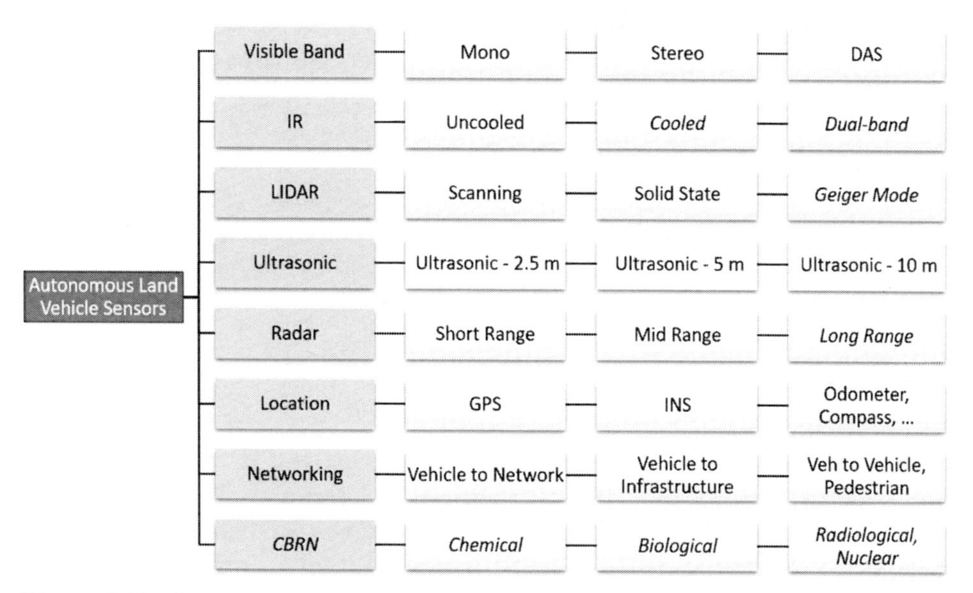

Figure 5.22 Types of sensors suitable for military ALVs. Sensors in plain text are appropriate for driverless cars. Other sensor types, such as acoustic and seismic, are suitable for ALVs when stopped.

Autonomy will be employed in different ways for different missions and mission phases. Different types of users will extend their reach into theater and facilities, adapting to surprise. ALV applications include:

- identification and neutralization of improvised explosives,
- in-depth reconnaissance (including beaches, caves, and base perimeters),
- performing kinetic operations within the rules of engagement,
- counter-mine warfare,
- cargo delivery (including leader/follower "mules"),
- chemical, biological, radiological, and nuclear (CBRN) missions, and
- participation in human–robotic (manned–unmanned) teams.

Test and evaluation of autonomous software that interacts with a dynamic environment in an adaptive manner is particularly challenging, more so with weapon delivery. The acronyms LAW (lethal autonomous weapon) and LAR (lethal autonomous robot) are used for a military robot that selects and attacks targets without human intervention. Under what circumstances LAWs are allowed is not a technical issue, but rather an issue of rules-of-engagement, treaty, and law.

Sensors for ALVs are essentially the same as for manned and teleoperated ground vehicles, and drones, except that they may differ in reliability, cost, ruggedness, and range. Some of the sensors, algorithms, and processing are the same as would be used in driverless cars and can benefit from advances and investments in this arena. General Robert Cone says that robots may replace one-quarter of ground soldiers by 2030.[36] A less-confident assessment is that there are still many unsolved problems and legal issues.

ATR-related sensor fusion for autonomous vehicles is essentially the same as for ATR in general. In addition, autonomous systems will require sensor fusion for route planning, navigation, health maintenance, and obstacle avoidance. When a human is required in-the-loop, hypothesized target images will be transmitted via satellite to a commander for permission to fire. If there is no human-in-the-loop for weapon delivery by a fully autonomous system, ATR will require a much lower false-alarm rate than when a human is the final decision maker.

5.7.4 Intelligent preparation of the battlespace

Intelligent preparation of the battlespace (IPB) is the process of fusing sources of technically derived information with human-gathered intelligence.[37] To be fused, the data is synchronized and coordinated in time and space. IPB produces awareness of the operational environment, including enemy situation and potential courses of action. IPB is broader in scope than ATR, but could benefit from ATR capabilities and output. IPB is a three-step process, implemented and updated in as close to real-time as possible.

Step 1: Evidence Collection and Fusion. Data is captured from a wide variety of sources. Information from each source is verified and corroborated by the other sources. This data helps form a dynamic model of the environment. The central data type is MASINT, which is scientific and technical intelligence information obtained by analysis of data. MASINT includes collection and assessment of emissions from enemy weapons and machines. Other relevant data types are geospatial intelligence (GEOINT), image intelligence (IMINT), SIGINT, human intelligence (HUMINT), weather data, etc., although the boundaries between the various categories of information are not always distinct. Hence, IPB fuses intelligence data collected by technical sensors with each other and human-gathered intelligence and analysis. The fused product indicates the location, composition, movement, and strength of enemy forces. The fusion process involves models and doctrines, and is less statistical in nature than fusion done within an ATR.

Step 2: Evaluating the Adversary. The adversary's intent and goals are inferred from an understanding of the adversary's training, motivations, and beliefs. This can involve religion, culture, tribal loyalties, payments, and willingness to die.

Step 3: Prediction of Adversary's Courses of Actions. Hypotheses are derived about the adversary's future actions based on the data fused in Step 1 and inferred intent and goals derived in Step 2. The prediction takes into account known future actions by friendly forces.

A number of issues remain unresolved. It is difficult to determine what the enemy will do next because there is so much uncertainty about the nature, disposition, and intent of enemy forces. Measurable indicators are sparse in complex urban environments filled with unorthodox adversaries and many more civilians. UAVs and satellite imagery might provide inadequate detail or information that is too untimely to describe a rapidly evolving urban situation. Sensitive information might not be allowed on the communication network.

The long-term solution is "every soldier (and robot) is a sensor" and intelligence asset, contributing information to a much smarter fusion engine than exists today. The IPB system will need to provide more than conclusions. It will need to explain how it reached its conclusions.

5.7.4.1 Dissemination and integration

The IPB provides components of a common operating picture (COP).[38] A COP is a visual display of relevant information shared by multiple commands. It provides a graphical depiction of fused information extracted from sensor data and human intelligence, built on a foundation of geospatial data. The COP depicts maneuvering troops, warfighting machines, and critical infrastructure. A COP is a tool for collaborative planning, decision making, and execution, assisting all echelons to obtain situational awareness.

The future COP will let the viewer switch between the baseline IPB-derived data and live video/data feeds from UAVs, UGVs, UGSs, manned vehicles, satellites, and soldier-borne sensors (Fig. 5.23). Other forms of data will be available, such as analyst comments, social media feeds, and cybermetrics.

5.7.5 Zero-shot learning

An ATR conveys information to humans in the form of track files, class labels, and probabilities. This is not how humans exchange information with each other. People pass along information by telling stories (Fig. 5.24). Humans are hardwired to process data in narrative format. Narratives can explain complex phenomena for which there are no pre-trained class labels. A good story captures attention. Our brains are more engaged by a story than by cold, hard facts. A narrative neatly packages the essential information needed by the intended audience for the relevant environment.

As seen in the cartoon in Fig. 5.24, the luckless person cannot name what he encountered. He cannot provide a class label for what bit him because he never saw that kind of thing before. But he can provide a vivid description of the mysterious thing. This is done in terms of semantic attributes, where "attribute" is a named property of an object or event. Attributes differ from features (e.g., wavelets) that don't have semantic meaning. The ill-fated

Figure 5.23 Notional depiction of a future common operating picture.[38] (Not subject to copyright.)

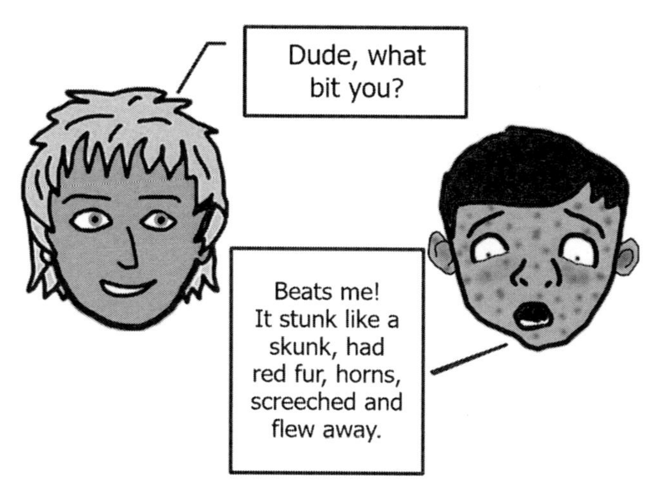

Figure 5.24 Humans pass along information in the form of a narrative.

person instantly recognized the creature's key attributes using his bodily sensors and cortical processing. Then his brain fused these attributes with other information to construct a narrative of what happened. He classified the target (by description) without ever before seeing a target like that. His description was explanatory.

The challenge of using deep learning in ATR is collecting the needed enormous, perfectly truthed, sufficiently comprehensive and representative training set for all target types of interest, under all possible conditions. The cost increases exponentially when collecting simultaneous data with multiple sensors. The problem is particularly acute when fusing sensor data at the feature level. Zero-shot learning (ZSL) circumvents this problem by characterizing objects or events as a combination of simple semantic primitives.[39] Target or activity classes are not explicitly learned, only their semantic primitives are learned. Semantic primitives are shared by multiple classes. Training requirements are reduced by an order of magnitude.

ZSL learns object or event primitives during off-line pre-training. More than one sensor type may be involved, but each primitive results from just a single sensor type. Subsequently, during real-time inference, a set of attributes is recognized in the incoming sensor data. A possibly never-before-seen object or activity is characterized by stringing together the recognized primitives, combined with track data, to form a narrative.

5.7.5.1 Manned–unmanned teaming

Over the next 25 years, Army operations will increasingly be carried out by manned–unmanned teams (MUM-Ts). An important issue is how humans and ATRs on robots will communicate with each other. One possibility is natural language communication using narratives. The narrative is fundamental to human cognition and understanding. That is, humans and robots will talk to each other in the "flesh" or over radio. But the form of the dialog will be more carefully composed than human-to-human communication. Narrative constructs can follow strict conventions. A well-defined construct is critical to man–machine communication. A strict narrative has five distinct elements:

1. Protagonist — Who or what is the entity? What are its sensor-derived dominant attributes?
2. Time — When did this take place?
3. Location — Where did this take place?
4. Plot — What happened? A plot is a summary of causally related events. It is the substance of the narrative. In our case, a plot can be a sequence of actions, a single action (like digging or raising missile launcher), or group of simultaneous actions (like walking while carrying large object).
   a. Causation — A plot that represents a progression of logically connected events can be considered from a Bayesian point of view: probability that action A caused action B in the specified context. Bayesian networks are causal networks with the strengths of the causal links represented as conditional probabilities. Bayesian causal inference provides a basis for a model for, or an analysis of, a plot involving related events that unfold over time.
5. Protagonist's motivation — Why did this happen? This covers intentionality, reasons, and goals.

Element 1 is inferred by the ATR's inference engine. Descriptive attributes can be determined through ZSL. Elements 2 and 3 are available from track files. A track file provides target location and velocity vector as a function of time. Element 4 can be determined from a recurrent neural network, wherein simpler action primitives are obtained from ZSL. Element 4a, causal links, connects a sequence of actions or event primitives, often modeled by a Bayesian network. Element 5 is beyond current ATR capability and needs to be determined by HUMINT.

With a simple *descriptive narrative*, sensor-derived information is used to portray a scene, thus leaving out elements 4 and 5. The descriptive narrative can be as simple as "large truck seen at this location at this time." This is analogous to the information conveyed machine-to-machine in various data exchange strategies for critical event information, such as MITRE's Cursor-on-Target (CoT) XML schema and other similar schemas [UCore, National Information Exchange Model (NIEM), common battlespace object (CBO), etc.].

Adding elements 4 and 5 provides a deeper understanding. Be aware of the difference between elements 4a and 5: causation and motivation. Each answers different types of *Why* questions. The motivation or reason someone did something is different from explaining it as an effect of some cause. Reasons can cite purposes. But a causal explanation cannot cite a purpose.

Motivation can be good or bad or unknown. A causation cannot be good or bad in the same way. If one says that the insurgent shot a child, causing him to die, this doesn't imply motivation, which could have been accidental or with evil intent. There is a temporal relationship between the separate events of being shot and dying. The cause of the child dying is that he was shot; the reason that he was shot may be known or unknown. The cause for the enemy tank being stationed at the end of the bridge is that it was driven there. The tracker can determine that. The motivation could be unknown, by happenstance, or to prevent anyone from crossing the bridge. An ATR may be able to determine a cause, but, at least for now, it is not smart enough to determine motivation.

Longer stories, covering larger scenarios, can be built up from a string of short narratives. A story may be used to communicate information about a bigger scenario or event, or even information derived through persistent surveillance. Humans can communicate with machines verbally, through short oral narratives or longer stories.

5.7.5.2 Expert systems

The biggest obstacle to ATR deployment is obtaining a sufficiently comprehensive and representative data set with which to train the ATR. The database should include adversarial targets and confusor objects at all angles and ranges, under all conditions and backgrounds. The problem is compounded when trying to do multisensor fusion. Sometimes there is just no way to get the necessary training data. Human observers and ZSL can circumvent this problem by reporting a set of semantic attributes in place of a class label. What else can be done? Let us look back into the history of ATR and AI—before the era of big data and fast computers. Expert systems (ESs) are a rich part of the ancient history of intelligent systems. ESs are computer programs that seek to mimic the way human experts solve problems. A group of experts does better than a single expert. For our focus, the experts are well-trained combat pilots, experienced image analysts, and skillful soldiers. An ES uses a knowledge base and rules to infer target class.[40] Rules are in the form of "IF ... THEN ..." logic. The IF part is the antecedent. It is the condition to be tested. The THEN part is the consequent. It is an action to be executed when a rule fires. The rules are formulated off-line from the knowledge base and are executed on-line by an inference engine (IE). The IE determines which rules, if any, are satisfied by the available facts. For our case, the facts are obtained by applying ZSL or similar approaches to incoming data from one or more sensors on one or more platforms.

The first step in developing the ES is knowledge acquisition. Target identification experts must be located and interrogated. Knowledge can also come from training manuals and other documentation. Knowledge falls into several categories:

- *Expert knowledge* is information widely available to subject matter experts (SMEs). For example, the engines, axles, and exhaust pipes of moving vehicles are hot.
- *Tacit knowledge* is used in practice. It is unconscious, internalized, and difficult to explain. It is learned through years of experience, and honed as a skill (e.g., a jazz jam session).
- *Domain knowledge* relates to a narrow area of expertise (e.g., expertise about jet engines, military vehicle paint, or submarine periscopes).
- *A priori knowledge* comes before and is independent of knowledge gained by processing sensor data. For example, 80% of the tanks in country X's inventory are T-72s.
- *Commonsense knowledge* is understood by nearly all adults as a result of a lifetime of observation and learning. Commonsense knowledge is difficult to catalog and encode. For example, tanks don't perch on tree tops, the moon and farm animals are not threatening objects, and jet planes don't stop midair.
- *Deep perceptual knowledge* involves complex spatial and temporal relationships. This type of knowledge is difficult to ascertain and encode (e.g., a group of persons acting suspiciously).

The second step in developing a knowledge base is to turn the acquired knowledge into a set of rules. We will illustrate this process with the fanciful example of Fig. 5.24.

Rule 1	IF		an animal has fur
	THEN		it is a mammal
Rule 2	IF		an animal has feathers
	THEN		it most likely flies
Rule 3	IF		an animal flies
		AND	it lays eggs
	THEN		it is a bird
Rule 4	IF		an animal can take off into the air
	THEN		it can fly
Rule 5	IF		a mammal flies
	THEN		it is a bat
Rule 6	IF		a bat has teeth (i.e., can bite)
		AND	has horns
	THEN		it is a Bulgarian horned bat

⋮

The incoming multisensor data is processed by ZSL (or human observer) to generate a set of facts. Again, using the curious example of Fig. 5.24:

FACT 1: Animal has fur.
FACT 2: Animal flies.
FACT 3: Animal bites.
FACT 4: Animal has horns.
FACT 5: Animal is red. (Disclaimer: This is a fictitious example; the bat species described is not actually red.)
FACT 6: Animal screeches.

The extracted facts are fed into the IE. The IE applies the rules stored in its knowledge base to reason its way to a target class decision, step by step, as shown in Fig. 5.25. However, there is no guarantee that sufficient knowledge has been accumulated, sufficient rules have been generated, or sufficient semantic attributes have been extracted from the sensor data, to reach a conclusion.

Unlike some other types of classifiers, an ES can justify its reasoning by tracing its inference steps to conclusion. This provides the desirable property of explainability, which is often missing from ATR and machine learning systems. The ES differs from deep-learning neural networks in that the ES is not easily crashed by future data that differs from training data in insubstantial ways.

5.8 Discussion

The natural environment produces multimodal information. Humans and other creatures have evolved sensory organs to perceive this information and brains for processing and fusion. ATRs need to do the same to understand the battlespace. This chapter covered various kinds of fusion: multi- {sensor, band, look, platform, classifier, function, attribute} fusion.

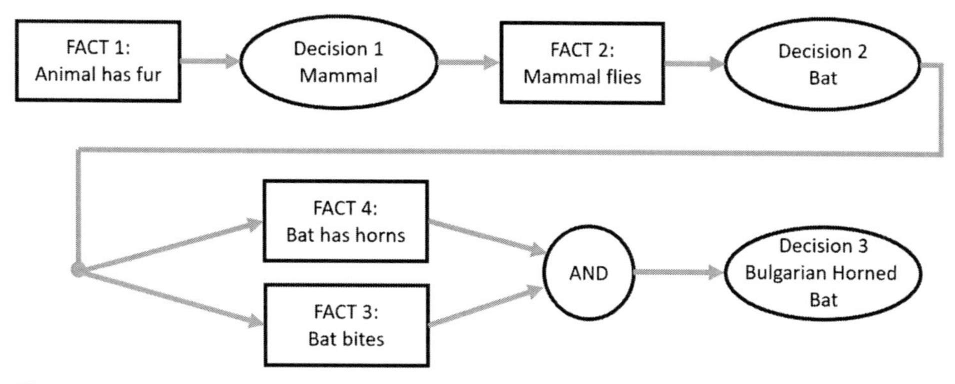

Figure 5.25 An inference engine, using forward chaining, starts with the known facts and asserts new facts until a conclusion is reached.

Combining multiple types of data provides insights that cannot be obtained from narrow, single-source data. This leads to less uncertainty, better accuracy, more reliability, and much deeper understanding of the environment. Intelligent creatures carry out these types of combinations to survive. From a military point of view, survival relates to situational understanding, area defense, search, and attack. ATRs incorporating fusion can be thought of as the survival nuclei of military weapons and operations.

The general recommendation is to always start with a detailed description of the problem to be solved. Know the way that the problem is currently being tackled, what is wrong with the current solutions, and who are the customers paying for a better solution. Then determine what resources are available to solve the problem. Resources include money, ConOps, platforms, sensors, algorithms, processors, human experts, T&E team, etc. Fusion is the act of combining the various constituents into a unified whole. These constituents can be sensor data, metadata, classifier outputs, tracker outputs, features extracted from different sensors, functional capabilities, etc. To establish success, one must know the key performance parameters, exit criteria, and acceptable test and evaluation regime. In the end, algorithmic results need to be conveyed to the humans-in-the-loop in an understandable form and one day perhaps also to the human's robotic partner.

References

1. A. Treisman, "The binding problem," *Current Opinions in Neurobiology* **6**(2), 171–178 (1996).
2. B. Wallace, "Multifuction RF (MFRF)," https://www.darpa.mil/program/multifunction-rf [accessed Nov. 24, 2017].
3. J. Esteban, A. Starr, R. Willetts, and P. Bryanston-Cross, "A review of data fusion models and architectures: towards engineering guidelines," *Neural Computing and Applications* **14**(4), 273–281 (2005) [accessed Nov. 25, 2017].
4. F. Castanedo, "A review of data fusion techniques," *The Scientific World Journal* (2013), https://www.hindawi.com/journals/tswj/2013/704504/ [accessed Nov. 25, 2017].
5. B. K. Gunturk, J. Glotzbach, Y. Altunbasak, R. W. Schafer, and R. M. Mersereau, "Demosaicking color filter array interpolation," *IEEE Signal Processing Magazine* **22**(1), 44–54 (2005).
6. B. J. Schachter, P. J. Vanmaasdam, and J. G. Riddle, "Converting an image from a dual-band sensor to a visible color image," U.S. Patent number 954456 (January 10, 2017).
7. R. F. Hess, L. To, and J. R. Cooperstock, "Stereo vision: The haves and have-nots," *i-Perception* **6**(3), 1–10 (2015).

8. M. Winterbottom, J. Gaska, S. Wright, S. Hadley, C. Lloyd, H. Gao, F. Tey, and J. McIntire, "Operational based vision assessment research: Depth perception," *Defense Technical Information Center*, ADA617034 (2014).
9. S. R. DeGraaf, "3-D fully polarimetric wide-angle superresolution-based SAR imaging for ATR," *Thirteenth Annual Adaptive Sensor Array Processing Workshop* (2005), https://www.ll.mit.edu/asap/asap_05/pdf/Papers/24_degraaf_Pa.pdf [accessed Nov. 25, 2017].
10. T. W. Cronin, S. Johnsen, N. J. Marshall, and E. J. Warrant, *Visual Ecology*, Princeton University Press, Princeton, New Jersey (2014).
11. F. A. Sadjadi and C. S. Chun, "Passive polarimetric IR target classification," *IEEE Transactions on Aerospace and Electronic Systems* **37**(2), 740–751 (2001).
12. F. Goudail and J. S. Tyo, "When is polarimetric imaging preferable to intensity imaging for target detection?" *Journal of Optical Society of America A* **28**(1), 46–53 (2011).
13. K. P. Gurton and M. Felton, "Remote detection of buried land-mines and IEDs using LWIR polarimetric imaging," *Optics Express* **20**(20), 22344–22359 (2012).
14. Y. Zhang and Q. Ji, "Efficient sensor selection for active information fusion," *IEEE Transactions on Systems, Man and Cybernetics – Part B Cybernetics*, **40**(3), 719–728 (2010).
15. J. Li, K. Cheng, S. Wang, F. Morstatter, R. P. Trevino, J. Tang, and H. Liu, "Feature selection: A data perspective," *arXiv* preprint:160. 07996 (2016), https://arxiv.org/abs/1601.07996 [accessed Nov. 25, 2017].
16. H. Li, X. Wu, and W. Ding, "Group feature selection with streaming features," in *Data Mining, IEEE 13th International Conference*, 109–114 (2013).
17. L. Meier, S. V. DeGeer, and P. Buhlmann, "The group lasso for logistic regression," *Journal of Royal Statistical Society: Series B, Statistical Methodology* **70**(1), 53–71 (2008).
18. X. S. Bahrampour, N. M. Nasrabadi, and A. Ray, "Sparse Representation for Time-Series Classification," Chapter 7 in *Pattern Recognition and Big Data*, A. Pal and S. K. Pal, Eds., World Scientific Publishing, Singapore, pp. 199–215 (2015).
19. X. Jin, S. Gupta, A. Ray, and T. Damarla, "Multimodal sensor fusion for personnel detection," *14th International Conference on Information Fusion*, Chicago, Illinois, pp. 437–444 (2011).
20. N. Virani, S. Marcks, S. Sarkar, K. Mukherjee, A. Ray, and S. Phoha, "Dynamic data driven sensor array fusion for target detection and classification," *International Conference on Computational Science* **18**, 2046–2055 (2013).

21. Y. Li, D. K. Jha, A. Ray, and T. A. Wettergren, "Information fusion for passive sensors for detection of moving targets in dynamic environments," *IEEE Trans. on Cybernetics* **47**(1), 93–104 (2017).
22. D. Ruta and B. Gabrys, "An overview of classifier fusion methods," *Computing and Information Systems* **7**, 1–10 (2000).
23. N. Najjar, "Information Fusion for Pattern Classification in Complex Interconnected Systems," Ph.D. Thesis, University of Connecticut (2016).
24. A. P. Dempster, "Upper and lower probabilities induced by a multivalued mapping," *The Annals of Mathematical Statistics* **38**(2), 325–339 (1967).
25. G. Shafer, *A Mathematical Theory of Evidence*, Princeton University Press, Princeton, New Jersey, (1976).
26. B. G. Foley, *A Dempster-Shafer Method for Multi-Sensor Fusion*, Thesis, AFIT/GAM/ENC/12-0, Air Force Institute of Technology, WPAFB, Ohio (2012).
27. E. El-Mahasini and K. White, *A Discussion of Dempster-Shafer Theory and Its Application to Identification Fusion*, DST-Group-TN-1443, Defense Science and Technology Group, Australia (2015).
28. D. Dubois and H. Prade, "A set-theoretic view of belief functions; logical operations and approximations by fuzzy sets," *International Journal of Gen. Syst.* **12**, 193–226 (1986).
29. Y. Bar-Shalom, P. K. Willett, and X. Tian, *Tracking and Data Fusion: A Handbook of Algorithms*, YBS Publishing, Storrs, Connecticut (2011).
30. C. Y. Chong, S. Mori, W. H. Barker, and K. C. Chang, "Architectures and algorithms for track association and fusion," *IEEE Aerospace and Electronic Systems Magazine* **15**(1), 5–13 (2000).
31. K.-C. Chang, S. K. Rajat, and Y. Bar-Shalom, "On optimal track to track fusion," *IEEE Transactions on Aerospace and Electronic Systems* **33**(4), 1271–1276 (1997).
32. J. L. Gertz and A. D. Kaminsky, *COTS fusion tracker evaluation*, Lincoln Lab Project Report ATC-302, (February 2002).
33. S. Haykin, *Cognitive Dynamic Systems: Perception-Action Cycle, Radar and Radio*, Cambridge University Press, Cambridge, Massachusetts (2012).
34. J. R. Guerei, *Cognitive Radar: The Knowledge Aided Fully Adaptive Approach*, Artech House, Norwood, Massachusetts (2010).
35. DoD Defense Science Board, Task Force Report: *The role of autonomy in DoD systems*, DTIC ADA566864 (July 2012). [Updated every few years.]
36. CBS News, "U.S. Army general says robots could replace one-fourth of combat soldiers by 2030," January 23, 2014. https://www.cbsnews.com/news/robotic-soldiers-by-2030-us-army-general-says-robots-may-replace-combat-soldiers/ [accessed Nov. 20, 2017].
37. *Intelligent Preparation of the Battlefield/Battlespace*, Army Publication ATP 2-01.3, Marine Corps Publication MCRP 2-3A (Nov. 2014).

38. D. Huyn and M. McDonald, "War TV: A Future for a Common Operating Picture," (Dec. 28, 2016) http://cyberdefensereview.army.mil/The-Journal/Article-Display/Article/1134603/wartv-a-future-vision-for-a-common-operating-picture/ [accessed Nov. 25, 2017].
39. M. Palatucci, D. Pomerleau, G. E. Hinton, and T. M. Mitchell, "Zero-shot learning with semantic output codes," *Advances in Neural Information Processing Systems* **22**(NIPS 2009), 1410–1418 (2009).
40. J. Roy, *A Knowledge-Centric View of Situation Analysis and Support Systems*, Technical Report DRDC Valcartier, 2005-419, Canada (January 2007).

Chapter 6
Next-Generation ATR

Acknowledgments: This chapter results in part from work done under the DARPA/MTO Cortical Processor project. Contributors were Paul Feinberg, Mike Fitelson, Alexander Grushin, Benjamin Bachrach, and Mike Novey; DARPA program managers were Dan Hammerstrom and Hava Siegelmann.

6.1 Introduction

The human brain receives, integrates, and processes sensor data and various forms of metadata.[1] It detects, recognizes, and tracks objects of interest. It communicates with other brains. The brain has motor control over its host body. On an abstract level, the brain and ATR have a lot in common. They have to solve similar computational tasks. This leads to similarities in design.

Any network whose neurons send feedback signals to each other is a recurrent neural network (RNN). The human brain is an RNN with many feedback loops. RNNs can learn to process sequential data not easily learned by other types of neural networks. A recurrent ATR is suitable for processing still frame, video, and various kinds of temporal signals.

Section 6.2 discusses brain versus ATR hardware design. Section 6.3 covers algorithm/software design. A strawman (reference) design is provided, but with no claim that this is the only way to construct a next-generation ATR. The strawman design should be thought of as a brainstormed, simple, draft proposal intended to generate discussion of its advantages and disadvantages, and to trigger the generation of new and better proposals.[2] The strawman is not expected to be the final creation. It should be kicked around and refined until a finished model is obtained that meets a project's key performance goals. The final ATR design can be very different from the strawman design.

6.2 Hardware Design

The basic constraints on the brain and ATR are the same (Fig. 6.1):

- size, weight,
- power (energy),
- speed (latency), and
- noise.

These constraints are not independent. Each constraint is linked to the other constraints. Constraints are multifaceted, involving many subtleties and environmental influences. An ATR satisfies its constraints differently from a data center or self-driving car, just as a bird satisfies its constraints differently from a whale.

ATR hardware is described by a specification ('spec') sheet. Table 6.1 provides a spec sheet for the human brain. It shows the evolutionary result of environmental influences and design constraints.

The brain's design is constricted in many ways to what is barely sufficient for the host body to survive and reproduce. Thus, the brain provides proof of what is achievable, but not bounds on what is possible. Let us consider how the brain's design results from its constraints.

Size and Weight Constraints: The brain's size and weight are limited by the muscular structure of the human body and the cost of collecting enough calories to keep the body going in both fat times and lean times. Head size is also limited by the size of the mother's birth canal. The neocortex, whose function is similar to an ATR, makes up about 76% of the human brain (Fig. 6.2).

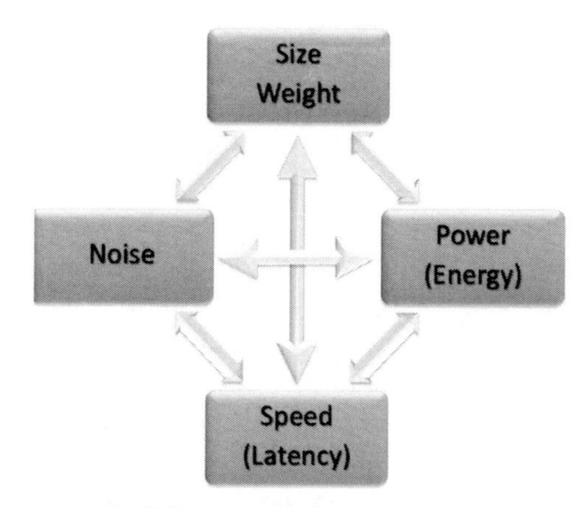

Figure 6.1 Brain design and ATR design are constricted by the same set of constraints.

Table 6.1 Spec sheet for the human brain.

Item	Approximate specification	Notes
Size	1100–1450 cm^3	
Weight	1300–1400 g	About 2% of adult body weight.
Power	10–20 W	Power use is little affected by the task performed.
Processing elements	80–100 billion neurons	
Data lines	1000–10,000 synapses per neuron	~1% substantially transmitting at any one point in time.
Total maximum bandwidth in cortex[4]	1 TB/sec	Maximum capacity never achieved because only a small percent of neurons fire simultaneously.
Latency	100 ms for object recognition	Latencies of 50, 100, 200, 300 ms reported; shorter times with cueing and for target versus nontarget.

Cortical components are arranged to minimize the total length of axons needed to join them (Fig. 6.3). Local connectivity of neurons is fairly sparse, but global connectivity is very much sparser, reducing the volume required for inter-processor communication. Structural connectivity implies functional connectivity.

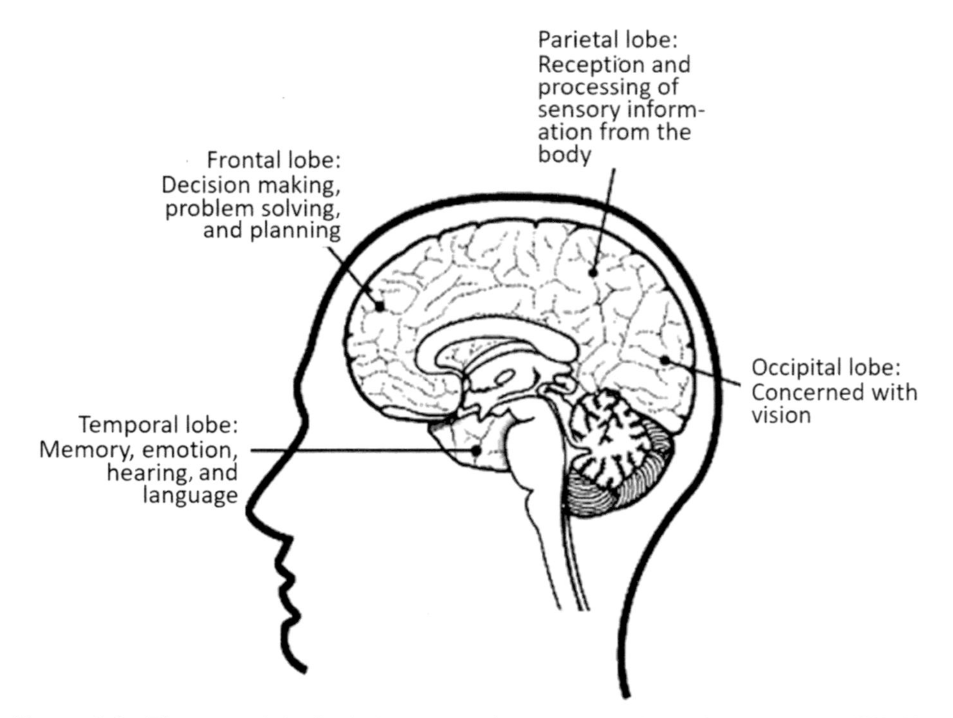

Figure 6.2 The part of the brain known as the neocortex is analogous to an ATR. The neocortex is involved in sensory perception, generation of motor commands, spatial reasoning, thinking, and external communication. It is divided into temporal, frontal, parietal, and occipital lobes. The visual cortex is located in the occipital lobe. (Public domain image from http://www.wpclipart.com.)

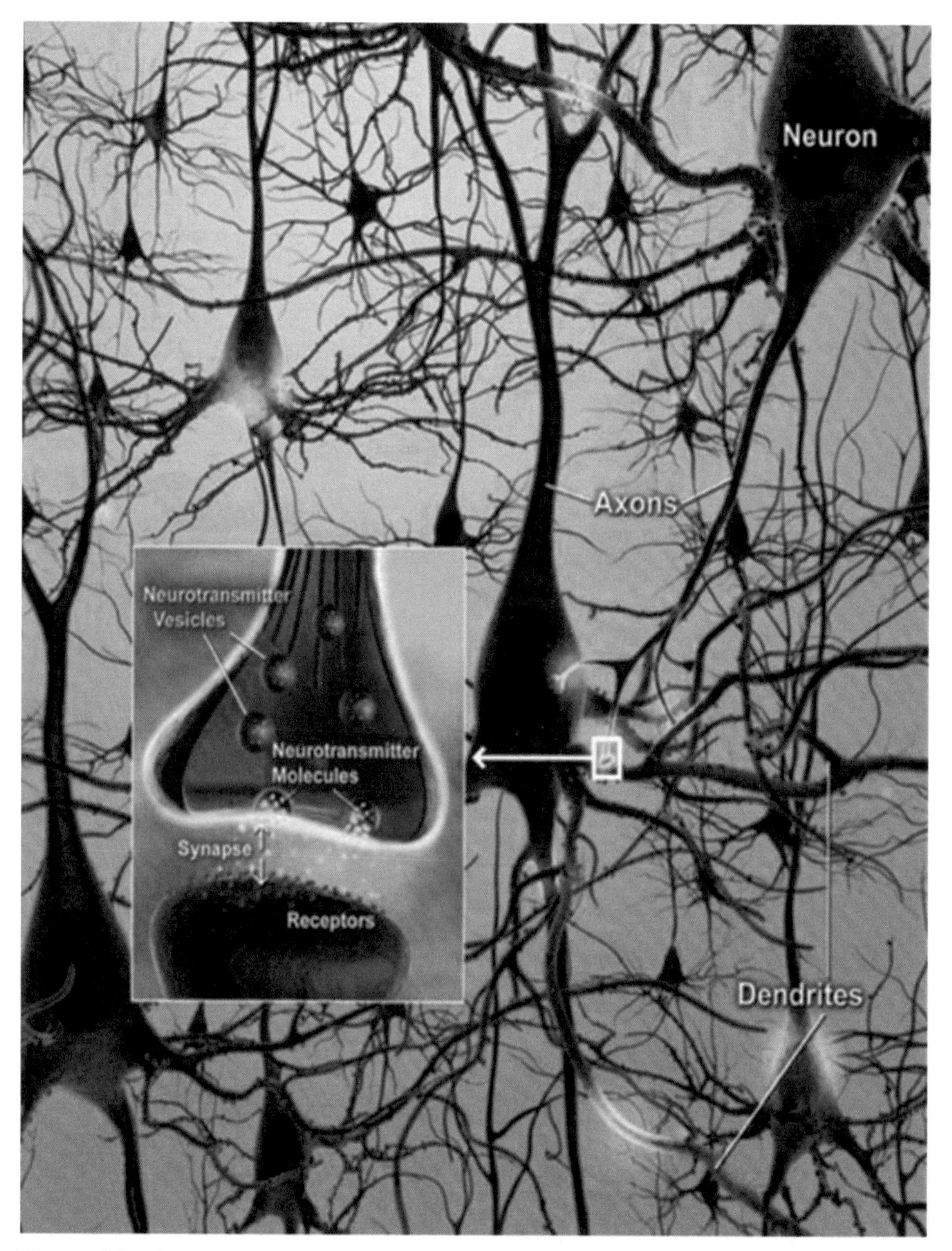

Figure 6.3 A neuron has three main parts: cell body (soma), dendrites, and axon. Dendrites are short branches that receive messages from other neurons and pass them on to the cell body. An axon is a long, single fiber ("wire") that transmits messages [in the form of action potentials (voltage spikes)] from the cell body to the dendrites of other cells (including muscles and glands). Myelin insulates some axons, analogous to the insulation of an electrical wire. A synapse is the junction across which a nerve impulse passes from an axon terminal to a neuron, thus permitting a sending neuron to signal a receiving neuron. The synaptic weight w_{ik} is the influence that neuron k has on neuron i, and is adjusted during learning. (Image from National Institutes of Health, www.nia.nih.gov.)

Wiring: The dominant method of transmission of information in the brain is in the form of voltage spikes along axons, through synaptic junctions, to other neurons (Fig. 6.3). Brain volume and energy requirements are reduced by an efficient wiring pattern.[3] Neurons have a direct connection to about 1 in 100 other nearby neurons, but to only about 1 in 1,000,000 distant neurons.[4] This drives down material cost, size, weight, power, and latency.

Energy: The brain meets its energy budget by the use of miniature components, elimination of unessential signals, and representation of information with sparse codes. Communications are coded by the strong activation of several neurons at once, in the form of sparse codes, which are represented as spike trains. The brain's energy budget places limits on how many neurons can code a particular data item. With just 3% of neurons concurrently firing in an active neocortex, about 75% of energy consumption is used for signaling.

Spikes are not the only way that information is transmitted in the brain. Fluctuations in relatively inactive neurons contribute to brain waves, which play an active role in information processing.[5]

The cost of processing information is eased by reducing the flow of predicable, redundant, and useless information. As information streams from stage to stage in the brain, it is condensed to what is essential for understanding and decision making. The visual machinery of animals that spend their lives in total darkness (like cave dwelling fish) have been reduced or eliminated through evolution. When a child loses one sensor modality (e.g., vision), cortical resources are re-directed to other sensor modalities (e.g., hearing, touch).[6] The cortex has the capacity for substantial long-term evolutionary plasticity, short-term environment-induced plasticity, as well as rapid goal-directed plasticity.

Power use is also limited by cooling requirements for the brain or electronic device. Cooling becomes the limiting factor as energy dissipation increases. Cooling requirements may also constrain the speed of learning, as learning requires energy. The clock rates of commercial chips, when used for military applications, are often reduced to meet cooling demands.

Some limitations imposed by the brain's energy budget are not limitations that we would want to impose on an ATR. It is difficult for a human to focus on or track more than one thing at a time. This makes multi-tasking, such as texting while driving or walking while talking on a cell phone, risky behaviors. Wolves take advantage of focus-of-attention limits by attacking their prey in packs.

With a slight simplification, we can characterize the cortex as an energy-efficient, hybrid device. Neurons mix digital and analog functionality. Signals and processing within a cell body are in analog. The results are converted to digital data (spikes) for transmission, which are then re-converted to analog by the receiving neuron. There are many reasons that this might not be the best design for a neuromorphic ATR.

Speed, Latency: The clock frequency of a processor chip is limited by heat dissipation and to a lesser extent gate delay, speed of electrical transmission, cross-talk, and noise. The practical limit with conventional cooling techniques is around 3 or 4 GHz, which has already been reached. This suggests that the way to improve chip performance is with a larger number of first-rate processors, more local memory, and better inter-processor communication.

Latency is a function of the speed of computational elements, speed of inter-processor communication, and number of stages in the processing pipeline. Low latency is necessary for animal survival and ATR efficacy. For the human visual system "... a single forward sweep as short as 13 ms is capable of extracting a picture's conceptual meaning without advance knowledge. ... durations as short as 13 ms are clearly sufficient, on a significant proportion of trials, to drive conscious detection, identification, and immediate recognition memory."[7] A small active population of neurons makes a best guess in a rapid forward pass using low-spatial-frequency information, while recurrent processing refines this to a detailed perception in roughly 100 ms. Visual noise adds to latency.[8] A typical target detection+ recognition latency requirement for an ATR is 100 ms (3 frame times) with imagery at 30 frames per second. Three frame times for a next-generation 120-frame-per-second camera is 25 ms. Target tracks typically take 3 frame times to initiate. If a target is already under track, then detection is not required, thus reducing the latency to achieve recognition. The ATR or brain's latency for recognizing a human activity (spatiotemporal data) is more difficult to quantify because it requires that the activity sufficiently unfold before it can be recognized. Recurrent loops keep data in local memory and appear to be essential for activity recognition.

Noise: Noise is a limiting factor in biological and electronic sensors and computational units. It is a constraint on energy, efficient coding, and minimization of wiring costs. Neurons are noise-limited devices of noise-restricted bandwidth.[6] A neuron doesn't do 64-bit, floating-point computations or even 16-bit precision arithmetic.

The brain adapts to changing environmental conditions.[6] Noise and scene variability are somewhat mitigated by plasticity—the ability of the neural network to continuously modify its properties.

The human visual system reduces a high-bit-rate streaming signal to a low-bit-rate signal, as also does an ATR. Each processing stage further compresses the streaming imagery and cognizes it a little bit better, while at the same time introducing noise, e.g., sensor noise, quantization noise, and synaptic noise (Table 6.2). In a parallel-pipelined manner, edge vectors and other features are extracted and bound together into shapes and textures, and eventually into semantic objects with names or descriptions.[5] The visual world gets ripped apart, only to be reassembled and then semanticized in succinct form.

Table 6.2 Visual information flow. (Data adapted from Ref. 9.)

Biology	Bandwidth (bits/s)	ATR analogy
Deposited on retina	10^{10}	2000 × 2000 pixel FLIR camera, 16-bits/pixel, 120 frames per second
Leaves retina	6×10^6	H.265 compressed FLIR image or stream of feature data
Arrives at layer IV of visual cortical area V1	10^4	128 × 128 feature region-of-interest image
Visual awareness	10^2	ATR report

Brain versus ATR: The human brain is an efficient computing and control device. It far surpasses current computers in many ways. Birds and mosquitos perform sophisticated tasks with much smaller brains than humans. The design of any brain results from what is biologically feasible. This doesn't mean that a computer or ATR can't be built that surpasses biology for more narrowly focused tasks. For example, current processing chips have much higher clock rates than brains. Brains don't benefit from clock-driven synchrony.

Computers and brains have multiple forms of memory (short term and long term). The brain does not store or retrieve memories in the same way as a computer. Memory is a decentralized, distributed process in the brain. A particular neuron can participate in the encoding of many different memories. The brain's circuits don't really even "store" data in precise slots as does a computer. Biological memory is better interpreted as adaptation to statistical associations among signals.

The brain's various limitations necessitate its attentional mechanisms.[10] The brain has difficulty focusing attention on or tracking more than one thing at a time. By contrast, ATRs need to track multiple targets at once.

ATR customers can't afford the twenty plus years of learning from birth to deployment, as for a soldier or pilot. If a recurrent ATR learns something in weeks, it outperforms a baby by months.[11] Gradient-descent optimization learning algorithms (like backpropagation of error) work well in artificial neural networks but probably aren't biologically plausible. For ATR design, "working well" is more important than being "biologically plausible."

Traditional ATRs are trained in advance. They are deployed only after passing rigorous test and evaluation trials. Current ATRs do not continue learning after deployment. If they did, then each of the ATRs on 100 different planes would perform differently. Military hardware isn't supposed to work that way. Real brains are always learning, always rewiring themselves, to meet environmental demands. Future, more brainy, ATRs (like current soldiers, Marines, pilots, and image analysts) will require continuous learning and adaptation to the battlespace. Military procurement, and test and evaluation procedures, will have to change to accommodate intelligent learning devices.

The cortex is a hybrid device. Low-precision, very slow, very noisy, analog modules process data. The results are converted to digital data (spikes) for transmission through the network. Spikes aren't the only way that information is passed in the brain. Quieter groups of neurons spawn brain waves to shunt some neural inputs and enhance others, alter spike timing, and consolidate memory.[5] The quieter neurons can also be a reserve pool for learning new things.[5] An ATR chip should have the equivalent of a reserve pool and not be utilized at 100% capacity.

Energy is used to transmit information within the brain. There is also a substantial, fixed cost for maintaining the brain, with active tasking only having a small effect on total energy used. Noise is a constraint on both energy-efficient coding and the minimization of wiring costs.[6] A brain is in a body—not in a box. Eye, head, and body movement are essential to vision, requiring muscular movement and consuming substantial energy. If we say that the brain consumes a certain amount of energy for visual processing, that amount ignores the support that it is getting from the rest of the body. Most ATRs are embodied in a platform. The nature of the platform (e.g., vibration, speed, altitude, mission) affect ATR design. Strict power limits dictate energy-efficient military system designs. ATR designers need to carefully examine energy–performance tradeoffs. Power consumption is a critical factor at the processor chip level. This involves not only evaluating the numerous architectural design choices but also optimizing constituent circuits (for new chip designs). Design decisions are judged according to cost–performance tradeoffs. Computation performance, input/output bandwidth, energy, size, price, weight, development time, programmability, logistics trail, reusability, reparability, long-term chip availability, upgradability, and risk, all change with design choices.

How closely should a neuromorphic ATR mimic biology? Maybe not that closely. Leopards could hunt better at night if their eyes were cryogenically cooled, infrared cameras. Birds could fly south faster if they had little jet engines. Nature can't create everything that humans can invent. Humans can't duplicate everything that nature took billions of years to invent. We should look to the brain for inspiration in designing an ATR, but not try to emulate all of its features because some of those features are shortcomings, while other features are too difficult to duplicate. Moreover, input/output (I/O) and tasks are not identical for ATRs and human brains. ATRs are fed data from military sensors and metadata sources. ATR tasks are narrowly focused. Brains are fed information from various biological sensors, throughout the day, over a lifetime. Brains are implemented in a liquid substrate. The brain has to rely on extremely robust signals distributed among a very large population of neurons. Integrated circuits use a uniform rigid substrate that allows a very fast clock and reliable synchronization of signals. Many digital chips can do precise computations, like 64-bit floating point. However, we have found that 16-bit floating point is just about perfect for implementing a recurrent ATR, resulting in energy savings. Lower precision cannot support the small, gradual, weight updates required for off-line training and on-line continuous learning. Analog computations seem to offer insufficient precision for the numerous feedback loops in an RNN chip designed to operate over difficult military data. Analog image processing is outside the mainstream of processor design. Digital processing requires many fewer computational units to implement an ATR than the tens of billions of neurons in the neocortex doing similar functions. The massive parallelism of the brain, in total, offers more processing power than would be available in a digital ATR, but it is not obvious that we could make a better performing ATR with say 10^{22} ops/s.

The brain's building blocks are fundamentally different from those of an electronic chip. The brain uses neurons and synapses as computational components, which are made of inorganic ions, proteins, fat and saltwater.[1] Water constitutes 70% or more of cell mass. The brain's efficiency stems from massively parallel networks and molecular components operating at nanometer scale. So how should we design a single-chip, recurrent ATR? We shouldn't start with the brain's design principles, but instead start with the building principles that work best with modern CMOS technology: 16-bit, digital floating-point processors (rather than a mixed bag of unreliable analog processors), precise digital communication (rather than spike trains), synchronization via global clock (rather than asynchronous operation), moderately massive parallelism (rather than extreme massive parallelism), and sufficient local memory per processor (rather than introducing a bottleneck of off-chip memory). The current state-of-the-art is 10- to 16-nm CMOS components, eventually going to 3–7 nm and stacked-die design.

Past ATR designs involved choosing from available military-compliant processor chips that include certain heterogeneous multicore processors, GPUs, and FPGAs. Neuromorphic ASICs are now being introduced by several companies for commercial applications. Nearly all of these are for feed-forward CNNs. They are not well suited for RNNs, which comprise our strawman ATR. Most neuromorphic chips are co-processors (accelerators). They require a host chip, which is generally an Intel heterogeneous multicore processor. Separating processing into a general-purpose host chip (controller) and neural accelerator is a reasonable way to go, since Intel has a huge budget to advance its general-purpose chips, offering new models each year at various design points.

6.2.1 Hardware recommendations for next-generation neuromorphic ATR

Neural processing and biological vision have been inspirations for ATR engineers since the start of the field in the 1960s. But, just as it hasn't proven fruitful to design flapping wing airplanes and walking cars, we do not want to mimic biology too closely. ATRs should take advantage of the march of technology in integrated circuits and software infrastructure. Several companies are spending billions of dollars on new chip foundries, intellectual property modules, and tools for chip design and programming. Approaches that don't make maximum use of mainstream advances will face many obstacles. Just as hydrogen-powered cars might at first seem like a good idea, one has to ask if the infrastructure is in place to support them. ATRs are fed by sensors not known to biology, such as cryogenically cooled infrared cameras (nonfoveated and monocular) and synthetic aperture radar, and operate at long range to targets. So there is an automatic mismatch between biological sensor data processing and processing required by neuromorphic ATRs. The strawman recurrent ATR is not a brain. A mechanism or property should never be included in an ATR just because it is present in the biological brain. An ATR is designed and programmed by brains. The results of its processing are usually presented to human brains. The ATR is a purposeful device. Its purpose differs somewhat from that of the highly evolved brains in human or animal bodies.

6.2.1.1 What *shouldn't* be copied from biology?

1. *The human brain's enormous complexity*: Different areas of the brain have different neuron types, synapse types, connectivity patterns, and supporting systems. There are many types of retinal ganglion feature extractors within the eyes, many different types of neurons and synapses within the brain, along with other biological components such as glial cells. The structure and function of most of these are little understood.
2. *Stereo vision*: Stereo vision is suitable for determining range at short distances. ATRs operate at long distances to targets.
3. *Retinotopic image processing:* Military imaging sensors have constant spacing between pixels and are usually monochrome, operating at a constant frame rate.
4. *Lifetime of learning*: A contract to design an ATR spans a few years at most. However, unsupervised continuing learning after deployment is a worthy goal.
5. *Wet substrate*: Transistors are built on a dry silicon (semiconductor) substrate. CMOS processes and variants dominate modern integrated circuit manufacturing. They provide very good performance per watt.
6. *Extreme massively parallel processing*: If an ATR possessed 100 billion processors and hundreds of trillions of connections between processors, it would be difficult to program, train, debug, regression test, configuration control, and understand. As complexity increases, detection and prevention of malicious hardware Trojans becomes more difficult.
7. *Huge training database*: Military data is expensive to collect.

8. *Single target tracking*: Some ATRs have to detect, recognize, and track several or even several hundred targets simultaneously.
9. *Communication by electrical spikes*: Spike trains, mimicking the way neurons communicate, are not a good way to transmit precise data. Neural code or codes are not yet deciphered, and may themselves be adaptive.
10. *Very low-precision analog arithmetic*: Our tests of (artificial) RNNs indicate that less than 16-bit floating-point precision reduces performance, particularly during training.
11. *Asynchronous clock*: The visual brain is a totally asynchronous organ. There is no central neural clock that synchronizes activity and communication. It is unclear whether asynchrony is a useful feature to be imitated or a defect to be overcome. Computer chips run on a high-speed clock; the higher the GHz value the more processing steps per second.
12. *Non-programmable*: The brain is teachable in supervised fashion and can continuously learn from its environment in unsupervised fashion, but it is not directly programmable. An ATR is not like an FFT. You can't just code it once and expect it to work forever after, or for it to learn whatever it needs to learn entirely on its own. ATR is an active area of research and development with new and improved algorithms always in the works. Particularly useless are "revolutionary" new processor chips that are not directly programmable or easily programmed by software engineers of average skill.
13. *Sleep*: Sleep helps consolidate memories. Trillions of the brain's synapses shrink by nearly 20% during sleep, forgetting unimportant information and renormalizing and resetting the brain for the next day when they will grow stronger while learning new things. No so-called neuromorphic chips follow this strategy called homeostatic scaling down.

6.2.1.2 What *should* be copied from biology?

1. *Low latency*: Several video frame times of latency seems about right.
2. *Low power*: The human brain consumes 10–20 W. An eagle's brain uses a fraction of a watt, which doesn't seem to hinder it from finding a rabbit at several kilometers range. For some ATR applications, like a very small drone, one watt is the limit. The power used by the host chip also has to be considered, but the host chip is often already on the platform.
3. *Aggressive power management*: Use of on-chip power management to dynamically shift processor speed puts some processors to sleep when not needed in order to maximize performance and temperature, while minimizing overall power consumption but still keeping within the total power budget.

4. *Low size, weight*: It is currently feasible to design a highly programmable, extremely high-performance, single-chip recurrent ATR. Consider the following alternatives:
   a. A nonprogrammable ASIC-based ATR would consume very little power. However, the fields of ATR, AI, computer vision, and computational neuroscience are relatively young and still evolving, so committing to a rigid model or algorithm up front would mean missing out on improvements for the life of the ATR.
   b. GPUs originally designed for wide vector-centric graphics are commonly used for off-line training of convolutional neural networks. GPUs are not particularly efficient for the training of any type of neural network and are definitely not well suited for low-power on-line military operation. Communication between GPU and memory or GPU and host processor is a bottleneck. With minuscule on-chip memory, weight values must be constantly reloaded and activations constantly saved and retrieved, using the GPU's latency- and bandwidth-constrained interface to very large off-chip memory. GPUs implement neural networks by partitioning the training data into typically 32 sample mini-batches, doing nearly all computations as very inefficient GEMM (general matrix-matrix multiplication) operations. A less than one-watt RNN ATR can't be a GPU.
   c. FPGAs are relatively difficult to program or update when the original programmers aren't available. They offer only moderate processing power per watt. Each new generation of FPGA often differs in design from the previous generation, requiring new software tools and techniques and dealing with newly found bugs. Impressively skillful FPGA programmers are remarkable but in short supply. That being said, future 7-nm FPGAs will be quite impressive and worthy of consideration for single-chip ATR.
   d. General-purpose heterogeneous multicore processors are appropriate as a controller for a neuromorphic processor but cannot by themselves support the extremely high-performance processing discussed in this chapter.
   e. More exotic approaches like memristors, quantum computers, spike-based neuromorphic processors, DNA-based computers, and cryogenically cooled processors are still in the research stage.
5. *Local connectivity of processing elements*: In an RNN chip, a processor should be able to efficiently communicate with its neighboring processors and less efficiently with more distant processors. Mimicking the brain's 100 billion processors, each wired to 10,000 other processors, is too complex to be practical.

6. *Local memory*: Keeping sufficient memory local to processing units, with no bottleneck of external memory, isn't exactly how neural circuits work, but is close enough.
7. *Recurrent processing*: ATRs, like brains, should be able to natively deal with temporal and spatiotemporal data. Even single-frame data, like SAR or step-stare IR, can be made spatiotemporal by performing saccades from target key point to key point.
8. *Massively parallel processing*: Moderate massively parallel processing is required for a very high-performance future ATR, but not extreme massively parallel processing. 10^3 first-rate processors on a chip seems sufficient for ATR, while 10^{11} (as in a brain) is way too many.
9. *Fairly low-precision arithmetic* (e.g., 16-bit floating point): Half-precision (16-bit) floating point is adequate for RNNs that are processing difficult video data. 5-bit arithmetic precision is considered adequate for feed-forward neural networks, but that isn't what is needed for the future ATR.
10. *Multiple forms of parallelism*: Real-time video processing requires several forms of parallelism. For example: a frame of image data is divided over multiple processors such as a grid of processors (geometric-parallelism/multiplexing); algorithms are partitioned over multiple processors (algorithmic parallelism); each of a sequence of images is dispatched to a separate processor (temporal multiplexing); the problem is addressed serially over multiple processors (pipelining); or two or more streaming signals are transmitted over a common channel (time-division multiplexing).[12]
11. *Fast learning, continuous learning*: Current ATRs are trained in advance and do not continue learning once deployed. Past attempts at on-the-fly learning have been largely unsuccessful. A higher-performing future ATR will need to adapt to the situations that it encounters and be able to rapidly reconfigure itself to follow new task orders.
12. *Self-repair and regulation*: Useful features for a future ATR include: built-in test, self-repair, powering down computational units when not needed, and making itself useless upon capture.
13. *Up and running from the start*: Surely not for a human baby, but many animals such as deer and horses are up and running shortly after being born. ATR algorithm/software designers need hardware that is up and running at the very start of a project. ATR engineers can't wait until the very end of a project when the contracted hardware is ready. Algorithm/software designers can make do with an older-generation chip, or an emulation on an FPGA or workstation.
14. *Multimodality*: The human brain performs multisensor fusion and makes continuous use of metadata, such as inertial data from a vestibular system and proprioceptive data indicating the relative

positions of the parts of the body. An ATR needs to process various forms of data and metadata. This may include video data, laser or radar range data, IMU data, GPS data, and data from additional platform instrumentation.

15. *3D design*: Through-silicon vias (TSVs) are electrical connections passing through a silicon die. TSVs reduce interconnect length, thus improving communication speed. True brain-like 3D design requires TSVs, which are not likely to be both reliable and affordable until after the year 2020.

6.3 Algorithm/Software Design

The next-generation strawman ATR is a biologically inspired single-chip RNN. Its capabilities will include vehicle and dismount detection/recognition and human activity detection/recognition. Thus, the strawman ATR is suitable for still-frame data (e.g., SAR, sonar) and, more challengingly, video data (e.g., FLIR, video SAR). The four unique features of the reference design are presented here.

1. Coupled model [$\mathbf{M}$ = ES-Pl-RNN($\mathbb{Q}_{16}$)] and controller $\mathbf{C}$: The model $\mathbf{M}$ is embodied and situated, adaptive and plastic, and based on RNNs, using 16-bit floating-point arithmetic [$\mathbb{Q}_{16}$]. Following the lead of noted computer scientist Jürgen Schmidhuber, a controller $\mathbf{C}$ is coupled to the model $\mathbf{M}$ to support abstract reasoning, planning, higher-level decision making, reinforcement learning, experimentation, and creativity.[13]
2. Embodied and situated (ES): All natural intelligence is embodied and situated. The strawman solution is embodied in a small drone with the autonomy to control flight to get a better look at what is going on. Each action (change in drone position and look angle) causes changes in the perceived environment (situation) that are analyzed bottom-up through the perceptual hierarchy, combining top-down influences that lead to processing for further action through the control circuitry toward motor effectors. Actions cause changes to the perceived world that are then analyzed, leading to new action, and so the cycle continues.
3. Adaptive and plastic (Pl): Adaptation, neuroplasticity, and continuous learning are hallmarks of bio-intelligence. The strawman design will adapt to the environment and adjust its architecture and weights (intrinsic program) at each time step. Thus, like the brain, the next-generation ATR will continually reorganize and "rewire" itself to adapt to new situations and shifts in the environment (e.g., nightfall).
4. Long short-term memory recurrent neural network (LSTM-RNN) block as basic computational unit: LSTM-RNN is arguably the best second-order RNN.[14] The LSTM-RNN treats memory as a dynamic process. Long and short-term memories are recurrently reconsolidated;

i.e., maintained and reinforced; or faded when no longer useful. Our version of the LSTM block is implemented as a distributed acyclical graph of strongly connected components, as such, providing arbitrary connectivity and easy parallelization.

6.3.1 Classifier architecture

The classifier is implemented as a taxonomy-based Bayesian decision tree. Each decision is made by a hierarchical temporal memory (HTM). Each HTM is constructed from LSTM-RNN memory blocks.

6.3.1.1 Decision tree

Several strategies are available for decomposing a K-class problem, as shown in Fig. 6.4. Each branch in the diagrams represents a decision, which in our case is made by an HTM. An all-versus-all (AvA) approach is appropriate if all classes are known in advance. Only one classifier needs to be trained for the K-class problem. However, if K is very large, the classifier will be difficult to train. An one-versus-all (OvA) approach is often used when classes are not known until the mission-of-the-day. K classifiers have to be trained for the K-class problem. Only those classifiers required for a mission are activated at the time of the mission. A one-versus-one (OvO) approach works well when the classes can be hierarchically arranged as a taxonomy. Each decision can be a simple binary decision (but doesn't have to be binary). A taxonomy-based decision tree is simple to understand and interpret. It mirrors human decision making better than other approaches.[15] The K-class problem is then reduced to $K(K-1)/2$ binary classifiers. With this approach, branches can be cut from the tree if not required for the mission-of-the-day. OvA and OvO schemes can add a class without retraining the entire classifier.

The strawman ATR's classifier is implemented as a decision tree in Fig. 6.5. Each decision is made by an LSTM-RNN-based HTM.

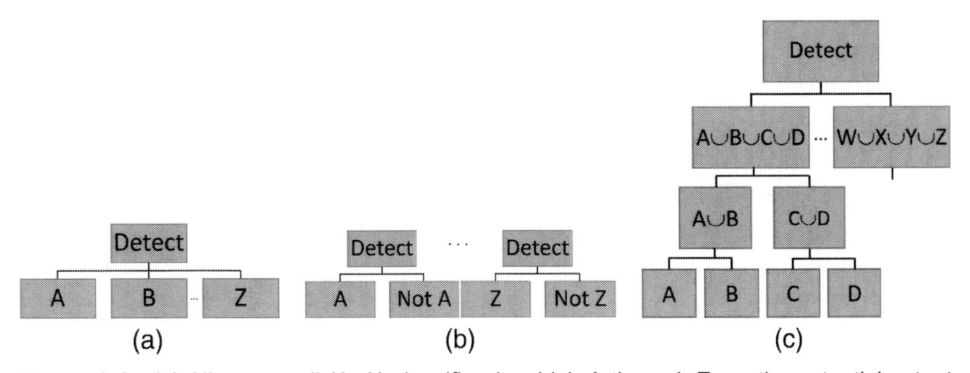

Figure 6.4 (a) All-versus-all (AvA) classifier, in which A through Z are the potential output classes. (b) One-versus-all (OvA) classifier. (c) One-versus-one (OvO) classifier as a decision tree.

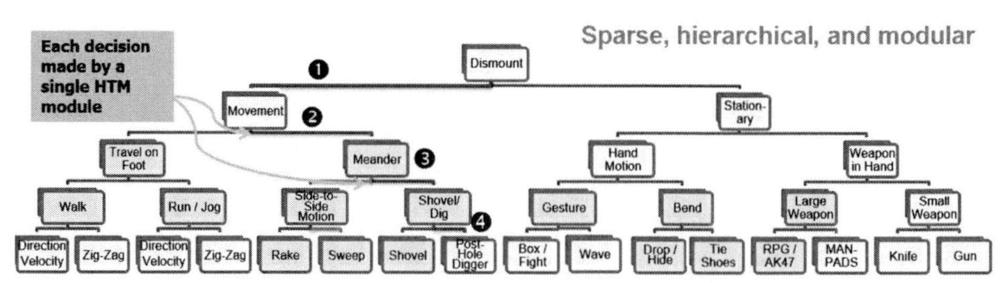

Figure 6.5 Example of Bayesian decision tree, where each decision (i.e., which branch to go down the tree) is made by an LSTM-RNN-based HTM module.

The decision tree is upside-down compared to a normal tree, with the root on the top and leaves on the bottom. As shown in Fig. 6.5, a decision tree is a visualization of the decision-making process, illustrating possible decisions and possible outcomes modeled on the hierarchy of the taxonomy. Thus, the decision tree illustrates not only all possible classification outcomes, but also the paths by which they can be reached.

At any level of the decision tree, the classifier can make a declaration, which is a decision to provide a label. The label corresponds to a node name within the taxonomy. The classifier might be able to declare an activity as "person digging," but might not be able to specify whether the person is digging with a shovel or a post-hole digger. In this case, all levels in the decision tree above "person digging (shovel/dig)" would be declared, but not those below "person digging." At longer ranges-to-target, or in adverse weather, it might not be possible to traverse the tree all the way down to the leaf nodes. This is not so bad for a military system, as any level of classification provides useful information. It may be sufficient evidence to act (actionable intelligence), knowing that a person is digging along a road to a tactical operation center (TOC) at midnight, not having finer-grained information on what type of instrument is being used for digging.

Assigning class priors and cost functions to each decision in the decision tree introduces Bayesian concepts. Then the tree is traversed going down the path of lowest loss or cost. The priors and cost functions can change at any time, for example, by a task order from the TOC or by the context of the situation. For example, if the TOC receives intelligence data that a bomb has exploded at a location, then the priors to find persons running away from that location would be increased. Priors can change at night (e.g., task of searching with a flashlight) or due to weather (e.g., person carrying an umbrella).

The equations for traversing the tree are presented in Table 6.3.

Each decision that the decision tree makes as it is traversed is made by a separately trained classifier. Our strawman classifier is an HTM, shown in Fig. 6.6. The HTM is constructed from LSTM-RNN memory blocks. One LSTM-RNN block is composed of first- and second-order computational

Table 6.3 Equations for traversing the decision tree.

Traversing the Decision Tree: [O(log N)] complexity

Go down the path with least loss or equivalently lowest cost:

- $loss(\mathbf{x}, S_k) = \sum_{i=1}^{K} C(S_k|S_i)p(\mathbf{x}|S_i)P(S_i)$.
- $p(\mathbf{x}|S_i)$ is output by HTM made of LSTM blocks.
- $\mathbf{x}$ is the new data (being considered at the moment).
- $p(S_i|\mathbf{x})$ is the posterior probability.
- $P(S_i)$ is the Bayesian prior for class S_i.
- $p(\mathbf{x}|S_i)$ is the likelihood of the data given the class.
- $C(S_k|S_i)$ is the cost or loss of mistakenly assigning an item to class S_k that is actually in class S_i.

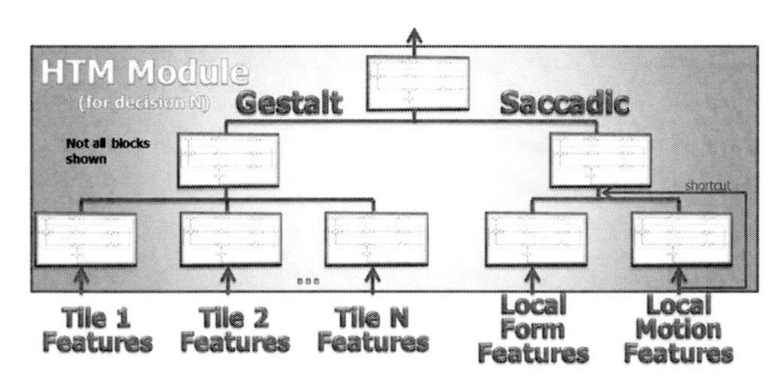

Figure 6.6 HTM module constructed from LSTM-dag blocks. Features on the right side of the diagram are extracted locally about the fixation points of saccades. Features on the left are extracted within a region-of-interest about the detected target.

units and many feedback loops (Fig. 6.7). The computational units are sometimes likened to neurons.

LSTM-RNN is able to solve many complex and difficult sequence modeling problems unsolved by other methods. By contrast, the widely popular feedforward CNN is intended for spatial patterns (i.e., still image frames). Due to its recurrent nature, LSTM-RNN is inherently deep in time, whereas the CNN has finite depth and can be intractable for indeterminately long sequences.

As noted previously, LSTM-RNN can also recognize still frame targets by crisscrossing the targets with saccades and extracting a small region-of-interest about the fixation point of each saccade, thus turning a still frame target image into video. This approach is analogous to the way a human recognizes an object (e.g., face) in a snapshot.

An LSTM-RNN memory block stores data across arbitrary time lags. This ability is key to recognizing events that unfold over long or short time scales. The original LSTM-RNN had difficulty with input data streams that were not segmented into self-contained temporal subpatterns. This problem was solved by adding "peephole" connections, which reveal current internal states at the LSTM-RNN nodes, providing visibility into the duration of events (peephole connections are shown as dashed lines in Fig. 6.7).

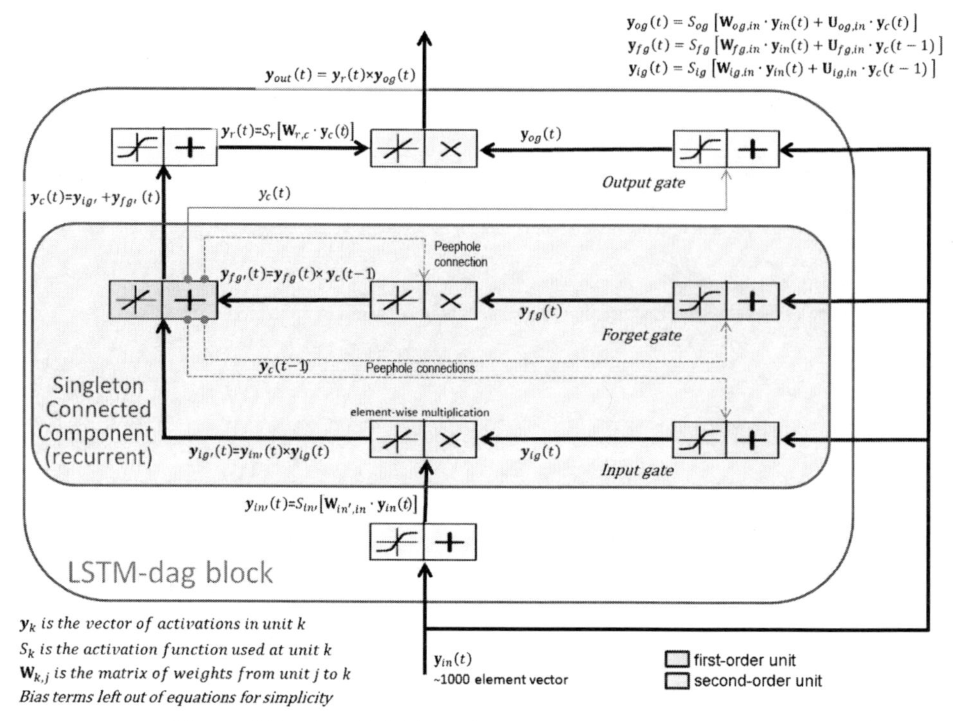

Figure 6.7 One LSTM-dag block. (Adapted from Ref. 16.)

Figure 6.7 portrays LSTM as a distributed acyclical graph (dag) of strongly connected components, referred to as LSTM-dag. The graph is based on the work of Hwang and Sung.[16] This more-generalized form of LSTM is suitable for real-time implementation on a graph-oriented massively parallel chip.

6.3.1.2 Embodied and situated (ES)

Knowing is inseparable from doing. Doing requires a body. Cognition and learning are bound to context and situation. Natural intelligence is thus both embodied and situated (ES). Our strawman ATR is embodied in a small drone (Fig. 6.8). It has the autonomy to control its situation with respect to

Figure 6.8 Our strawman model [M = ES-PI-RNN($\mathbb{Q}$)] is *embodied* in a small drone that can autonomously change its situation, e.g., fly to a different location to get a better look at an activity or target. *Situatedness* is then the dependence of meaning (semantics) on the situation, context, and task. The situated being (drone) uses self-initiated agency to obtain meaning.

the environment. Each action (change in drone position and look angle) causes changes in the perceived environment (situation) that are analyzed, leading to further action. Thus, response to a stimulus produces a new stimulus, which changes the probability of subsequent responses. Each subsequent response produces a new stimulus, which changes the probability of following responses, and so on. This process is called autochaining.

6.3.1.3 Adaptivity and plasticity (Pl)

The proposed strawman architecture is *adaptive* in that it continually adapts to its environment. It is *plastic* because its neural weights and network architecture are continually changing. Neuroplasticity is an active area of research and debate.

Architectural Plasticity: In our strawman design, transitory situational knowledge (in the form of Bayesian priors and cost functions) is shielded from deep knowledge (in the form of neural weights). The ATR's architecture is continually reorganized and "rewired" to adapt to new task orders, new situations, and shifts in the environment (e.g., nightfall, person seen waving). If a particular class prior goes to zero or is set to zero, the class is effectively pruned from the decision-making architecture.

Synaptic Plasticity: Reinforcement learning (RL) allows an agent in an environment to learn a policy to maximize reward: **C**: S × A → R, where S denotes situation, A denotes action, and R denotes reward. This is illustrated in Fig. 6.9 Reinforcing feedback R modifies **M**'s synaptic weights according to a bio-inspired or gradient descent learning rule. As we shall cover next, adding a controller **C** helps evolve the model **M** and establish actions to maximize reward.

6.3.2 Embodied, situated, plastic RNN [M = ES-Pl-RNN($\mathbb{Q}$)] coupled with a controller C

The strawman model [**M** = ES-Pl-RNN($\mathbb{Q}$)] is put into Prof. Schmidhuber's revolutionary, new, recurrent neural network artificial intelligence (RNNAI)

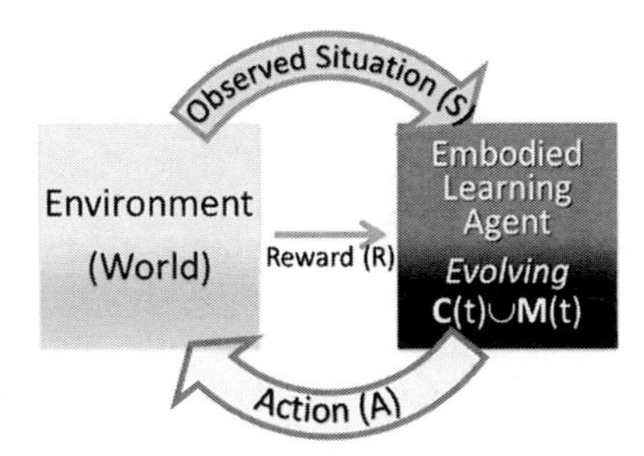

Figure 6.9 Reinforcement learning is learning how to act to maximize a numerical reward. An agent selects an action, the action causes changes to the perceived world (environment) that are then analyzed; a reward is returned to the agent, which then leads to a new action, and so the cycle continues.

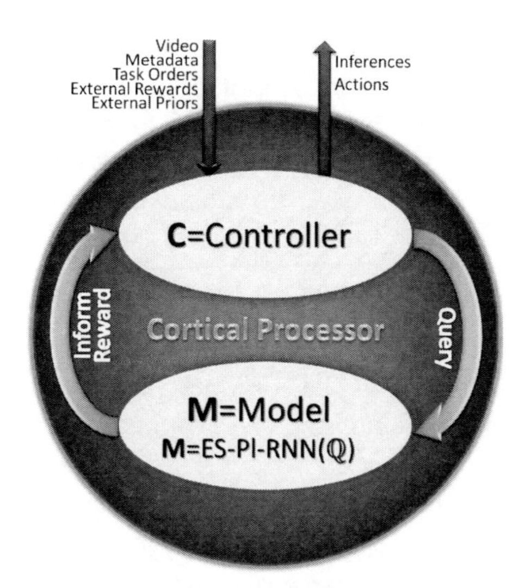

Figure 6.10 Tightly coupled model and controller: the RNN is trained to become a predictive model of the world. It uses that predictive model to train a separate controller, which acts as a reward maximizer.

framework (Fig. 6.10).[13] The model **M** and controller **C** form a coupled RNN, where **M**'s outputs become inputs for **C**, and where **C**'s outputs in turn become inputs to **M**. The goal is to train **C**'s parameters to help accomplish a new or better inference task whose solution shares mutual algorithmic information with **M**'s task. To facilitate this, **C** is allowed to learn to actively inspect and reuse the algorithmic components of **M**. Once a new inference task is learned, it is added to **M**. The search space of the learning algorithm is much smaller than that of a possible competing system that has no opportunity to query the model, but has to learn the new task from scratch without forgetting what it has already learned. Thus, the strawman ATR continuously uses all available information to modify the current **M** to produce a new **M** that can infer all previously learned activities or targets plus the new one; or it simplifies, compresses, improves, or speeds-up previous solutions without forgetting any solution.

Let us consider several concrete examples of what can be accomplished by the tightly coupled **C** ∪ **M**.

Example 1: Complex Scenarios: To address complex activities or scenarios, **C** will improve the performance of **M** by recognizing patterns of complex activity that encompass the previously learned simpler activities. Specifically, while **M** will recognize the primitive activity that is currently observed, **C** will infer the underlying complex activity that is most likely occurring, given the sequence of observed more-primitive activities. **M** will provide this sequence of recognized primitive activity labels to **C** as input, along with other data. In turn, **C** will modify the priors of **M** based on the inferred complex activity in order to improve the prediction accuracy for the next observation. For example, suppose that **M** has output the sequence of activities {*"walking to car," "entering driver's side of car"*}. Then, **C** will temporarily increase the prior for what may come next, which is *"driving car."* It will also notify the tracker of its supposition.

Example 2: Abridged Activity Model: When learning a new activity involving atomic actions *abc*, **C** can use **M** to test the efficacy of shorter subpatterns, such as *ab*, *bc*, or *b*. For example, it may turn out that a previously learned activity involving "walking ... dropping object ... walking" can be replaced by the simpler core activity of "dropping object."

Example 3: Generative adversarial network (GAN): The controller implements a GAN.[17,18] The GAN's goal is to create a training set that conforms to current environmental conditions. The GAN consists of a discriminator and a generator, each of which is implemented as a neural network. The task of the generator is to learn to modify the initial video (or still frame) training images to look like they are sampled from the current operational environment. The task of the discriminator is to determine if a sample is from current conditions (which we shall refer to as real) or is an initial off-line training image that has been modified by the generator (which we shall refer to as fake). To this end, the system collects a set of real, but unlabeled, object images from the current environment. The discriminator and generator are each trained with backpropagation of error. The discriminator tries to distinguish newly observed unlabeled object images (the reals) from the modified training images (the fakes), while the generator is trying to modify the original training images to make the discriminator think that they are real. At the end of this on-line process, the outdated initial training images will have been modified to appear as if from current conditions. The controller then retrains the model's neural networks with the modified training set.

Example 4: Algorithmic Transfer Learning: **M** is extended to recognize a new activity by reusing its existing components to facilitate training, thus increasing the speed and effectiveness of learning. This is in contrast to most existing approaches, where a new class requires the complete retraining of an existing classifier. For example, **C** asks **M** to pass the data for a new activity of interest through its decision tree. The new activity gets to a certain node down the tree, but can't go any further. The node that it gets stuck at represents a parent category for the new activity. The HTM for that node N of the decision tree is then retrained using training data for N's previous daughter classes plus the new class. After retraining, the parent node N may have three daughter nodes instead of two, but the other decision-making neural networks in the tree will not be altered. The nodes above the parent node N have transferred what they have previously learned to help recognize the new class. The bottom layer of node N's HTM can also be reused to learn the new class.

Example 5: The World as the Teacher: On-line reinforcement learning differs from off-line supervised learning in that correct input/output pairs are rarely available in the former. Suppose that the ATR on a small drone gets a single look at an activity or target and makes a classification decision based on the single look. It has no way of knowing if its decision is correct or not, so it can't feedback an error signal to improve the training of its neural networks. Now suppose that the controller directs the drone to take several looks at the activity or target from several vantage points. It now has trustworthy consensus information that can be fed back to improve the on-line training of the neural networks. Thus, the system has collected information about the environment by interacting with it. It then uses this information to improve its model.

6.3.2.1 Training the controller C

The controller **C** and model **M** are trained in a mutual but alternating fashion over many cycles. In other words, when **C** is trained, the weights of **M**'s LSTM-based HTM networks are frozen; subsequently, **M** is trained while **C**'s weights are frozen. The process is then repeated until some performance criterion is met. First, we describe the training of **C**, which is accomplished through reinforcement learning. Specifically, **C**'s observations include primitive activity labels with track and object attributes. Its actions include signals that are sent to **M** to upgrade it and signals sent to motor effectors to change the observed situation. **C**'s rewards are determined by the accuracy with which **M** makes its predictions. In other words, if **C** is able to improve the performance of **M** through modifications to it or its situation, it receives a higher reward. The reinforcement learning problem is to develop a *policy* (mapping from observations to actions) such that reward is maximized. **C**'s policy can be in the form of another LSTM network that is trained via the recurrent policy gradient method.[19] Importantly, the search space of the learning algorithm is much smaller than that of a possible competing system that has no opportunity to query **M** but has to learn the policy from scratch.

In summary, **C** learns to access, query, and exploit in arbitrary computable fashion the "program" of the much larger **M**. The learning progress of **M** is the intrinsic reward for **C** that motivates **C** to come up with additional promising experiments, learning to shape the observation stream through action sequences or experiments that help the learning agent figure out how the world works and what can be done in it. **C** ∪ **M** is creative. It makes mistakes and tries new things.

6.3.3 Software infrastructure

In ATR development, the software environment is just as important as the hardware environment. Compatibility with normal software design tools and standards is necessary for a reasonably priced, repairable, and upgradeable military product. The strawman single-chip ATR will be readily programmable using standard languages such as C/C++ and Python. The chip's programming framework will seamlessly interface with one or more of the leading neural network frameworks and computer vision libraries such as Caffe, CNTK, Deeplearning4j, dlib, Keras, Lasagne, MXNet, Neon, OpenCV, PaddlePaddle, TensorFlow™, Theano, or Torch.

6.3.4 Test results

Testing was over video data with eight decision tree leaf classes: rake, sweep, shovel, post-hole dig, drop/hide, wave, box/fight, and tie shoes. Six of the classes were from Army-provided data, and two of the classes were from KTH database videos. Training and testing data were collected by different organizations, against different test subjects wearing different clothing (e.g., hats, camouflage, sunglasses, gloves) at different locations and times, and were taken with different cameras. That is, there was absolutely no intermingling of training and testing data, or an attempt to make them look alike.

Results are shown in Fig. 6.11(a) for both 8-leaf-node activity classes, and 15 nodes encompassing the whole decision tree. This makes sense for the military problem, since either video quality won't always support traversing the tree all the way down to the leaf nodes, or a nonleaf decision may suffice for the mission. Results are also reported for three looks from different vantage points [Fig. 6.11(b)], as if taken from a drone.

6.4 Potential Impact

The goal is to make targeting systems smarter and more autonomous, continuously adapting to new situations and conditions. A next-generation ATR hardware/algorithm/software concept was presented as a biologically inspired strawman design. The design integrates a model [**M** = ES-Pl-RNN($\mathbb{Q}$)] and a reinforcement learning controller [**C**]. Both **M** and **C** are RNNs or composite designs incorporating RNNs. **C** is like an artificial ATR engineer (homunculus) operating on-line to direct the functioning of the brain-like **M**.

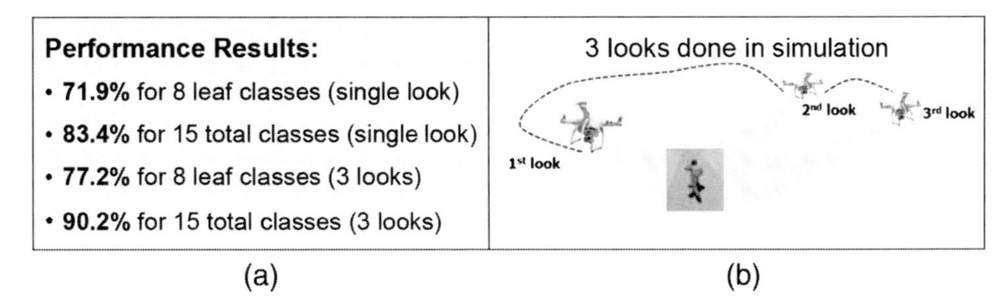

Figure 6.11 (a) Performance results and (b) three looks from different vantage points.

C adds to its external reinforcement learning reward an intrinsic reward for performing experiments that improve **M**.[13] Borrowing concepts from biology, **C** ∪ **M** achieves its goals by being embodied and situated (ES), adaptive and plastic (Pl).

The foundation of **M** is a classifier that is implemented as a Bayesian decision tree, where each decision is made by an HTM, where HTMs are constructed from LSTM-RNN blocks. Test results on Army data are encouraging.

Coupling a controller **C** to the model **M** forms the complete strawman system (**C** ∪ **M**), which is more powerful, in many ways, than a standard ATR. **C** ∪ **M** can learn a never-ending sequence of tasks, operate in unknown environments, and realize abstract planning and reasoning. This next-generation ATR is suitable for implementation on two chips: a single, custom, low-power chip (<1 W) for implementing **M**, hosted by a standard processor serving as the controller **C**. This ATR will be appropriate for various military systems, including those with extreme size, weight, and power constraints.

The motivations of neuroscientists and ATR engineers are synergistic. Neuroscientists provide information on circuits and algorithms implemented in the brain. ATR engineers use these clues to develop artificial vision systems. One way to determine if the visual system is understood is whether an artificial system can be constructed based on broad neuromorphic principles that can match human performance on a set of tasks. This leads to a type of Turing test for measuring the intelligence of an ATR, which is the subject of the next chapter. Potential impacts of the smarter ATR are better adherence to the rules of engagement, and robotic systems that keep friendly forces out of harm's way.

References

1. A. A. Faisal and A. Neishabouri, "Fundamental Constraints on the Evolution of Neurons," in *The Wiley Handbook of Evolutionary Neuroscience*, S. V. Shepard, Ed., John Wiley & Sons, Ltd. pp. 153–172 (2016).
2. Straw man proposal, Wikipedia, https://en.wikipedia.org/wiki/Straw_man_proposal
3. S. B. Laughlin, "Energy as a constraint on the coding and processing of sensory information," *Sensory Systems* **11**(4), 475–480 (2001).
4. S. B. Laughlin and T. J. Senjowski, "Communication in neuronal networks," *Science* **301**(5641), 1870–1874 (2003).
5. K. Clancy, "Here's why your brain seems mostly dormant," *Nautilus* **2015**(27), 6 August 2015, http://nautil.us//issue/27/darl-matter/heres-why-your-brain-seems-mostly-dormant

6. J. E. Niven and S. B. Laughlin, "Energy limitation as a selective pressure on the evolution of sensory systems," *Journal of Experimental Biology* **2008**(211) 1792–1804 (2008).
7. C. Potter, B. Wyble, C. E. Hagmann, and E. S. McCourt, "Detecting meaning in RSVP at 13 ms per picture," *Attention, Perception and Psychophysics* **76**(2), 270–279 (2014).
8. R. Kuboki, Y. Sugase-Miyamoto, N. Matsumoto, B. J. Richmond, and M. Shidara, "Information accumulation over time in monkey inferior temporal cortex neurons explains pattern reaction time under visual noise," *Frontiers of Integrated Neuroscience* **10**, 43 (2017).
9. M. E. Raichle, "Two views of brain function," *Trends in Cognitive Sciences* **14**(4), 180–190 (2010).
10. P. Lennie, "The cost of cortical computation," *Current Biology* **13**(6), 493–497 (2003).
11. A. Sandberg, "Energetics of the brain and AI," Technical Report STR 2016-2, *arXiv*:1602.04019 (2016).
12. M. Fleury and A. C. Downton, *Pipelined Processor Farms: Structured Design for Embedded Parallel Systems*, John Wiley & Sons, Inc., New York (2001).
13. J. Schmidhuber, "On learning to think: Algorithmic information theory for novel combinations of reinforcement learning controllers and recurrent neural network models," *arXiv*:1511.09249 (2015).
14. J. Schmidhuber, "Deep learning in neural networks: An overview," *Neural Networks* **61**, 85–117 (2015).
15. J. Gareth, D. Witten, T. Hastie, and R. Tibshirani, *An Introduction to Statistical Learning with Applications in R*, Springer, New York (2015).
16. K. Hwang and W. Sung, "Single stream parallelization of generalized LSTM-like RNNs on a GPU," *40th IEEE International Conference on Acoustics, Speech, and Signal Processing (ICASSP) 2015*, pp. 1047–1051 (2015).
17. I. Goodfellow, J. Pouget-Abadie, M. Mirza, B. Xu, D. Warde-Farley, S. Ozair, A. Courville, and Y. Bengio, "Generative adversarial nets," *Advances in Neural Information Processing Systems* **27** (NIPS 2014), pp. 2672–2680 (2014).
18. A. Shrivastava, T. Pfister, O. Tuzel, J. Susskind, W. Wang, and R. Webb, "Learning from simulated and unsupervised images through adversarial training," *arXiv*:1612.07828 (2016).
19. D. Wierstra, A. Foerster, J. Peters, and J. Schmidhuber, "Recurrent policy gradients," *Logic Journal of the IGPL* **18**(2), 620–634 (2010).

Chapter 7
How Smart is Your Automatic Target Recognizer?

7.1 Introduction

The human brain has about 90 billion neurons, each with roughly one thousand synaptic connections to other neurons. However, it does not follow that if we build a computer with equivalent processing power and connectivity, it would match human functionality as an emergent property. We do not now know how the brain does the vast majority of what it does. The long-term goal of neuromorphic engineering is to design artificial systems patterned after both the design and functionality of biological neural systems (not necessarily human). Much of the human brain is used for scene understanding, object detection, recognition, tracking, multisensor fusion, and motor control. So, in a sense, its function is similar to that of an ATR. The ATR can be viewed as a substitute, or at least a workload reducer, for the warfighter's brain. For the purpose of this discussion, we will consider the neuromorphic ATR to be a black box. We will not be as passionately concerned as to whether the inner workings of the black box are true to brain biology in all possible respects, as is, for example, Henry Markram, head of the European Human Brain Project.[1] As ATR engineers, we just want the black box to transform its inputs to the required outputs. We want the black box to meet certain key performance requirements involving size, weight, power, cost, latency, mean time between failure, and logistics trail. The black box must demonstrate capabilities that are needed in combat. It must become fully operational in a military environment, having passed a difficult operational test and evaluation process.[2] That is, it must be more than "just research." It should be more rugged and reliable than a comparable commercial product.

The human brain has evolved solely for survival of the species. It is "designed" to work as part of a system, which includes various sensors, vestibular (IMU) data, positioning system, and articulated parts and processes that it controls. The brain learns over a lifetime in both supervised and unsupervised modes. It never stops changing its wiring. It's hard to view the brain as disembodied from the rest of the system. Visual perception works with a pair of eyes, feeding retinal code (not video) to the brain. The eyes are always in motion. Human vision often functions as part of a multisensory fusion system. Except for relatively recent passive pastimes, such as reading and watching TV, visual perception's main function is to initiate and guide motor control. A person walks to an unknown object to get a closer look, might first push it so see how it reacts, then might pick it up, touch it, smell it, shake it, and might even take a bite. Humans are highly social animals, with actions in collaboration with or in reaction to those of other persons. There are good reasons to study biological systems. Multisensor fusion, networked processing, and robotic self-controlled platforms are of interest to civilian and military system designers. Biological systems provide models known to work. They help spark the engineer's imagination.

When we ask: "How smart is your ATR?" we mean: "How well can this machine perform some of the tasks performed by the well-trained human pilot, soldier, sailor, Marine or analyst?" If we arrive at the point where a machine can do all of the tasks of humans, then what are the implications? These types of questions have a surprisingly long history of discussion and analysis. In Jewish legend, Adam of the Biblical creation story was created as a golem, or unfinished human being, until the point where he was given a soul. "Can we then say: God is to Golem as man is to machines?"—asks legendary MIT cyberneticist, mathematician, and communications pioneer Norbert Wiener (1964).[3] A Talmudic tract states that something that looks somewhat human and acts human has to be considered a member of the community and given human rights. Later stories about golems describe animated anthropomorphic beings made out of inanimate matter, imbued with a sense of life when a specific series of letters is programmed into them.[4] The concept of intelligence was studied, discussed, and hotly debated in stories of golems dating from early Judaism (1500s–1800s and before) and until the present, now in the context of artificial intelligence (AI) and robotics. Clockwork automatons in the 17th century provided further basis for imagining an anthropomorphic robot. These early stories were remarkably insightful in that they understood that natural language communication is critical to human intelligence. They couldn't envision an artificial being that possessed human language. The omission was remedied by the Turing test (developed in 1950 by Alan Turing), which measures the ability of a machine to exhibit intelligent behavior indistinguishable from that of a human. Intelligence is verified through natural language interaction with the unseen machine. Rabbi Dr. Rosenfeld referred to any computing machine that can pass the Turing test as a "robot."[5] He stated that "if intelligent golems could be created in the laboratory, it should not matter if they are biological ("androids') or mechanical ("robots") ... it should not even matter whether these golems have human form. ... Thus, it is conceivable that even an intelligent computer could be" considered human (according to religious law).[6] In a military context, whether an ATR or robot can substitute for human judgement has to be considered in terms of adherence to the "rules of engagement."

In a *New York Times* opinion piece, Stanford University computer scientist Jerry Kaplan points out that nuclear missiles and chemical and biological weapons, are "dumb" and difficult or impossible to control once deployed.[7] The lowly landmine kills indiscriminately, often continuing to do so long after it is deployed. "A.I.-based weapons, in contrast, offer the possibility of selectively sparing the lives of noncombatants, limiting their use to precise geographical boundaries or times, or ceasing operation on command (or the lack of a command to continue.)" "... a machine won't grow impatient or scared, be swayed by prejudice or hate, willfully ignore orders or be motivated by an instinct for self-preservation."[7] But, as will be discussed later, some technologists, along with science fiction movie producers, to be sure, do not share this hopeful outlook.

7.2 Test for Determining the Intelligence of an ATR

As Reggia et al. note: "Significant progress has been made in artificial intelligence without having a generally accepted definition of intelligence."[8] Rather than precisely defining what constitutes a smart ATR, we will provide a list of capabilities that an intelligent ATR should possess. Our version of a Turing-like test is posed as a list of eleven questions, narrowly focused on ATR. We will briefly review the questions and offer some thoughts on where we now stand. This approach, along with the provided conclusions on the state-of-the-art, represent a significant departure from the conventional wisdom and will likely be met with some suspicion. If this approach is deemed reasonable, a test and evaluation (T&E) organization will have to transform the eleven questions into an actual scorable test that can be competitively taken by ATRs considered for procurement. The test will have to be tailored to specific sensors, platforms, and missions. It will have to be taken over sufficiently varied data to reach meaningful conclusions.

The golem stories and the Turing test view intelligence with anthropomorphic bias (Table 7.1). Much of the human brain is devoted to processing data from a matched pair of foveated eyes, interpreting spoken and written language, with important issues being food, shelter, and social engagement. An ATR usually receives data from radar, LADAR, sonar, infrared, and/or precise positional sensors and communicates by ones and zeros. Therefore, the ATR requires types of processing that are somewhat different from human brain processing. Furthermore, there is no reason to believe that the human model is the only viable model for intelligence. Human intelligence, residing in a three-pound brain, can't exhaust all possibilities. Future ATRs, or creatures/robots

Table 7.1 Anthropomorphic bias.

Although futurists and marketeers might think otherwise, chances are that most intelligent machines will not:

- look like humans,
- think like humans,
- perform like humans,
- have human-like eyes,
- communicate verbally to other machines,
- have human longings and emotions,
- sleep each night and daydream when staring out the window,
- have familial, religious, and ethnic ties, or
- replicate like humans.

in distant galaxies, might have drastically different forms of intelligence. Even on Earth, communities of social insects and marine mammals are intelligent in ways that differ from human intelligence. Due to lack of imagination, the test items discussed in the following eleven subsections admittedly suffer from anthropomorphic bias. The items also have temporal bias, focusing on the issues of the day. An ATR that passes a more concrete version of this test could substitute for a human but will not have more advanced "superintelligence."

7.2.1 Does the ATR understand human culture?

Human activities vary by time of day, day of the week, season, on holidays, by country, and in relationship to buildings and roads. People come to and go from schools, factories, market places, and religious institutions in somewhat periodic fashion. Different activities take place on the battlefield, sports field, farm field, and in relation to water bodies. Culture, custom, and organizational behavior are specific to location. Costume is specific to location, gender, age, occupation, and weather. The ATR is often faced with insufficient data to make a decision. An understanding of human culture would help the ATR distinguish a weapon from a farm implement, a farm truck from a terrorist's truck, a schoolgirl carrying a lacrosse stick from a combatant carrying a rifle, and so on. Can the ATR assign meaning to a person pointing an object at other people raising their hands into the air? Does the ATR know that a terrorist is less likely to be on a sailboat than a speedboat? Does the ATR know that a 90-year-old woman on a rocking chair or a child skipping rope is less likely to be an insurgent than a 22-year-old man on a motorbike? Can the ATR reach a conclusion about the men bunched on the bed of a pickup truck wearing black face masks?

7.2.2 Can the ATR deduce the gist of a scene?

Scene understanding is central to the objective of ATR. A human observer recognizes the gist of a scene in a single glance. That will be our assumption, although like most topics in brain research, there is more than one school of thought. For example, Wu et al. conclude that "semantic guidance is not entirely due to either the effect of scene gist or spatial dependency among objects."[9]

Scene gist is a global first impression.[10] Perceptual gist refers to the coarse structure of a scene. Conceptual gist refers to semantic understanding. We will expand the concept of scene gist to include the most salient activity taking place in a spatiotemporal scene. Humans seem to comprehend a scene at a coarse level almost effortlessly. Scene gist is rapidly inferred without the need to recognize each object and structure in a scene, which is opposite to the way that ATRs now operate.

Professional image analysts are very good at their job. Image analysts rapidly determine the gist of a scene and use this information to guide their attention. They do not spend equal amounts of time and energy on all parts of a scene. Of course, prior knowledge and goals help. Any person can determine the gist of an IR scene without ever having seen an IR image. An IR image is more literal than a synthetic aperture radar (SAR) image. However, even novices can deduce the gist of a SAR image of a city.

Understanding scene gist is a step in the direction of smarter ATRs. Precise metadata is generally available to help. Goals are known. Nevertheless, determining the gist of a scene and doing so as well as an experienced image analyst is proving difficult to automate.

7.2.3 Does the ATR understand physics?

Physics tells us that the concept of ATR makes sense. We live in a universe that is describable and understandable using physics. With a sufficiently good model of how the world works and the constraints it imposes, it is possible to remotely recover information about the world and then process the information by computer.

There is seldom sufficient low-level information in imagery to discern what is really going on. Physics provides important priors to help understand a scene. Physics is knowledge about matter and motion, and includes such concepts as energy, force, and electromagnetic radiation and its propagation through the atmosphere. Physics describes how objects behave. Physics explains a scene through velocity, momentum, gravity, thermodynamics, electricity, magnetism, waves, and optical properties.

Each sensor type is based on certain physics. There are many types of imaging devices and, more generally, many types of sensors and metadata sources. Most provide data that can be used with an ATR. Some sensors actively probe the environment. Radar, sonar, LADAR, gravitometers, vibrometers, INS/GPS, thermal sensors, among others, are designed with a deep understanding of physics. Imagers can use such properties as frequency, polarization, quantum spin Hall effect, multiple apertures, and Doppler. We will limit the following discussion to more conventional scene physics, focusing on the visible and thermal bands.

Physics describes how a 2D image is formed from a 3D scene. Lenses and/or mirrors are used to help form the image. The physics of light, its propagation through the atmosphere, and properties of scene surfaces determine the brightness of a point in the image plane. Geometry determines the location in the image plane of the projection of a point in the scene. Adjacent points on an object map to adjacent points in the image. The basis of ATR is that an object is recognizable from the image, although the image may be distorted by the atmosphere, optics, focal plane array, and various forms of noise.

Visible light propagates through the atmosphere, is attenuated, reflects off of surfaces, and forms highlights and shadows. The higher attenuation of red compared to blue is a clue to range and hence scale. Scale is critical to scene understanding. Shadows move with an object in the visible band but not in the LWIR band. Shadow direction can be determined from location on the Earth, time of day, and day of the year. Burning objects are brighter than their backgrounds and release smoke that rises into the atmosphere. Some vehicles release diesel smoke. Flames and smoke are important indications and should be included in an ATR's analysis.

The ATR can use physics to help interpret a scene and track objects. Physics tell us not to look for a tank perched on top of a tree or floating on a lake. Physics tells us that vehicles are not likely to be found on the sides of steep mountains. Mass determines a body's resistance to acceleration. When tracking a vehicle, we know that momentum helps determine its course. Fixed-wing aircraft and munitions cannot stop in midflight. Bodies in motion that contact each other sustain damage. The gravity vector is a critical axis of symmetry and hence an important prior. Gravity and air resistance determine the trajectories of munitions. Water applies more damping force than air. Water flows downhill. This is useful information for tracking river boats. An understanding of friction would tell the IR ATR that the wheels, tread, and axles of a vehicle in motion will get hot. If a tank's barrel is hot, it has recently been firing rounds. Wind moves smoke, stirs up dust, and cools warm surfaces. A long dust trail or contrail indicates a vehicle at its leading edge. Puddles reflect light and introduce clutter. Clouds cast shadows. Windshields and mirrors are highly reflective of sunlight. The ATR should not point the EO/IR sensor at the sun. The sound of a blast travels slowly through the atmosphere, while the flash travels much faster, a clue to the range of the source of the gunfire. Shape from shading and motion as well as stereo are based on scene physics and geometry. Is the ATR smart enough not to look for an armored vehicle on the roof of a house, floating on a lake, above the skyline, or on the side of a cliff? Can the tracker predict when a vehicle passing behind an obstacle will emerge? Can the ATR use the sun, moon, stars, shadows, and landmarks to help determine position?

7.2.4 Can the ATR participate in a pre-mission briefing?

The pre-mission briefing is the final intelligence briefing to an aircrew or squadron before they embark on a mission [strike mission or sometimes intelligence, surveillance, and reconnaissance (ISR) mission]. These briefings provide essential information and instructions to enhance mission success. They are interactive and tailored to user needs. Modifications would be needed to address the particular requirements of an ATR or fully robotic system that was capable of actively participating in a pre-mission briefing.

The pre-mission briefing illustrates the battle situation by means of maps, graphics, and photographs. It summarizes the activities of enemy and friendly forces. It reviews enemy threats at the battle zone as well as along ingress and egress routes. For an air mission, threats are covered in sequence from takeoff to landing. Threats include the enemy's early warning radar, surface-to-air missiles, and anti-aircraft artillery. The briefing provides the latest information on enemy tactics, force structures, weapon ranges, jamming capability, and electronic countermeasures.

The briefing goes into great detail about the nature and purpose of the mission and what is being targeted. The aircrew is shown photographs of the exact targets or target types, and is provided the targets' last known locations. The briefing covers rules of engagement and special instructions, such as to avoid hitting certain sites containing civilians or having religious or historical significance, or critical infrastructure. The briefing covers methods to enhance aircrew safety such as evasive tactics and the locations of search and rescue forces. The briefing provides instructions for radio communication, data transmission, and what should later be covered in a post-mission briefing.

7.2.5 Does the ATR possess deep conceptual understanding?

Physics makes the world understandable and predictable. Physics doesn't explain why objects are designed the way they are or what they are supposed to do. A person looks at the world with a basic understanding of how things work. A tractor-trailer has wheels that move it along. A trailer without a tractor isn't going anywhere. When a person enters the driver's side of a vehicle, the vehicle may soon move in the forward or possibly backward direction. A truck doesn't move sideways or straight up. A helicopter has different possibilities. This is useful information for a tracker. The object leaving the door of a house is likely to be a person and occasionally could be a dog, but is unlikely to be a deer. This is useful information for a classifier. A house and a barn have different uses. A fence or river might serve as a barrier. A train is less likely to leave its tracks than a tank is to leave a paved road. Stop signs and traffic lights influence object motion, but less so for a tank than a car. Smoke from a chimney indicates that a house is occupied. Smoke billowing out of a front door means that the house is on fire.

More abstractly, "how things work" relates to conceptual knowledge. Conceptual knowledge is general knowledge of concepts and is not tied to a specific problem or situation. Deep conceptual understanding involves comprehension of ideas, relationships, connections, and their applications to various situations, including situations not previously encountered. Concepts often reside in networked or hierarchical structures. Conceptual knowledge can be learned by direct observation, information received from external sources, and thoughtful reflection. Conceptual understanding provides a basis for explanation and justification of one's actions. Consider the following scenario:

> A strange vehicle is spotted. It appears to have tubular projections mounted to its chassis. By comparison with previously observed vehicles, this could be a double-barrel cannon. The vehicle is seen firing on friendly forces that are at a certain distance from it. The vehicle dispenses rounds while on the move. The vehicle is observed to have a certain speed and turn radius. Friendly troops are seen lying on the ground.

Some of the concepts involved in conceptual knowledge include: the world and ourselves, how observed elements are related, the fact that something noteworthy has been witnessed, the order in which events take place, what action leads to what result, what is a vehicle, what does it mean to be in motion, how ground vehicles maneuver, what constitutes the enemy, what constitutes friendly forces, what it means to be injured and die, and how forces can interact. A concept might also be what is known as a composite concept, such as the understanding that the enemy is probably driving the vehicle. Conceptual knowledge can be the basis of procedural knowledge, such as don't get within a certain distance of this newly observed vehicle. Conceptual knowledge solves ambiguities. It sometimes involves the ability to think backwards:

> A person is seen running down the road. Shortly thereafter, a report received by radio indicates that an IED has exploded nearby.

For the ATR to think backwards, it has to store a number of minutes of past data. Deep conceptual understanding will be much more difficult to implant into an ATR than deep learning from labeled samples.

7.2.6 Can the ATR adapt to the situation, learn on-the-fly, and make analogies?

In biological organisms, adaptation is a form of insurance that offers resilience to varying environments. It is a mechanism for becoming more suitable for new or changing conditions. One reason that humans seem smart and current ATRs seem dumb is that humans can readily adapt to the situation at hand. Human eyes adapt to the brightness of a room and the color of its lights. A pilot can learn the situation as he is flying to the mission area. He can react to conditions that were unforeseen and unplanned. He can learn that the enemy has placed its heavy weapons into revetments and that decoys are in the open. Can the ATR learn on-the-fly the lay of the land and unthreatening activities taking place? Can it include observed weather in its calculation of likely activities or target appearance? Can it discount lightning, sun dogs, sunset colors, large snowflakes, swirling dust, reflections from mud puddles, shadows from clouds, blowing leaves, or large flocks of migrating birds? Can it follow tank tracks on the ground? Can the undersea ATR ignore the fish and focus on the targets? Can the SAR ATR learn that target shadows are well-defined and hence very informative for one locality, but are fragmented and uninformative for another locale? A person who is shown a single picture of a chair can then identify more chairs of that type in a large room, even though he won't be viewing the chairs from the same aspect angle, depression angle, lighting, and scale as in the picture. A pilot might be shown just a single picture of the target of the day. Can the ATR execute "one-shot learning?" That is, can it learn to recognize a target from a single instance? Can the ATR make comparisons between an object at one location known to be a target and a similar but more ambiguous object at another location? Can the ATR perform transfer learning from one problem to an analogous problem? For example, can learning the human pose for firing rocket-propelled grenades (RPGs) be transferred to predicting the human pose for firing man-portable air defense systems (MANPADS)?

7.2.7 Does the ATR understand the rules of engagement?

The report by Scharre and Horowitz describes different categories of autonomy.[11] Once activated, a fully autonomous system operates on its own, detecting targets and attacking them. Loitering munitions fly a search pattern, swooping down on the target (e.g., radar transmitter) once detected. Land and ocean mines are generally autonomous. For some defensive systems, the reaction time to incoming missiles and munitions is so short that human control is limited by necessity. Many weapon systems are semi-autonomous, in that autonomy is restricted to some particular function. Each weapon system differs in the details of what functions are autonomous. Some guided munitions can control, abort, or retarget after the human operator initially locks onto a specific target (i.e., lock-on-before-launch). It is the human, possibly aided by a machine, who chooses targets to engage. Cruise missiles and GPS-guided bombs attack particular locations chosen by a person. Lock-on-after-launch air missiles operate differently. For example, a radar might detect a target beyond the visible range. The pilot decides whether to launch a missile without visually confirming the target.

Although ATRs can be a component of a weapon system, from the point of view of ATR designers, nearly all ATRs are fully autonomous image or signal processing systems. A typical ATR performs its job without any input or correction from a human. It just chugs along processing the data that is fed to it. The human can turn it on, shut it off, or ignore its conclusions. The ATR feeds its output to a larger system, which can even be a robotic system. There is generally a human somewhere in the loop to decide the action to be taken. The action can be shooting a missile, firing a gun, or getting out of harm's way. If the ATR is used with a purely reconnaissance system, it is just a gatherer of information for possible human action in the distant future. The human action could be to submit a report to the U.N. rather than take lethal action. The point here is important in that much confusion surrounds the concepts of "automatic" and "autonomous." Engineers generally design ATRs to operate as automatic or autonomous systems. However, in the larger context of current military operations, the ATR is not the "guy" in charge.

If we assume that a particular robotic system will never be trusted to take lethal action on its own, we may ask why it has to understand the rules of engagement. After all, there will always be humans-in-the-loop to decide the course of action. However, consider the ATR as a filter. It takes in a huge amount of information and outputs very little information. It outputs only the information deemed important for a mission. What is important is governed in part by the rules of engagement. The ATR might report that men holding RPGs are gathering at a spot. In order for the human recipients of this information to make informed judgment, the ATR might also report that a group of boys is playing soccer just north of this spot, and that a family is watching a man changing a tire just south of the spot, and that a wedding is taking place just east of the spot, and that a natural gas facility was detected just west of the spot. Acting as a filter, the ATR should supply critical context but shouldn't overburden human operators with unimportant information, such as a long list of bright blobs if there is insufficient detail on the blobs to make the high-confidence classification decisions required by the rules of engagement.

Military operations are governed by international law, including the law of armed conflict, national law, and national policy. Rules of engagement are issued by governments and commanders to limit the circumstances under which military action can be employed. Rules of engagement are issued in a variety of forms, including national military doctrine, deployment orders, operational plans, and standing directives.[12] When the use of force is not justified by self-defense but is required to accomplish a mission, reasonable force may be exercised within the constraints of the rules of engagement. Thus, the rules of engagement both authorize and limit the use of force.

Rules of engagement are reviewed to ensure that they are clear and lawful, are sufficient to address the requirement of the mission, and can effectively deal with situations likely to be encountered. Targeting directives for a mission may set out limitations, such as restricted target list or restricted zones of operation. An attack is not permitted when the expected incidental injury and death to civilians and critical infrastructure is excessive in relationship to the military advantage gained by the attack. Conversely, commanders, planners, legal advisors, and armed forces must also be aware that not all adversaries will abide by recognized laws and rules. The truly smart ATR will use the rules of engagement to help focus its efforts and determine what information to report.

The rules of engagement can be difficult to apply in the fog of war. Implementation involves subtleties and fine distinctions. Is the truck heading toward the tactical operations center a threat, or is the driver lost? Are the enemy forces waving white flags setting a trap, or are they surrendering? Is the kid pointing a real gun or a toy gun? Is the nervous woman approaching the checkpoint wearing a bomb, or is she just scared of foreigners? Are the teenagers throwing rocks a threat or just an annoyance? Looking much farther into the future, the ATR will need deep social intelligence to make critical decisions. This will involve understanding and communicating in natural language; understanding natural life patterns and anomalous situations; and being able to "read" facial expressions, body language, tone of voice, and driver behavior. Robotic soldiers and aircraft will be in a complex social relationship with their human counterparts. They will need to cooperate, working together as a team toward a common goal. They will have to understand each other's capabilities, needs, and limitations. One can hope that the rules of engagement will be applied more precisely than is possible today.

7.2.8 Does the ATR understand order of battle and force structure?

Military personnel are taught order of battle and force structure.[13] The order of battle is the hierarchical organization or command structure; it includes the disposition of equipment, units, and subunits, as well as their size, activity, location, tactics, past history, and future possibilities. A force structure describes how military personnel and equipment are organized for operations.

Military operations take into account the operating environment and its impact on the order of battle. It is impossible to predict the exact nature of a conflict. Conditions and circumstances create a fluid situation. Nevertheless, much is known about the operating environment and the nature of opposing forces. Military personnel use this information to interpret the observed situation.

The equipment and force structures of nation states are more formal than those of non-state actors, which have no fixed table of organization and equipment. Insurgent organizations might not have heavy weapons and more sophisticated armaments. They tend to rely more on pickup trucks, small arms,

IEDs, and antitank grenade launchers. Accordingly, even local insurgent organizations have typical lists of equipment. Some local insurgent organizations might have unclear or shifting allegiances that are motivated by tribal or religious affiliations, family protection, or criminal activity. Military personnel analyze the entire spectrum of conditions, circumstances, and influences affecting the operational situation.

The intelligent ATR should understand these matters as well as they are understood by trained military personnel. Force structure provides examples of equipment and expected numbers of each type found in a specific organizational unit, such as an antitank brigade, combat helicopter brigade, UAV squadron, tank division, maritime unit, artillery brigade, or surface-to-air missile (SAM) site. This information could help the ATR set *a priori* probabilities. If the ATR is directed to find an S-125 SAM site, it should know that the launch positions are arranged in a parallelogram around a radar, with certain support vehicles nearby. The ATR should transmit an alert when dangerous and surprising maneuvers are encountered. It should report strategic events, such as convoys moving, forces gathering, tanks leaving their compound, ships leaving port, etc.

7.2.9 Can the ATR control platform motion?

Any creature's brain has three basic functions: finding food to eat, avoiding being eaten, and obtaining a mate. The creature accomplishes these tasks with motor control. Fortunately, the ATR has to deal with the equivalent of only the first two functions: finding targets to attack and locating threats to avoid. Except for missiles, current ATRs have limited or no control of platform motion.

Our brains evolved with our bodies. The human brain engages in a number of motor control functions. Motor control areas receive motivation and data from parts of the brain dealing with sensor processing, cognition, and memory. The motor cortex has several main parts: primary motor cortex, supplementary motor areas, and posterior parietal cortex. Other brain areas contribute to motor control through strategy, tactics, timing and coordination, as well as by controlling eye motion and speech.

The primary motor cortex encodes the direction, speed, and force of movement. The premotor cortex and supplementary motor areas make preparations for movement, often guided by sensor input. These areas of the brain select motor plans by encoding complex sequences of motor output and taking context into account. A father would not throw a ball to a 5 year old the same way he would to a 15 year old.

The posterior parietal and prefrontal cortex receive multisensor input. They ensure that motor commands are translated into accurate motions in relationship to objects and structures in the spatiotemporal environment. Motor programs are subroutine-like pre-structured motor-activation commands.

Motor programs are executed in open loop (feed-forward) fashion using sensor input to determine the current state and help plan and meet goals—as in bowling. Closed-loop motor control continuously and precisely adjusts muscle movement using feedback—as in threading a needle. Reflexes are hardwired and automatic—as in ducking a punch.

Can the ATR issue motor commands to achieve goals? Onboard a UAV, it would use open-loop control to avoid obstacles during takeoff and landing, swoop down on an object to get a better look at it, control sensor scan patterns for area coverage, set way points to avoid known anti-aircraft batteries, and perform cargo drop during a resupply mission. The ATR would use closed-loop control to follow a target. For a platform with robot arms, the ATR would direct the robot arms toward an object, grasp it, rotate it until it is identified, and then determine what to do next. The equivalence of reflexes would be used for collision avoidance and response to incoming fire.

7.2.10 Can the ATR fuse information from a wide variety of sources?

Soldiers are constantly communicating with each other on the battlefield. They receive information from other soldiers and various sensors, and combine this information to understand the situation and prepare for action. They occasionally have to communicate with persons speaking another language. Can the ATR help by performing real-time natural language translation? Can the ATR read a foreign language sign? Verbal language translators and sign readers now exist, but are not yet embedded into ATRs. When a person hears a loud sound, he instinctively turns his head toward it to enable visual processing of the event. This is a form of multisensor fusion. Many aircraft have a visible-band sensor, a radar system, an infrared sensor, a laser, and an INS/GPS system. Other information and intelligence are available but are not currently fed into the ATR. The smart ATR will fuse all available information from on and off platform to make its decision. This will not be simplistic feed-forward of sensor data into a fusion box but will consist of feedback loops controlling sensor parameters and modes, tasking one sensor to obtain data needed to aid another sensor to do its job better.

In some cases, sensor data might contain very little useful information. However, prior knowledge and contextual awareness can be considerable. The task might not be to recognize a target from scratch, but just to answer the question of whether the target is still at the location it was when last seen. If the target is a terrorist pointing a rifle out of a window, the next glimpse of movement at the window might be sufficient information for action. The use of prior information, or priors, can strongly bias a decision. Priors can be inferred from a deep understanding of the situation, or they can be received as instructions from a commander, from other platforms in the area, and from other sensor types. Additional priors can be received from intelligence units such as electronic warfare (EW), signal intelligence (SIGINT), communications intelligence (COMINT), electronic intelligence (ELINT), human intelligence (HUMINT), image intelligence (IMINT), and Blue Force Tracker.

7.2.11 Does the ATR possess metacognition?

Reggia et al. talk about the *computational explanatory gap*,[8] which refers to researchers' inability to determine the way high-level cognitive functions map onto low-level neural computations. They specify high-level cognitive information processing as "aspects of cognition such as goal-directed problem solving, executive decision making, planning, language, and metacognition." We will focus on metacognition.

Metacognition is about knowledge and control of cognition. It is one of the hallmarks of general intelligence. A metacognitive system has knowledge about itself. It knows what it knows and what it doesn't know. When it gets an answer, it knows why and how it got the answer. It can develop strategies to learn the missing information needed to get an answer and perform a task. It understands the tools that it has available to gather the required information. It has strategies for connecting the newly gathered information to its stored knowledge base. It can make generalizations and analogies to transfer knowledge from one subject area to another.

The hypothetical metacognitive ATR understands its own capabilities. It understands its limitations in the context of the situation, availability, and quality of input data, and understands the problem that it is trying to solve. It can place confidence bounds about its conclusions. The metacognitive ATR continues to monitor its own health. It detects the failure of one of its computational or memory units and takes it offline. It can then eliminate a less important function to keep up with the data rate. This gives the ATR a degree of self-repair and fault tolerance.

The metacognitive ATR performs self-regulation. It adjusts its internal parameters if it determines that it is producing too many false alarms, or too much data for the transmission bandwidth or the storage capacity, or too much data for the operator to comprehend. It continuously monitors the quality of the input data as well as the weather, and reduces reliance on a particular sensor (e.g., FLIR) if it determines that the weather (e.g., rain) is reducing its quality, then switches reliance to another sensor (e.g., radar). When cognition fails, metacognition kicks in. The metacognitive ATR continuously monitors the resources available to it. It turns off certain operations if available electrical power is getting low. It may decide to shut itself down *en route* to a target area and turn itself back on as it enters the area. The metacognitive ATR erases its memory when captured, an action that is referred to as "making itself useless."

Certain ATRs have some aspects of metacognition, but none so far have a comprehensive ability to strategize, plan, monitor, evaluate, repair, and control itself and its performance.

7.3 Sentient Versus Sapient ATR

A sentient being is capable of perception and awareness. It has insight into the world about it. Dogs are sentient. Dogs rely on olfactory perception to interpret the world in much the same way that humans rely on visual perception. Sensed information helps guide their future actions. Dogs can make basic decisions from what they perceive. They have a minimal level of consciousness. Dogs experience a range of sensations and emotions. They feel pain, suffering, anxiousness, and joy. They communicate with us through body language and actions, such as barking or jumping up and down when we return from a long trip.

A sapient being exhibits higher forms of intelligence, understands right from wrong, and as a result (hopefully) exhibits moral behavior. A sapient machine acts with judgment, intellect, rational thought, and reason. The sapient machine understands complex situations. This allows it to seek outcomes that are optimized for time, energy, and consequence. The sapient machine has roughly human-level intelligence. The sapient entity has insights into how humans interact, how the world works, the aftermaths of actions, etc. It understands the rules of engagement. A dog may be taught not to bark at the mailman. But it probably does not understand why it shouldn't bark at the apparent intruder. A dog may be loyal and loving, but doesn't exhibit careful planning, reflection, wisdom, and discernment. It probably can't experience beauty, and definitely can't solve differential equations or drive a car. It is sentient but not sapient. Sapient implies being sentient, but sentient does not imply being sapient.

Can an ATR be sentient? Its function is to process sensory information, which is the basis of sentience. The ATR (or robotic platform), however, can't experience pain, suffering, or joy. Still, it is reasonable to say that the ATR is sentient since its job is to process sensor data. Can the ATR become sapient? We wouldn't expect or want the ATR to exhibit some of the qualities of sapience, such as free will and consciousness. It is even hard to define such terms or know if they really exist. We just want the ATR to implement certain narrowly defined tasks. Some of these tasks require stronger intelligence than that possessed by current ATRs, but what does that mean? According to the rules of the Turing test, a machine that passes the test passes for a human but doesn't necessarily exhibit stronger or more general intelligence. Strong AI implies some level of autonomy and self-guidance. A system designed to just pass the Turing test cannot exhibit the motor control to guide its host platform to get a better look at a target, cannot understand human scenarios taking place on the ground, cannot send out alerts about dangerous or suspicious activities, and cannot direct the system to move out of harm's way. Beyond strong intelligence lies super-intelligence, which is now, and may forever be, in the realm of science fiction and venture-fund-seeking startup companies. A super-intelligent ATR will have more creative problem solving

Table 7.2 Levels of intelligence. (SWaP is size, weight, and power.)

Type of AI	Weak (narrow)	Strong (general)	Super-intelligent
Capability	Single area of expertise. Programmed by people using purely statistical machine learning from very limited human-labeled training set.	Capability comparable to that of a very well-trained human.	Much smarter than a human in all possible ways, including break-through conceptual creativity, interactivity with other super-intelligent entities, and general knowledge. Ability to learn from observation and communication.
Example	Current ATRs: Target detection, recognition, and tracking.	ATR controlling future fully autonomous robotic system; very low SWaP, fault tolerant with self-repair.	Alien biological/robot hybrid with millions of years of evolution beyond that of humans. (More dubiously, runaway AI on Earth advancing at an exponential rate.)

capacity than any group of humans (Table 7.2). It will take advantage of higher processing speed to operate a million times faster than its biological counterpart. Its memory banks will include all recorded knowledge. The super-intelligent robot will transcend human intellect in all possible aspects, as well as have a capacity to act independently in its own self-interest. The eleven questions provided in this chapter are the criteria for judging the intelligence or neuromorphism of the ATR up to the point of strong (general) intelligence.

7.4 Discussion: Where is ATR Headed?

To predict the future of ATR, one must consider its enabling technologies such as AI, brain models, computers, and robotics. News stories, books, science fiction movies, and ramblings by boutique futurists predict that AI will soon surpass human intelligence (Fig. 7.1). Then, as the stories go, AI will

Figure 7.1 Recent news headlines: Fear mongering about artificial intelligence. Or to put it another way: Will your future washing machine be smarter than you and order you around?

inevitably turn malevolent and try to wipe humans off the face of the earth (like the humanoid monsters of movies and video games). Of course, predictions of incremental advances in AI won't make anyone rich or famous. The flamboyant "experts" base their claims on the premise that advances in AI are foreseeable and unquestionably exponential. These "experts" see beyond the foreseeable future. A seldom-heard contrarian viewpoint is that AI advances are decelerating. Perhaps the easy problems and more predictable environments have already been tackled, like chess and mainstream UAVs and UUVs. Designing a robot to cook a meal, set and clear a table, do the dishes, and send the kids to bed is more difficult than designing a UAV to follow way points. Much more difficult are robots traversing cluttered cityscapes, understanding complex scenarios of human activity, and then deciding what to shoot at. All in all, a bit of concern is warranted for runaway AI someday in the future, but perhaps less so than for nuclear war, asteroid/comet/meteorite impacts, hyper-novas, super-volcanoes, mega-thrust earthquakes, epidemics, climate change, mega-tsunamis, engineered viruses, overpopulation, and a long list of other potential calamities.

The original golem stories couldn't envision artificial men (robots) that could communicate in natural language. Engaging in intelligent dialogue and understanding semantic content are proving difficult, but definite progress is being made in the commercial sector. A truly intelligent ATR would have to verbally communicate with people back and forth, repeatedly, until a true understanding and plan of action are reached.

No one knows the final form that AI will take. It is clear that productivity will continue to improve. Charts exist plotting improvements in productivity over the last several hundred years. Improvement is steady, but there is not a recent exponential uptick in productivity due to AI. Employment in the farming sector has declined from more than 90% of the workforce to just a few percent, but this is over a 300-year period. Robots now assemble most of a car. Much of the machinist's job has been automated. Employment opportunities in certain narrower sectors of the economy will decline precipitously in the near future due to improved machinery or algorithms. System autonomy is gradually reducing the number of soldiers and pilots required for military operations. Computer code and improved machinery could conceivably, over time, perform a wider variety of tasks now done by humans. However, very long-term predictions of the effect of AI on the economy or military forces, based on current models and current thinking, are likely to be wrong. The "singularity," evoking a sudden accelerating pace of smarter and smarter machines outpacing human capabilities, then replacing human workers, then replacing human soldiers, then replacing humans altogether, may never come about.

Advances in ATR can be viewed mainly as a byproduct of advances in computers, sensors, and database size, rather than advances in computational neuroscience or machine cognition. Raw processing power and memory density continue to make remarkable strides. This means that ATRs can be made smaller, faster, and less power hungry; however, increased processing power by itself doesn't make ATRs smarter. Unsupervised, fully automated learning is advancing at a slow pace. "One-shot" learning is challenging. ATRs with millions of processing cores, heterogeneous mixes of analog/digital processors, or billions of highly interconnected artificial neurons are likely to be difficult to program, debug, upgrade, and modify once fielded. Self-learning ATRs will be difficult to keep under configuration control.

One can draw comparisons between the ATR, similar systems from the commercial sector, and the human brain (Table 7.3). Some goals and functions

Table 7.3 ATR versus other intelligent systems.

Item	ATR	Commercial AI (image-based Internet search as an example)	Human brain
Processing	Military-grade processors (e.g., FPGAs), not fastest clock rates; long time between design and deployment; restricted SWaP. Current systems have ability to detect, recognize, and track moving and stationary objects matching those in a training set.	Latest commercial-grade processors (e.g., GPUs), fastest clock rates; massive parallelism. Human-labeled training set. Ability to search for object types cued by humans.	Massive parallelism and connectivity, feedback loops everywhere; slow clock rate; low SWaP. Brain controls and receives feedback from platform (body). Ability to detect, recognize, and track objects, as well as reason, plan, solve problems, make analogies, think abstractly, comprehend complex ideas, and learn quickly. Often works in groups. Some fault tolerance and self-repair.
Sensors	FLIR, radar, sonar, LADAR; day/night operation.	Visible-band cameras.	Matched pair of foveated eyes, visible band, parallel streams of compressed data from 40 types of retinal ganglion neurons; also hearing, touch, taste, smell, etc.
Metadata	Precise military-grade IMUs, INS/GPS, soon to be operational in GPS-denied environments; time season, altitude, etc.	None used with search engines.	Vestibular system, time, season, proprioceptive feedback.
Latency	30 to 100 to 200 ms	Improving.	Latencies of 50, 100, 200, 300 ms reported.
Learning	Training set limited by high cost of data collections. Limited closed-class list.	10^6–10^8 human-truthed data samples. Large, generally closed class list.	Lifetime of supervised and unsupervised learning. 10,000 classes learnt. Open class list.
Setting	Deployed, military operations. Non-cooperative adversary.	Computer network.	Independent agent in control of platform and in communication with other independent agents.

are comparable, but there are also notable differences. The human brain is extraordinarily complex. Neuroscience research is wide scale and well-funded. Thousands of technical papers are published each year, augmenting the engineering literature. This offers a great body of work for ATR engineers to dive into. Developments in neuroscience have sparked the imagination of ATR engineers and of those working in its enabling technologies—from the early days of ATR development to the present. However, a good whole brain model is many years away. A good neuron model is perhaps a decade away. Even then, abstract brain models, journal articles and well-funded research projects will not provide a blueprint for making an "operational/deployable system." Employing deep-learning techniques, Internet search companies are spending bundles of money on machine learning. However, they are not up against an adversary trying to prevent detection and recognition. Their sensors do not visualize at the longest possible ranges. Self-driving cars, if they actually come about, will be for on-road, not off-road use, and will not often be shot at.

Predictions of the future have a better chance of being correct if they are extrapolations. Computers are bound to be more capable in the future, and man will one day colonize Mars. But, even mundane extrapolations don't always come true, not necessarily because of technological impediments, but because society heads in a different direction. Consider flying cars, household nuclear power plants, undersea cities, over-the-air TV, low-speed landline Internet, direct-current power grid, mainframe computers, etc. Predictions are challenging and untestable until their time comes, whether based on mundane extrapolation, overly optimistic hyperbole, or self-promotion. The super-intelligent agent or robot, which will supersede ATR, is not yet on the horizon.

Consider an ATR group founded in the 1960s. A 1% improvement (e.g., reduction in classification error) per year would have been a remarkable achievement. A continued 1% improvement per year for the next 100 years will result in an ATR surpassing human capabilities. But, not so fast; as the ATR improves, the demands on it are likely to broaden. It will be expected to take over more functions now done by humans—possibly even control a robotic craft. Humans and robots will have to learn how to operate in proximity, to cooperate, and to collaborate. ATRs may learn from humans as apprentices and assistants. Human–robot teams will develop over the long run. Civil societies will require both humans and autonomous weapons to strictly adhere to the rules of engagement and laws of armed conflict. In the foreseeable future, Terminators and rampaging golems will remain the stuff of movies and video games.

It will be a long time before the ATR designer can answer all of this chapter's eleven questions to the affirmative. In the meantime, the percent of affirmative answers can be used to answer the question: "How smart is your ATR?"

References

1. H. Markram, "The Human Brain Project," *Scientific American* **306**(6), 50–55 (2012).
2. *Air Force Test and Evaluation Guide Book*, AFI 99-103, HQ USAF/TE, United States Air Force, Washington, D.C. (2013).
3. N. Wiener, *God & Golem, Inc.: A Comment on Certain Points Where Cybernetics Impinges on Religion, MIT Press* **42**, Massachusetts Institute of Technology, Cambridge, Massachusetts (1964). (Winner of second annual U.S. National Book Award in the category Science, Philosophy and Religion.)
4. M. Idel, *Golem: Jewish Magical and Mystical Traditions on the Artificial Anthropoid*, State University of New York Press, Albany, New York (1990).
5. A. Rosenfeld, "Religion and the robot," *Tradition: A Journal of Orthodox Thought* **8**, 15–26 (1966).
6. A. Rosenfeld, "Human identity: Hulakhic issues," *Tradition: A Journal of Orthodox Thought* **16**(3), 58–74 (1977).
7. J. Kaplan, "Robot weapons: What's the harm?" *New York* Times (August 17, 2015).
8. J. A. Reggia, D. Monner, and J. Sylvester, "The computational explanatory gap," *Journal of Consciousness Studies* **21**(9–10), 153–178 (2014).
9. C.-C. Wu, H.-C. Wang, and M. Pomplun, "The contribution of scene gist and spatial dependency of objects to semantic guidance of attention," *Journal of Vision* **13**(9), 1617–1622 (2013).
10. A. Oliva, "Gist of the scene," in *Neurobiology of Attention*, L. Itti, G. Rees, and J. K. Tsotsos, Eds., Elsevier, San Diego, pp. 251–257 (2005).
11. P. Scharre and M. C. Horowitz, *An Introduction to Autonomy in Weapon Systems*, Center for a New American Security, Washington, D.C. (February 2015).
12. A. Cole et al., *Rules of Engagement Handbook*, International Institute of Humanitarian Law, Sanremo, Italy (November 2009).
13. *Opposing Force Organization Guide*, FM7-100.4, Headquarters, Department of the Army, Washington, D.C. (May 2007).

Appendix 1
Resources

This appendix is a selected set of resources related to ATR, covering government sponsors, data sets, and software libraries. Much of the information is copied with minor editing from the referenced web pages of the various organizations.

Contents

A1.1 Air Force Research Lab COMPASE Center
A1.2 Army Aviation and Missile Research Development and Eng. Center
A1.3 Army Night Vision and Electronics Sensors Directorate
A1.4 Army Research Laboratory
A1.5 Automatic Target Recognition Working Group
A1.6 Chicken Little
A1.7 Defense Advanced Research Projects Agency
A1.8 Defense Technical Information Center
A1.9 Federal Business Opportunities
A1.10 Institute of Electrical and Electronics Engineers
A1.11 Intel®
A1.12 Military Sensing Information Analysis Center
A1.13 Military Sensing Symposia
A1.14 Motion Imagery Standards Board
A1.15 National Geospatial-Intelligence Agency
A1.16 National Oceanic and Atmospheric Administration
A1.17 Naval Research Laboratory
A1.18 North Atlantic Treaty Organization
A1.19 Open Source Computer Vision
A1.20 SPIE

A1.1 Air Force Research Lab COMPASE Center

http://www.wpafb.af.mil/library/factsheets/factsheet.asp?ID=17903

The COMPASE (COMPrehensive Assessment of Sensor Exploitation) Center is the technical program of AFRL/RYAA. For layered-sensing technology, the COMPASE Center provides services that develop, maintain, and communicate an understanding of the state-of-the-art and art-of-the-possible. The services include collaboration (web-based project coordination and file sharing utilizing a Virtual Distributed Laboratory), testbeds, data (collection, warehousing, and dissemination), independent test and evaluation of sensor exploitation systems, modeling, and simulation. With an emphasis on independence and technical specialization, the COMPASE center serves customers across the U.S. Department of Defense (DoD) and other national agencies.

A1.1.1 Modeling and simulation (M&S)

Modeling and simulation is offered by the AFRL COMPASE Center as an affordable way to test layered and automated sensor exploitation technology equations, theories, and concepts prior to development and/or acquisition. In many cases it is impractical or impossible to examine or assess the benefits of layered sensing or automated sensor exploitation technologies by building and flying (sub)systems. M&S tools and processes support general and program-specific efforts, providing researchers flexible and powerful decision-making tools for exercising system design and operations trade studies. The benefits of doing M&S in the laboratory are:

- Early feedback to researchers on effectiveness of various technologies and concepts
- Rapid and early proof of technologies and concepts prior to design and build
- Better input from users, affecting concept design decisions.

A1.1.1.1 M&S details

M&S can be viewed as either a disjointed collection of tools and procedures or as a coherent set of techniques with unique problem domains. These techniques are generally developed by experts in the fields of mathematics, systems engineering, and computer science. Practitioners in various disciplines use these methods to conduct research in various application areas. These tools and techniques are often used to aid in prediction when empirical means are not available or are prohibitively expensive. In all cases, M&S is used to achieve a particular goal. The focus is generally not on developing new M&S theories. On the other hand, M&S is also becoming its own discipline with unique problems such as interoperability and composability.

Layered Sensing is characterized by the appropriate combination of sensors and platforms (including those for persistent sensing), infrastructure, and exploitation capabilities to enable synergistic awareness. To achieve the layered sensing vision, AFRL is pursuing an M&S strategy through a Layered Sensing Operations Center (LSOC) (Fig. A1.1). An experimental intelligence,

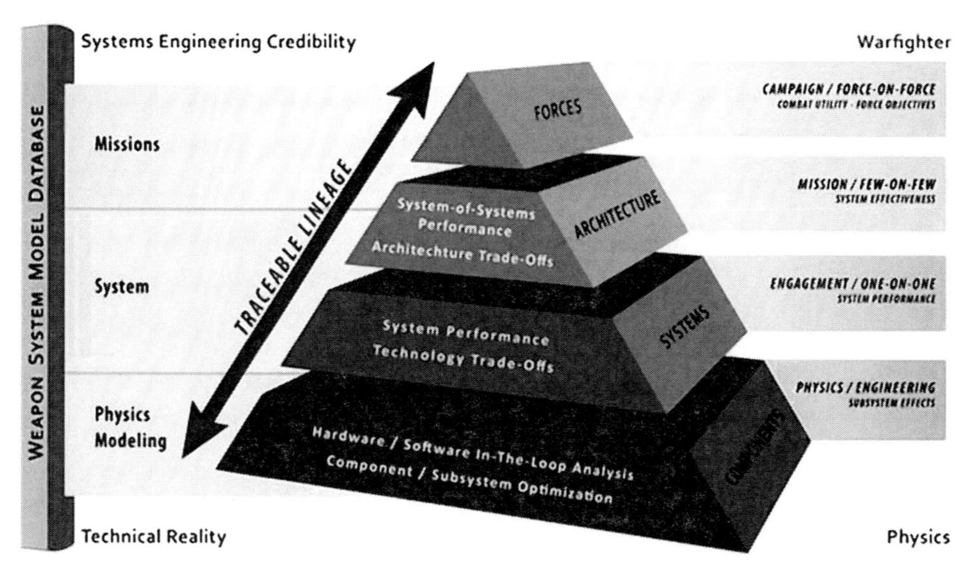

Figure A1.1 AFRL is pursuing an M&S strategy through its Layered Sensing Operations Center.

surveillance, and reconnaissance (ISR) system-of-systems testbed, the LSOC integrates DoD standard simulation tools with commercial, off-the-shelf video game technology for rapid scenario development and visualization. These tools help facilitate sensor management, performance characterization, system development, and operator behavioral analysis. LSOC goals are to:

- Quantify the benefits of Layered Sensing
- Provide a simulation environment for Layered Sensing architectural trade studies
- Develop a framework to effectively model and compare Layered Sensing components
- Discover Layered Sensing technology gaps through virtual scenarios and simulations.

In order to aid in the planning and programming of future research, AFRL's Gaming Lab utilizes both constructive simulations with traditional models and virtual simulations with operator perspective. This provides a comprehensive analysis of the Layered Sensing construct. The Gaming Lab leverages academic research from the Air Force Institute of Technology (AFIT) to define metrics for both real-time and post-processing analysis. Measures of effectiveness (MOE) and Measures of performance (MOP) are under development and validation. Future ISR data collection centers, along with visualization metrics for real-time analysis, are being explored through cooperative experiments with Ohio State University (OSU). These experiments are a "first pass" in evaluating ISR operator interfaces and teaming efficiencies.

The Gaming Lab focuses on leveraging commercially available gaming technology to advance M&S for defense applications. This is a shift from traditional DoD M&S. The Gaming Lab monitors progresses in the gaming industry as one of its major thrusts. This impacts the direction and analysis of all of the Gaming Lab's efforts. Objectives are twofold: (1) to keep up with the state of the art in gaming and (2) to develop AFRL-specific technologies to leverage these advancements. The Gaming Lab concentrates on market-moving products while avoiding boxed solutions that have a tendency to stagnate or become less relevant over time. Gaming technology affords the greatest visual fidelity for M&S. In the Sensors Directorate, fidelity matters. This fidelity allows the Gaming Lab to perform analysis on systems designed for use in complex dynamic worlds such as urban environments.

A1.1.2 Sensor Data Management System (SMDS)

The SDMS is a self-contained sensor data processing, archiving, and dissemination center providing centralized data management support to the sensor exploitation research and development communities. SDMS supports

- Data collections
- Development of problem specific datasets
- Data distribution.

A1.1.2.1 SMDS details

For over ten years, the AFRL-sponsored SDMS program has been committed to facilitating data sharing and collaboration across research and development programs and agencies (Fig. A1.2). Objectives are to

Figure A1.2 SMDS provides centralized data management support.

- Support data collections (test ranges, exercises)
- Acquire legacy data sets from various government labs and agencies
- Archive sensor data at various classification levels for the long term
- Process, reduce, reformat, and characterize sensor data
- Develop customer- and problem-specific datasets
- Distribute data (tape, CD/DVD, network, hard drive).

The SDMS offers the following benefits:

- Secure, long-term data archiving at a controlled government facility
- Easy, 24/7 web-based access to sensor data, organized and queryable
- High-speed, high-volume processing, calibration, truthing, and distribution of sensor data
- Data set and problem set development to support algorithm developments, analysis, testing, and evaluation
- Secure and rigorously enforced data distribution approval process to ensure data is only viewed by authorized users
- Database-driven request system that tracks data requests from the request through delivery to users
- Classified and unclassified archives with high-speed access to archived sensor data; archives contain research, development, acquisition and operational data
- Staff expertise in handling and processing radar, EO/IR, video, HSI/MSI, MASINT, LADAR/LIDAR
- High volume, high quality, short turnaround
- Data viewers.

A1.1.3 Test and evaluation (T&E)

COMPASE Center T&E aids DoD decision makers by analyzing emerging sensor exploitation technology and determining their potential usefulness. Unbiased and thorough, COMPASE T&E provides independent ATR and sensor exploitation technology assessment. COMPASE-T&E coordinates layered-sensing technology evaluations and related activities, such as data collection, experimental design, metric definition, tool development, and objective assessments. The COMPASE-T&E mission is to help the sensor exploitation community understand where they are and where they're heading. COMPASE-T&E provides

- Independent assessment
- Feedback to tech programs
- Maturity assessment
- Mission assessment
- Community standards
- Data collection support.

A1.1.3.1 T&E details

The COMPASE Center has provided independent evaluation support to over 50 programs since 1999. The COMPASE Center T&E determines the potential usefulness of emerging sensor exploitation technology. Resources include access to large databases of operational and laboratory data and the computational means to perform statistically significant experiments (Fig. A1.3).

A1.1.4 Virtual Distributed Laboratory (VDL)

VDL is a web-based portal facilitating cooperative research, development, and algorithm evaluation. It provides communication and information sharing services for the entire DoD-wide sensor exploitation research and development T&E community. VDL benefits and capabilities include:

- Government-controlled infrastructure
- Ability to share ITAR-level information
- High-speed networking
- Web-based tools for controlling access.

A1.1.4.1 VDL details

The VDL facilitates cooperative research, development, and algorithm evaluation. Since its inception in 1997, the VDL has been tying together algorithm developers, algorithm evaluators, and DoD simulation environments through VDL resources and DoD high-speed data networks (Fig. A1.4). The VDL is a joint-services, shared, virtual-workspace and web-based knowledge management system designed to facilitate collaboration, information sharing,

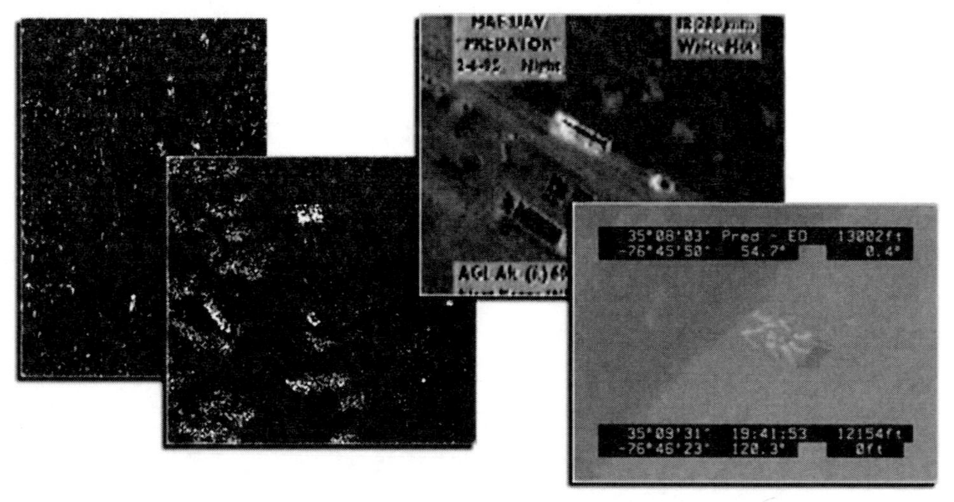

Figure A1.3 COMPASE-T&E provides independent sensor exploitation technology assessment for hyperspectral, SAR, IR, and other forms of imagery and signals.

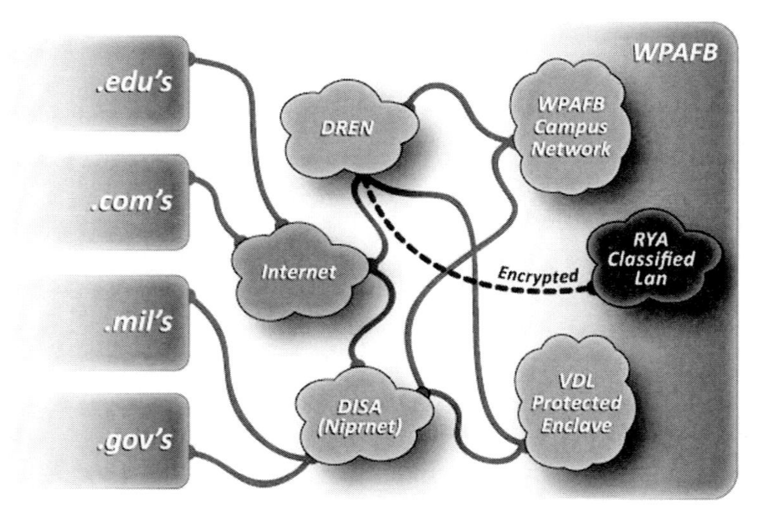

Figure A1.4 The VDL ties together algorithm developers, algorithm evaluators, and DoD simulation environments.

and cooperative research, development, test and evaluation. The VDL offers the following benefits:

Security:

- Government controlled infrastructure
- Unclassified/restricted project websites approved for sharing up to ITAR-level (export controlled) information
- Single security token (username/password) for accessing VDL resources.

Network:

- High-speed connectivity to secure Defense Research Engineering Network (DREN).

Hardware:

- Highly fault-tolerant resources
- High-speed networking.

Recovery:

- Daily back-ups of all content with off-site storage available.

Content management support:

- Project managers control who has access to their material and to specific content
- Centrally maintained email distribution lists; list servers with web-viewable archives
- Archives for completed programs
- Community event calendars.

Supported programs:

- ATR Center
- ATRPedia
- Clean Sweep
- COCO
- COMPASE Center
- Gotcha
- RadarVision
- NetTrack Phase 2
- RYA
- RYD
- RYR
- Signature Center
- SAFEGARD
- Xpatch User's Group.

For further information, contact
COMPASE Center: compase-center@lyris.vdl.afrl.af.mil
Modeling and Simulation: compase-mns@lyris.vdl.afrl.af.mil
Sensor Data Management System: sdms_help@mbvlab.wpafb.af.mil

A1.2 Army Aviation and Missile Research Development and Engineering Center

www.amrdec.army.mil/

The U.S. Army Aviation and Missile Research Development and Engineering Center (AMRDEC), a subordinate organization to the Research, Development and Engineering Command (RDECOM), is the Army's focal point for providing research, development, and engineering technology and services for aviation and missile platforms across their lifecycle. AMRDEC is the focal point for ATR technology related to missiles. AMRDEC's Huntsville facility is often used for ATR test and evaluation. It also supports data collection for ATR development.

AMRDEC traces its origins back to October 1948 when the Chief of Ordnance designated Redstone Arsenal (Huntsville, Alabama) as the center for research and development in the field of rockets. One year later, the Secretary of the Army approved the transfer of the Ordnance Research and Development Division Sub-Office (Rocket) at Fort Bliss, Texas, to Redstone Arsenal. Among those transferred were Dr. Wernher von Braun and his team of German scientists and technicians who had come to the United States after World War II. The Von Braun Team is most noted for its pioneering efforts in helping the Army at Redstone lay the foundation for U.S. space exploration.

With the transfer of the Von Braun team to NASA in 1960, research and development activities by the Army at Redstone turned to integrating space-age technology into weapons for the soldier in the field.

AMRDEC's core technical competencies depend on a preeminent, multidisciplinary, adaptive workforce that conducts leading-edge research, development, and life cycle engineering, while promoting discovery and innovation across government, academia, and industry.

AMRDEC's core technical competencies include

Aviation systems technologies

- Aerodynamics/aeromechanics (structures, flight control, crew station, survivability)
- Weapons and sensor integration (avionics)
- Propulsion
- Aviation autonomy and teaming (manned and unmanned)
- Aviation design/modification/integration/testing/qualification

Missile/rocket technologies

- Missile image processing
- Structures (propulsion, energetics, lethal Mechanisms, flight control)
- Guidance/navigation (embedded electronics and computers, infrared sensor/seekers)
- Missile weapons and platform integration
- Missile radio frequency (RF) technology
- Missile fire control radar technology

Cross-command engineering specialties

- Systems engineering (anthropocentric systems engineering, systems integration)
- System/subsystem concept design and assessment
- Software engineering
- Reliability engineering
- Sustainment engineering/industrial base analysis/obsolescence management
- Prototyping, modeling/simulation
- Quality engineering and management
- System safety
- Human factors engineering
- Manufacturing/production support (product/technical data)
- Survivability, lethality, vulnerability analysis and assessment
- Acquisition support

AMRDEC's mission is to "deliver collaborative and innovative technical capabilities for responsive and cost-effective research, product development, and life cycle systems engineering solutions" in order to equip the warfighter with the best technology today and tomorrow.

A1.2.1 AMRDEC Weapons Development & Integration Directorate

www.amrdec.army.mil/AMRDEC/Directorates/WDI.aspx

The capabilities of AMRDEC's Weapons Development and Integration Directorate include

- Automatic target recognition
- Image and signal processing
- Real-time embedded hardware and software
- Guidance, navigation, and control solutions
- Infrared and RF sensors and seekers
- Inertial and global positioning systems
- Hardware and software for fire control and platform integration
- Support and improvement for fielded systems
- Development and demonstration of new weapon systems

Facilities include

- ATR/Tracker Laboratory
- Towers (often used for ATR testing and data collection)
- Automated Infrared Sensor Test Facility
- Embedded Processor Lab
- LASER Countermeasures Lab
- Automated Laser Seeker Performance Evaluation System (ALSPES)
- Fiber Optics/MEMS Laboratory
- Radar Operations Facility
- Inertial Laboratory

A1.3 Army Night Vision and Electronics Sensors Directorate

http://www.nvl.army.mil/

The Army's Communications-Electronics Research, Development and Engineering Center is known as CERDEC. CERDEC's Night Vision and Electronic Sensors Directorate (known as NVESD or less formerly as NVL for Night Vision Lab) researches and develops sensor and sensor suite technologies for air and ground intelligence, surveillance, reconnaissance, and target acquisition under adverse battlefield conditions in day and night-time environments. NVESD is located in Fort Belvoir, Virginia with a large satellite facility at Fort AP Hill, Virginia. Data collections and field tests are often conducted at Fort AP Hill or the desert site of Yuma, Arizona, as well as at numerous other government test sites.

NVESD is the main focal point for ATR-related activities within the Army. It has world-renowned subject matter experts (SMEs) on ATR, sensor technology and T&E. NVESD has a long and distinguished history of data collections, flight tests, and funding of programs related to ATR. It is also the leader in technology related to the mitigation of various kinds of degraded visual environments. While AFRL tends to focus on radar, NVESD focuses on infrared (IR) sensors (Fig. A1.5). Radar sensors are active, while infrared sensors are often passive, literal and spatiotemporal, making each type of sensor suitable for different problem domains.

NVESD is "The Army's Sensor Developer," conducting research and development that provides U.S. forces with advanced sensor technology to dominate the 21st-century digital battlefield. NVESD exploits sensor and sensor suite technologies to

- See, acquire, and target opposing forces, day or night, under adverse battlefield environments
- Deny the enemy the same capabilities through electro-optic means and/or camouflage, concealment, and deception
- Provide capabilities for night driving and pilotage
- Detect, neutralize, clear, and mark explosive hazards including minefields and unexploded ordnance
- Protect forward troops, fixed installations, and rear echelons from enemy intrusion.

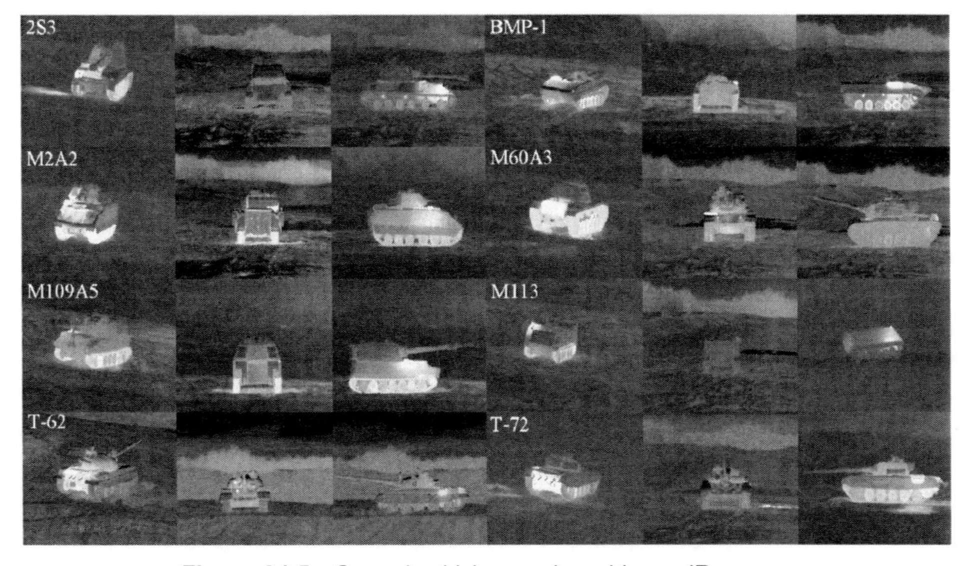

Figure A1.5 Ground vehicles as viewed by an IR sensor.

A1.3.1 History of Army night vision

NVESD has a long and distinguished history in ATR and sensor technology, punctuated by major breakthroughs in both disciplines.

A1.3.1.1 World War II and early attempts at night vision technology

Historically, military tacticians knew the benefits of maneuvering under the cover of darkness. However, in early days, the military rarely conducted maneuvers at night because of the risk.

During World War II, the United States, Britain, and Germany developed a rudimentary IR Sniperscope that used near-IR cathodes coupled with visible phosphors. The result provided a near-IR image converter to begin night fighting efforts. Although the military shipped approximately 300 of these Sniperscopes across the Pacific in 1945, few were used. With a range of less than 100 yards,-Sniperscopes could only aid in perimeter defense. These limited-range, rifle-mounted scopes were powered by cumbersome batteries. They required active IR searchlights so large that soldiers had to mount them on flatbed trucks. Enemy forces with similar equipment could readily detect the searchlights.

Despite its inadequacies, the IR Sniperscope initiated investigation into advanced night vision technology. Military leaders foresaw many other uses for such technology. Night vision goggles and weapon sights had the potential to equip armies to operate 24 hours a day. The next challenge in night vision technology was developing passive systems without the IR searchlights that betrayed a soldier's position to the enemy.

A1.3.1.2 Establishing NVESD

NVESD dates back to 1954 with the founding of the Research and Photometric Section of the U.S. Army Corps of Engineers Engineering Research and Development Laboratories (ERDL). ERDL began with minimal funding and no laboratory facilities. The Research and Photometric section of EDRL began developing personalized night vision equipment intended for use by individual soldiers in the field. This technology carved a unique niche for ERDL. Many similar organizations focused on developing large weapons systems.

NVESD's initial mission was "the Conquest of Darkness so that the individual can observe, move, fight, and work at night by using an image that he can interpret without specialized training and to which he can immediately respond." As NVESD expanded into new areas and across Army platforms, the mission also expanded to include new applications for sensor technologies.

A1.3.1.3 The 1940s and 1950s

Through the 1940s and 1950s, NVESD focused on improving the cascade image tube developed in Germany during World War II. NVESD contracted scientists from the Radio Corporation of America to research and develop a near-IR, two-stage cascading image tube. This new cascade image tube used a multi-alkali photocathode. It exceeded researchers' expectations. The Image Intensification system collected and intensified ambient light from the night sky. It was hindered by limited light gain and inverted images. To remedy these issues, NVESD added a third electrostatic stage to enhance light gain and to re-invert the image. With this, however, the tube grew to 17 inches in length and 3.5 inches in diameter, making it too large for practical applications.

From 1957 to 1958, NVESD scientist John Johnson worked to develop methods of predicting target detection, orientation, recognition, and identification. Johnson worked with volunteer observers to test each individual's ability to identify targets through image-intensifier equipment under various conditions. This time marked significant development in the performance modeling of night vision imaging systems. In October 1958, Johnson presented his findings in the paper, "Analysis of Image Forming Systems" at the first NVESD Image Intensifier Symposium. The paper described image and frequency domain approaches to analyzing the ability of observers to perform visual tasks using image intensification. These findings became known as the *Johnson Criteria* and proved important in understanding the performance of night vision devices and systems, and helped guide further developments. To this day, the Johnson Criteria are frequently referred to in ATR circles.

Night vision technology fielded in the 1940s and 1950s included

- **1940s** – Sniperscope
- **1940s** – Metascope
- **1955** – First NIR mapper
- **1958** – First IR linescanner.

A1.3.1.4 The 1960s

During the mid-1960s, NVESD scientists and engineers fielded the first generation of passive night vision devices for U.S. troops, including a Small Starlight Scope. These systems were referred to as first-generation image intensifiers. Second and third generations have since evolved.

Also during this decade, NVESD worked and contracted with scientists and engineers from other organizations to pursue research and development objectives. NVESD advanced beyond acting solely as a research institution to coordinating and managing further shared research initiatives in many fields, including astronomy, nuclear physics, and radiology, and continued to work with research personnel from leading commercial organizations. NVESD established fundamental strategies for collaborating with private industry in technology development and advancement.

Night vision equipment fielded in the 1960s included

- **1964** – Starlight Scope
- **1965** – AN/TVS–4 Night Observation Device
- **1967** – Low-Light-Level Television; Pulse-Gated I2-AN/TVS–2 Crew Served Weapon Night Sight
- **1969** – First laser rangefinder (Ruby); AN/PSS–11 Handheld Metallic Mine Detector.

A1.3.1.5 The 1970s

Based on the far-IR spectrum, thermal imaging forms an image of a target by sensing the differences between heat radiated by the target and its surrounding environment, as well as thermal differences between parts of the target. Before the 1970s, prototypes using this technology were very expensive.

While NVESD research and development focused on developing practical night vision equipment based on near-IR technology, scientists also worked toward technological advancements that would pave the way to far-IR night vision equipment. The advent of linear scanning imagers, consisting of multiple-element detector arrays, led NVESD to develop thermal imaging systems in the 1970s.

Multiple-element arrays provided a high-performance, real-time framing imager that could be practically applied to military use. This technology led to targeting and navigation systems known as forward-looking infrared (FLIR) systems. FLIR systems provide the advantage of 'seeing' at night as well as through smoke, fog, and other obscuring conditions.

FLIR imaging was in high demand for all weapons system platforms, spurring a proliferation of designs and prototypes. To satisfy this demand, a group of experts from NVESD designed a Universal Viewer for Far Infrared in 1973 that led to the family of Common Modules. Thousands of Common Modules were fielded across multiple platforms. The Common-Module-based FLIR systems were less expensive to purchase and produce than previous designs.

Night vision accomplishments of the 1970s included

- **1971** – Handheld thermal viewer; FLIR production; AN/PRS-7 Handheld non-metallic mine detector
- **1975** – AN/PVS–4 Individual weapon sight; NVESD Thermal Model publication
- **1976** – Common Module FLIR production
- **1977** – AN/PVS–5 Night vision goggles (NVGs).

A1.3.1.6 The 1980s

In the 1980s, NVESD began improving its image intensification systems. The Army fielded the third generation of night vision based on image intensification technology, composed of the AN/AVS-6 Aviators Night Vision Imaging System (ANVIS) and the AN/PVS-7 night vision goggles.

In 1980, NVESD developed the AN/GVS-5 Laser Infrared Observation Set, which significantly increased the probability of hitting stationary or airborne targets with the first round fired. Fielded in 1988, the AN/AAS-32 Airborne Laser Tracker greatly improved the offensive capability of Army helicopters.

NVESD also developed Generation II FLIR systems with improved thermal imaging technology. Improved sensor resolution and sensitivity was coupled with reduced exposure time through signal processing for aided target detection and recognition. This led to Generation II FLIR with greater standoff range.

Also in the 1980s, NVESD pioneered the revolution in ATR. While U.S. forces now had the ability to see in the dark, technology still needed improvement to help soldiers distinguish between friend and enemy. By uniting with private industry scientists, NVESD helped develop algorithms for more effectively detecting targets while minimizing false alarms. Real-time ATR processor chassis were first field tested and flown in the 1980s under NVESD supervision.

Accomplishments during the 1980s included

- **1981** – AN/VGS–2 Tank thermal sight
- **1982** – AN/TAS–6 Night observation device, long range
- **1984** – AN/AVS–6 ANVIS goggles, third-generation I2
- **1986** – AN/GVS–5 Laser rangefinder
- **1987** – AN/PVS–7 Night vision goggles
- **1988** – AN/AAS–32 Airborne laser tracker.

A1.3.1.7 The 1990s

In the 1990s, NVESD developed an eye-safe laser rangefinder, the AN/PVS-6 Mini Eyesafe Laser Infrared Observation Set (MELIOS). The 1990s also saw the next generation of Aviators Night Vision Imaging System with Heads-Up Display (AN/AVS-7) and an improved group of lightweight thermal weapon sights for ground troops.

During this time, NVESD pioneered the concept for Horizontal Technology Integration (HTI), a new method of developing and acquiring equipment for the U.S. Army. This HTI system focused on developing equipment that integrates FLIR subsystems from a single Project Manager across several weapon systems. This use of common hardware reduced equipment procurement costs.

By the end of the 20th century, NVESD had provided the Army the legacy to "Own the Night." NVESD has transitioned unique sensor technologies that have resulted in fielding over 400,000 image intensifier systems, 60,000 thermal systems, 40,000 laser systems, and 15,000 countermine systems.

A1.3.1.8 21st Century and beyond

CERDEC NVESD is strongly focused on the Army vision for the transformation of the current to the future force. In order to provide the technology transition resulting in superior tactical sensors for tomorrow's warfighter, NVESD's mission is to

- Conduct research and development to provide U.S. forces with advanced sensor technology to dominate the 21st century digital battlefield
- Acquire and target enemy forces in battlefield environments
- Detect and neutralize mines, minefields, and unexploded ordnances
- Develop humanitarian demining technology
- Deny enemy surveillance & acquisition through electro-optic, camouflage, concealment, and deception techniques
- Provide for night driving and pilotage
- Protect forward troops, fixed installations, and rear echelons from enemy intrusion.

A1.3.2 Intelligence, Surveillance, Reconnaissance and Targeting (ISR&T)

ISR&T includes the coordination of multiple sensor functions and the management and application of sensor information. ISR&T improves situational awareness and decision making, for the purposes of enabling decisive action.

The spectrum of threats encountered by U.S. soldiers is varied and complex. Degraded visual environments, demanding weather conditions, urban environments, and adversary asymmetric warfare strategies are serious risks to soldier safety and mission execution.

ISR&T technologies provide actionable information that enables soldiers to be aware of their surroundings and perform their duties with increased safety and effectiveness. This operational insight provides tactical and strategic alternatives for mission success and mitigates the risk of surprise attacks.

A1.3.3 IED, Mine, and Minefield Detection and Defeat

The detection and defeat of explosives must occur before the explosives cause harm to soldiers and innocent civilians. This core area involves identifying and combating explosive hazards that include improvised explosive devices (IEDs) and mines employed individually or in minefields, buried under ground, on the surface or otherwise camouflaged, and set and triggered by a variety of tactics. The DoD predicts that the threat from explosive hazards is likely to remain high in the coming decade, as abilities to counter them continue to evolve.

NVESD identifies, investigates, develops, matures, evaluates, and demonstrates technology and system-level prototypes for the U.S. Army that are used in the detection and neutralization (including electronic countermeasures) of explosive hazards that include IEDs and mines, whether employed individually or in minefields, buried underground, camouflaged, or on the ground's surface.

Superior explosive-hazard detection and neutralization technologies coupled with an inclusive government approach that integrates federal, state, local, tribal, territorial, private sector, and global participation in countermine/counter-IED activities best position the United States to detect and neutralize mines/IEDs in the nation or abroad.

Along with its mission to develop systems for military application that detect and neutralize mines, minefields, and unexploded ordnance, NVESD also applies these technologies to humanitarian operations.

A1.3.4 Degraded visual environments

When a helicopter lands in arid terrain, the entire aircraft can become engulfed in a dense particle cloud. *Brownout* refers to the condition of reduced in-flight visibility due to blowing/swirling dust and sand, often kicked up by downwash from the helicopter's rotor blades (Fig. A1.6). The pilot has difficulty seeing the landing zone and nearby objects. This can

Figure A1.6 NVESD develops advanced camera capabilities and algorithms to increase flight safety in degraded visual environments, such as brownout. Visibility to the pilot is significantly improved in the processed FLIR imagery; see lower right insert. (U.S. Army photo.)

cause loss of spatial orientation and loss of situational awareness, possibly leading to a crash. The problem is worse when landing in an enclosed area or an area with scattered junk, fences, wires, dunes, large rocks, crannies, hostile fire; or with people, vehicles, or large animals moving inside the brownout area.

The physics of the brownout problem is extraordinarily complex. The state-of-the art is advancing under a diverse set of programs throughout the world, with NVESD, often working with DARPA, playing a leading role. Novel algorithms are being developed and tested in the lab and in flight. The fluid dynamics and sedimentology are studied, modeled, and simulated for insight into particle transport leading to brownout. New rotor geometries, landing gear, landing trajectories, and flight-training regimes are under development and test.

NVESD continues developing and demonstrating advanced IR cameras, some of which directly address the brownout problem. Flight experiments are conducted with a Blackhawk helicopter under brownout conditions. Feedback from the Blackhawk pilots indicates that the new camera technology improves situational awareness under a majority of brownout conditions.

More generally, degraded visual environments (DVEs) refers to visibility conditions that are reduced to the point where situational awareness and aircraft control cannot be maintained as comprehensively as in normal visual meteorological conditions. Categories of DVE include not only brownout, but also smoke, fog, rain, snow, and turbulence. DVEs are a leading contributor to rotorcraft accidents and reduction of operational effectiveness.

The Army and its contractors are investigating DVE mitigation solutions. Complete solutions consist of three elements: improved flight controls, sensors/algorithms, and cueing symbology. DVE mitigation algorithms are a subcategory of ATR and are a specific type of image enhancement. They benefit both target detection and pilotage under adverse conditions. (The same types of algorithms apply to ground vehicles maneuvering in the desert and undersea vehicles maneuvering through marine snow and kicked up sediment.)

A1.4 Army Research Laboratory

www.arl.army.mil/

Although there were Army research facilities as early as 1820, The U.S. Army Research Laboratory (ARL) was created in 1989 to integrate the Army's corporate research labs. ARL is headquartered in Adelphi, Maryland. It has sites at Aberdeen Proving Grounds, Research Triangle Park, White Sands Missile Range, NASA Glenn Research Center, and Langley Research Center.

ARL consists of six directorates and the Army Research Office. These organizations focus on technology areas critical to strategic dominance across the entire spectrum of operations:

Army Research Office (ARO) - initiates the scientific and far-reaching technological discoveries in extramural organizations: educational institutions, nonprofit organizations, and private industry.

Computational & Information Sciences Directorate (CISD) - Scientific research and technology focused on information processing, network and communication sciences, information assurance, and battlespace environments; and advanced computing that creates, exploits, and harvests innovative technologies to enable knowledge superiority for the warfighter. CISD's technologies provide the strategic, operational, and tactical information dominance across the spectrum of operations.

Human Research & Engineering Directorate (HRED) - Scientific research and technology directed toward optimizing soldier performance and soldier–machine interactions to maximize battlefield effectiveness, and to ensure that soldier performance requirements are adequately considered in technology development and system design.

Sensors & Electron Devices Directorate (SEDD) - Scientific research and technology in electro-optic smart sensors, multifunction radio frequency (RF), autonomous sensing, power and energy, signature management, directed towards reconnaissance, intelligence, surveillance, and target acquisition (RISTA), fire control, guidance, fuzing, survivability, mobility and lethality.

Survivability/Lethality Analysis Directorate (SLAD) - Integrated survivability and lethality analysis of Army systems and technologies across the full spectrum of battlefield threats and environments as well as analysis tools, techniques, and methodologies.

Vehicle Technology Directorate (VTD) - Scientific research and technology addressing propulsion, transmission, aeromechanics, structural engineering, and robotics technologies for both air and ground vehicles.

Weapons & Materials Research Directorate (WMRD) - Scientific research and technology in the areas of weapons, protection, and materials to enhance the lethality and survivability of the nation's ground forces.

ARL provides enabling technologies for many of the Army's most important weapons systems. Technology and analysis products are moved into Army Research, Development, and Engineering Centers (RDECs) and to other Army, Department of Defense, government, and industry customers.

ARL does research on its own and funds research through Broad Agency Announcements. It also funds and participates in government/industry/university consortiums. ATR research covers a number of sensor types: FLIR, acoustic, LADAR, seismic, radar, and hyperspectral. ATR technologies under investigation include: neural networks, computational imaging, quantum imaging, eigenspace analysis, template matching, clustering, clutter models, fusion, and man–machine collaboration.

A1.5 Automatic Target Recognition Working Group

https://www.vdl.afrl.af.mil/atrwg/

The Automatic Target Recognition Working Group's (ATRWG's) charter is to advance the state-of-the-art in ATR. ATRWG (pronounced ahtrig) conducts workshops on ATR on a regular basis. ATR workshops cover not only ATR theory, algorithms, hardware, evaluation, and programs, but also more general military image/signal processing and related topics.

A1.5.1 ATRWG charter

ATRWG is a consortium of government and industry organizations dedicated to the advancement of ATR tools, technologies, methodologies, and operation. The ATRWG charter established two primary goals for the organization:

1. To standardize imagery data formats
2. To establish uniform criteria for the evaluation of ATR performance.

ATRWG fosters collaborative partnerships and provides workshops each year for technical interaction between researchers, developers, integrators, and testers. ATRWG also provides the DoD with information on ATR activities and progress.

A1.5.2 Requirement for registering for ATRWG

The requirements for being involved with ATRWG are as follows:

- Must be a U.S. citizen
- Must be working on a government contract
- Must have a valid security clearance.

Restricted access: Access to the ATRWG restricted website requires a Virtual Distributed Library (VDL) account. Access can be requested at the following link: https://www.vdl.afrl.af.mil/

Administration: The ATRWG administrators can be reached by email at: atrwgwebmaster@vdl.afrl.af.mil

A1.6 Chicken Little

The Air Force Development Test Center, Joint Munitions Test & Evaluation Program Office, more colorfully known as Chicken Little, conducts seeker/sensor captive flight and signature data collections on mobile ground targets. Dozens of organizations gather at the Eglin Air Force Base for data collections at scheduled events known as Sensor Week. Upcoming events are announced in Federal Business Opportunities.

A1.7 Defense Advanced Research Projects Agency

http://www.darpa.mil/

Much of the leading-edge research into sensors, ATR, and ATR-related technologies is, and has been, funded by the Defense Advanced Research Projects Agency (DARPA). DARPA holds workshops and pre-proposal briefings. DARPA accepts "seedling" white papers and proposals for far-out, high-risk new ideas. DARPA-funded ATR programs are often jointly sponsored and administrated by a government lab. DARPA is not itself a lab conducting research as depicted in Hollywood movies, but rather consists of a group of program managers, each leaders in their field, who develop and then manage programs conducted by proposal winners. DARPA program managers have a short stay at DARPA by policy. Their stay is typically limited to a term of three years, renewable once, occasionally twice.

DARPA was established in 1958 to prevent strategic surprise from negatively impacting U.S. national security, and to create strategic surprise for U.S. adversaries by maintaining the technological superiority of the U.S. military. To fulfill its mission, the Agency relies on diverse performers to apply multidisciplinary approaches to both advance knowledge through basic research and to create innovative technologies that address current practical problems through applied research. DARPA's scientific investigations span the gamut from laboratory efforts to the creation of full-scale technology demonstrations. As the DoD's primary innovation engine, DARPA undertakes projects that are finite in duration but that create lasting revolutionary change.

DARPA sponsors fundamental and applied research in a variety of areas. This research leads to experimental results and reusable technology that benefit multiple governmental entities, academia, and the private sector. The Internet is a noted example. The DARPA Open Catalog organizes publicly releasable material from DARPA programs. The Open Catalog contains a curated list of DARPA-sponsored software and peer-reviewed publications:

http://www.darpa.mil/opencatalog/

A1.8 Defense Technical Information Center

http://www.dtic.mil/

The Defense Technical Information Center (DTIC®) serves the DoD community as the largest central repository for DoD and government-funded scientific-, technical-, engineering-, and business-related information available today. A large variety of publications on ATR and related technologies are available at no cost.

A1.9 Federal Business Opportunities

https://www.fbo.gov/

A good place to search for U.S. government requests for proposals, announcements, workshops, and contract awards related to ATR can be found on-line at Federal Business Opportunities, commonly known as FedBizOpps or FBO. FBO is a free, web-based portal that allows vendors to review Federal Procurement Opportunities over $25,000.

FBO Daily™ (http://www.fbodaily.com/) was developed by Loren Data Corp. for contractors and agencies who desire a traditional Commerce Business Daily (CBD)-style daily listing of all notices posted to the FedBizOpps web site. Loren Data makes available a complete, daily publication, customized e-mail search services, and its much acclaimed free website.

A1.10 Institute of Electrical and Electronics Engineers

https://www.ieee.org

The Institute of Electrical and Electronics Engineers (IEEE) is the world's largest professional association dedicated to advancing technological innovation and excellence. IEEE publishes numerous journals and magazines. The IEEE conducts conferences on topics related to ATR. Most articles are available on-line. Most technical libraries offer access to IEEE publications. A number of IEEE journals contain articles related to ATR:

- *Aerospace and Electronic Systems Magazine*
- *IEEE Transactions on Aerospace and Electronic Systems*
- *IEEE Transactions on Computational Imaging*
- *IEEE Transactions on Image Processing*
- *IEEE Transactions on Learning Technologies*
- *IEEE Transactions on Neural Networks and Learning Systems*
- *IEEE Transactions on Pattern Analysis and Machine Intelligence* (Online Plus).
- *Intelligent Systems Magazine*

Also of interest, is the IEEE Applied Imagery Pattern Recognition (AIPR) workshop held each October in the historic Cosmos Club, in

Washington, D.C. Numerous other IEEE conferences and workshops are held worldwide each year.

A1.11 Intel®

https://software.intel.com/en-us/intel-ipp

Many ATR systems are built using Intel® heterogeneous multicore processing chips. The Intel® Integrated Performance Primitives (IPP) library is an extensive set of highly efficient software functions, including many directly applicable to ATR.

A1.12 Military Sensing Information Analysis Center (SENSIAC)

https://www.sensiac.org

SENSIAC is the DoD Sensing Information Analysis Center operated by the Georgia Institute of Technology, under contract to DTIC®. Its objective is to facilitate the use of scientific and technical information specifically Military Sensing Technology (MST) for the design, development, testing, evaluation, operation, and maintenance of DoD systems, military systems operated by allied and friendly nations, and the industrial and research base that provides and supports such systems.

SENSIAC fosters communications within the MST community; creates standards; and collects, analyzes, synthesizes, maintains, and distributes critical information within the field. It provides information products and services to the U.S. government, organizations performing government contracts or subcontracts, educational institutions, and infrastructure/tech base organizations involved directly and indirectly in the application of sensing technologies to the defense of the United States. SENSIAC provides an ATR algorithm development database: https://www.sensiac.org/external/about/mission.jsf

A1.13 Military Sensing Symposia

https://www.sensiac.org/

The Military Sensing Symposia (MSS) is a set of conferences dedicated to military sensing technologies. The current set of MSS specialty committees is:

- Tri-Service Radar
- Electro-Optical and Infrared Counter Measures
- Battlefield Acoustic and Magnetic
- Missile Defense Sensors Environment and Algorithms
- Battlespace Surveillance and Discrimination
- Passive Electro-Optical Sensors

- Active Electro-Optical Sensors
- Electro-Optical Detectors and Materials
- Sensor Fusion
- National Committee

plus the Executive Committee, which provides oversight over the various other committees from a top level. Conference schedules are posted on-line.

U.S. Government agencies and qualifying contractors can attend the MSS conferences. For more information, contact MSS (SENSIAC) at (404) 407-7367 or mss@gtri.gatech.edu.To find MSS papers approved for public release, visit the DTIC® website to perform a search.

A1.14 Motion Imagery Standards Board

http://www.gwg.nga.mil/

ATR design and performance analysis requires standards. ATR results must be viewed in relation to the quality of the data processed. Image and video quality measurement follow standards. Standards are also necessary for passing data between systems and platforms. Data formats follow accepted standards and guidelines. Image compression follows standards. The Motion Imagery Standards Board (MISB) holds meetings to review and develop standards dealing with:

- Advanced Sensors
- Advanced Compression
- Metadata
- Transport
- Interpretability, Quality, and Metrics
- Interoperability and Conformance.

MISB documents can be found at: http://www.gwg.nga.mil/misb/stdpubs.html

MISB document 0901 covers the Video-National Imagery Interpretability Rating Scale (V-NIIRS). This is a spatiotemporal relative of the still image rating scale known as NIIRS, developed over the years (for visible, IR, and SAR) by John Irvine and others, in conjunction with the intelligence and ATR communities.

A1.14.1 MISB history

In 2000, the National Geospatial-Intelligence Agency (NGA) Innovation Directorate founded the MISB, effectively establishing an official standards body responsible for reviewing and recommending standards for motion imagery, associated metadata, audio, and other related systems for use within the DoD, the Intelligence Community, and the National System for Geospatial Intelligence (DoD/IC/NSG).

A1.14.2 Standards-based motion imagery workflow

The mission of the MISB is to ensure the development, application, and implementation of standards that maintain interoperability, integrity, and quality of motion imagery, associated metadata, audio, and other related systems in the DoD/IC/NSG.

As a leading advocate for interoperability, the MISB monitors and participates in the development of, and changes to, adopted standards. It assesses their impacts on systems and DoD/IC/NSG architectures. The MISB, through its seven working groups and three conferences per year, strives to establish universal standards for the intelligence community (IC) and DoD. It does this through ongoing research into current commercial off-the-shelf tools and technologies, outreach to stakeholders, and community input and discussion.

The MISB also participates in the North Atlantic Treaty Organization Standards Agreement process for coalition forces interoperability. This includes working with U.S. and international standards bodies to monitor, advocate, and represent DoD/IC/NSG interests for motion imagery, metadata, audio, and related systems to support global interoperability.

Standards greatly increase the value of information. By providing an underlying "common language" for the sharing of information, standards foster breadth in knowledge and depth in intelligence. Nowhere are standards more crucial in realizing this added value than within the acquisition, processing, exploitation, and dissemination workflow processes for motion imagery rich-media assets.

MISB establishes standards for motion imagery encoding, metadata schemas, and dissemination protocols in conjunction with compliance enforcement and testing. Standards help prevent the proliferation of proprietary, stovepipe systems that are not interoperable. Stovepipe solutions impede the intelligence processing, exploitation, and dissemination (PED) process and minimize the intelligence value derived from motion imagery (MI) assets. MISB's mission is to unify the motion imagery workflow, effectively maximizing the value of MI assets for all stakeholders. Architecting the PED workflow within a standards-based foundation and guiding the development, acquisition, and implementation of tools, technologies, and processes creates solutions that have immense value for the warfighter.

A1.15 National Geospatial-Intelligence Agency

https://www.nga.mil/

Intelligence based on the earth's physical and man-made attributes—and the art and science of interpreting that information—began to change well before the tragedy of September 11, 2001. By combining America's most advanced imagery and geospatial assets within the National Imagery and Mapping

Agency (NIMA) in 1996, the U.S. created a critical mass of skills and technologies under a single- mission umbrella. As a result, the U.S. intelligence community was able to take its geospatial products to a new level. With the creation of the National Geospatial-Intelligence Agency (NGA) in 2003, this area of intelligence took a leap forward, allowing the integration of multiple sources of information, intelligence, and tradecrafts to produce an innovative and sophisticated new discipline that then NGA director James Clapper formally christened as geospatial intelligence, or GEOINT.

GEOINT is made up of three key elements: geospatial information, imagery, and imagery intelligence. This specialized discipline encompasses all activities involved in the planning, collection, processing, analysis, exploitation, and dissemination of spatial information. The purpose is to gain intelligence, visually depict knowledge, and fuse the acquired knowledge with other information through analysis and visualization processes. Geospatial analytics applied to very large data sets (big data) can extract actionable intelligence with visual depictions intelligence information that is not otherwise discernable. Thus, GEOINT is closely aligned with ATR. NGA products, research, and standards come into play in ATR programs.

GEOINT standards ensure interoperability across sensors, platforms, systems, and communities. An extensive tutorial on GEOINT standards can be found at:

http://www.gwg.nga.mil/documents/gwg/GEOINT%20Standard%20The%20Basics_Part%201.ppt

http://www.gwg.nga.mil/documents/gwg/GEOINT%20Standard%20The%20Basics_Part%202.ppt

http://www.gwg.nga.mil/documents/gwg/GEOINT%20Standard%20The%20Basics_Part%203.ppt

A1.15.1 NGA products

NGA products are often used in ATR research, development, and production programs. NGA products and services include:

- Aeronautical charts and publications
- CIA maps and publications available to the general public
- Custom-media-generation team
- Falcon View (not actually an NGA product)
- Historical maps and charts
- Imagery
- Net-centric Geospatial Intelligence Discovery Services
- Maritime safety products and services
- Military ordering of NGA products and services
- Nautical publications

- Topographic maps, publications, and digital products
- U.S. Board on Geographic Names
- GPS and earth-orientation products.

A1.15.2 Topographic maps, publications, and digital products

The Department of the Interior, U.S. Geological Survey (USGS) is the distributor of public-sale NGA topographic maps, publications, and digital products. To order, contact:

USGS Branch of Information Services, Map and Book Sales
Federal Center, Building 810
P.O. Box 25286
Denver, CO 80225
USA
Phone: 888-ASK-USGS or 303-202-4700
Internet: http://www.usgs.gov

A1.15.2.1 Imagery

Declassified satellite imagery (e.g., Corona, Argon, and Lanyard) used in early mapping programs can be obtained from the USGS EROS Data Center at:

605-594-6151 (e-mail at custserv@usgs.gov) or from the National Archives at 301-837-1926 (e-mail at carto@nara.gov).

A1.15.2.2 Aeronautical charts and publications

The Aero Browser–ACES (Aeronautical Content Exploitation System) is a map-based website that provides enhanced web technology for users to access multiple geospatial intelligence and aeronautical information databases and to package the information in user-specified formats. This capability provides access to data that comprises the many NGA aeronautical products and crosses over the traditional lines of AAFIF, DAFIF, FLIP, or Intel Imagery and makes all available as needed, with just a few clicks of the mouse. This site is open only to U.S. military and government employees with CAC/PKI credentials. To access: https://aerodata.nga.mil/AeroBrowser/

The Defense Logistics Agency (DLA) Aviation Division is responsible for the distribution of NGA aeronautical charts and Flight Information Publications (FLIP). Contact:

Defense Logistics Agency for Aviation
Mapping Customer Operations (DLA AVN/QAM)
8000 Jefferson Davis Highway
Richmond, VA 23297-5339
USA
Phone: 804-279-6500 or DSN 695-6500.

A1.15.2.3 Nautical publications

The U.S. Government Printing Office manages the public sale of NGA navigation publications. Contact:

U.S. Government Printing Office
Superintendent of Documents
P.O. Box 371954
Pittsburgh, PA 15250-1954
USA
Tel: 202-512-1800
Internet: http://bookstore.gpo.gov

A1.15.2.4 Maritime safety products and services

The Maritime Safety Office collects, evaluates, and compiles worldwide marine navigation products and databases. It is responsible for maritime safety and hydrographic activities, including support for the worldwide portfolio of NGA and National Oceanic and Atmospheric Administration standard nautical charts and hardcopy and digital publications.

Publications are available in digital format and include the U.S. Notice to Mariners, Sailing Directions, NGA List of Lights, U.S. Coast Guard Light Lists, American Practical Navigator (Bowditch), and other navigation science publications. The office coordinates the worldwide Navigational Warning Service's NAVAREA IV and NAVAREA XII safety messages, an essential part of the Global Maritime Distress and Safety System.

Electronic access to databases and products is provided at http://msi.nga.mil/NGAPortal/MSI.portal. The site includes U.S. Notice to Mariners and other selected publications in PDF format; a marine navigation calculator; and corrections to NGA, NOS, and U.S. Coast Guard hydrographic products. To find more information or to submit updated information for navigation publications or charts, e-mail webmaster_nss@nga.mil, or write to:

National Geospatial-Intelligence Agency
Maritime Safety Office
Mail Stop N64 SH
7500 GEOINT Drive
Springfield, VA 22150-7500
USA.

A1.15.2.5 Historical maps and charts

Historical maps and charts available to the public can be obtained from:

Library of Congress
Geography and Map Division
101 Independence Avenue, SE
Washington, D.C. 20540-4650

USA
Phone: 202-707-6277
Internet: www.loc.gov/rr/geogmap
(Place requests through the Ask a Librarian Service.)

National Archives Cartographic and Architectural Branch
8601 Adelphi Road
College Park, MD 20740-6001
USA
Phone: 301-837-1926, E-mail: carto@nara.gov
Internet: http://www.nara.gov

A1.15.3 CIA maps and publications available to the general public

Maps and publications released through the Library of Congress from 1971 and through the National Technical Information Service (NTIS) since 1980 can be purchased from NTIS. *The World Factbook* in hardcover and on CD-ROM can be purchased from the Government Printing Office. To purchase maps and publications in print after 1 January 1980, contact:

National Technical Information Service
US Department of Commerce
5285 Port Royal Road
Springfield, VA 22161
USA
NTIS Order Desk: (703) 605-6000
Internet: http://www.ntis.gov/

To purchase *The World Fact Book* (printed after 1980) in hardcover or CD-ROM, contact:

Government Printing Office
Superintendent of Documents
Washington, DC 20402
USA
(202) 512-1800
Internet: http://www.access.gpo.gov/

Publications in print before 1980 and those published through the present (excluding maps) are available in photocopy or microfiche from the Library of Congress. Contact:

Library of Congress
Photoduplications Service
Washington, DC 20540
USA
(202) 707-5650
Fax (202) 707-1771.

Free CIA products: The public may request the CIA Factbook on Intelligence and the CIA Agency brochure free of charge. Write to:

Central Intelligence Agency
Public Affairs Staff
Washington, DC 20505
USA
Internet: http://www.cia.gov/

A1.15.4 FalconView®

FalconView has been used in systems encompassing ATR. FalconView is a Windows® mapping system that displays various types of maps and geographically referenced overlays. Many types of maps are supported, but the primary ones of interest to most users are aeronautical charts, satellite images, and elevation maps. FalconView also supports a large number of overlay types that can be displayed over any map background. The current overlay set is targeted toward military mission planning users.

FalconView is an integral part of the Portable Flight Planning Software (PFPS). This software suite includes FalconView, Combat Flight Planning Software (CFPS), Combat Weapon Delivery Software (CWDS), Combat Air Drop Planning Software (CAPS), and several other software packages built by various software contractors.

FalconView is not a NGA product. All support and version issues must be addressed by a military command or with the developers of the FalconView program. Find information on Falconview at: http://www.FalconView.org

A1.15.5 Military ordering of NGA products and services

All orders of NGA products and services from military units must be requested through the Defense Supply Center in Richmond, Virginia.

A1.15.6 GPS and earth-orientation products

The NGA GPS Division ensures that the targeting and navigation grid (WGS84) is constantly realized in GPS for all users and meets National, DoD, and IC requirements. It provides timely, accurate, leading-edge GPS content, technical support, and situational awareness to the DoD, IC, and scientific community to support precise positioning, navigation, and targeting. Electronic access to the products is provided at:

http://earth-info.nga.mil/GandG/sathtml/ or direct ftp:ftp://ftp.nga.mil/pub2/gps/

A1.15.7 Net-Centric Geospatial-Intelligence Discovery Services

The user interface for geospatial exploration, discovery, and access is provided using the commercial software component, Net-Centric

Geospatial-Intelligence Discovery Services (NGDS) Client. NGDS is only available to individuals possessing DoD PKI (CAC access). NGDS can be found at: https://ngds.nga.mil or https://ngds.nga.mil/wes/Lite/WESLite.jsp. See also NGDS' Intellipedia page at https://intellipedia.intelink.gov/wiki/NGDS

A1.16 National Oceanic and Atmospheric Administration

http://www.noaa.gov/

The National Oceanic and Atmospheric Administration (NOAA) has many still-image and video databases. See:

http://oceanservice.noaa.gov/video/archive/videoarchive-page1.html/ and http://www/lib.noaa.gov/

A1.17 Naval Research Laboratory

www.nrl.navy.mil/

In 1992, the Secretary of the Navy consolidated existing Navy research, development, test, and evaluation facilities and fleet support to form The U.S. Naval Research Laboratory (NRL). NRL is the corporate research laboratory for the U.S. Navy and Marine Corps. NRL is aligned with the Office of Naval Research (ONR) and four warfare-oriented centers:

- Naval Air Warfare Center
- Naval Command, Control and Ocean Surveillance Center
- Naval Surface Warfare Center
- Naval Undersea Warfare Center.

NRL conducts a broad range of scientific research and technology development directed toward maritime applications. In fulfillment of this mission, NRL:

- Initiates and conducts broad scientific research of a basic and long-range nature in scientific areas of interest to the Navy.
- Conducts exploratory and advanced technological development deriving from or appropriate to the scientific program areas.
- Within areas of technological expertise, develops prototype systems applicable to specific projects.
- Assumes responsibility as the Navy's principal R&D activity in areas of unique professional competence upon designation from appropriate Navy or DOD authority.
- Performs scientific R&D for other Navy activities and, where specifically qualified, for other agencies of the Department of Defense and, in defense-related efforts, for other Government agencies.
- Serves as the lead Navy activity for space technology and space systems development and support.

- Serves as the lead Navy activity for mapping, charting, and geodesy (MC&G) R&D for the National Geospatial-Intelligence Agency (NGA).

NRL provides the Navy with a broad foundation of in-house expertise from scientific through advanced development activity. Specific leadership responsibilities are assigned in the following areas:

- Primary in-house research in the physical, engineering, space, and environmental sciences.
- Broadly based applied research and advanced technology development program in response to identified and anticipated Navy and Marine Corps needs.
- Broad multidisciplinary support to the Naval Warfare Centers.
- Space and space systems technology, development, and support.

NRL conducts ATR research and funds programs in many of the same areas as the Army and Air Force. In addition, it conducts ATR research in areas of specific interest to the Navy involving maritime operations: undersea, shoreline, and sea surface.

NRL encourages industry, educational institutions, small business, small/disadvantaged business concerns, historically black colleges and universities, and minority institutions to submit proposals in response Broad Agency Announcements (BAAs), which provide for the competitive selection of research proposals. See more at: http://www.nrl.navy.mil/doing-business/#sthash.vo7DbOoI.dpuf

A1.17.1 Naval Air Warfare Center Weapons Division (NAWCWD)

http:www.navair.navy.mil/

NAWCWD maintains a center of excellence in weapons development for the department of the Navy. Its Target Recognition Branch does considerable research, development, and testing related to ATR.

A1.18 North Atlantic Treaty Organization

https:// www.cso.nato.int and http://www.sto.nato.int/

The North Atlantic Treaty Organization is commonly referred to by its acronym NATO. NATO holds conferences on ATR, sensors, systems, and basic research. Conferences are held in Europe and the United States.

NATO's Collaboration Support Office (CSO) supports the collaborative business model of its Science and Technology Organization (STO). NATO nations and partner nations contribute their national resources to define, conduct, and promote cooperative research and information exchange. NATO has publications, seminars, lectures, security guidelines, data standards, and databases related to sensors, ATR, and degraded visual environments.

The total spectrum of the scientific collaborative effort is addressed by six Technical Panels that manage a wide range of scientific research activities, a group specializing in M&S, plus a committee dedicated to supporting the information management needs of the organization.

A NATO Standards Agreement (STANAG) defines processes, procedures, terms, and conditions for common military or technical procedures between member and allied countries. STANAGs form the basis of technical interoperability and data exchange. Deployed ATR systems that communicate with other systems must comply with STANAGs.

A1.19 Open Source Computer Vision

http://opencv.org/

Open Source Computer Vision (OpenCV) is a library of computer programming functions, free for use under the open-source BSD license. The library accelerates processing by using Intel® IPP when available on a computer system.

A1.20 SPIE

http://spie.org/

SPIE is the international society for optics and photonics. SPIE publications and conferences are often geared to industry rather than academia. SPIE holds unclassified sessions on ATR each year, recently at the Defense and Commercial and Sensing (DCS) conference, at the Baltimore Convention Center. A huge array of exhibitors show off their latest products. Other SPIE conferences also include talks and exhibits related to ATR. Most SPIE publications are available on-line to members, by pay-per-view, and through technical libraries on the SPIE Digital Library:

http://spiedigitallibrary.org/

Publications related to ATR include the following:

- *Journal of Electronic Imaging*
- *Optical Engineering*
- Various SPIE Conference Proceedings.

Appendix 2
Questions to Pose to the ATR Customer

This appendix summarizes some of the points made in the body of the book. A successful project starts with a clear description of the problem to be solved. However, well-defined ATR programs and tasks are the exception rather than the rule. One way to get a project going is by posing questions to the customer. The customer can be internal to one's own organization or an external prime contractor or governmental organization. To pose a question is to bring attention to the true nature of the problem. The questions do not have to be directed to a particular person and do not have to be answered immediately. Here are some basic questions.

Contractor Q1: What is the problem to be solved?

The customer should be able to describe the problem with a simple block diagram. If not, the problem is ill-defined and the project is ill-conceived.

Contractor Q2: What resources are available to solve the problem?

Resources could include: budget, staff, integrated product team, sensors, processors, test aircraft, ConOps, access to end-users, human perception lab, simulator, institutional review board, training data, testing data, etc.

Contractor Q3: How is the problem now solved?

The rationale for a new project must be that current solutions either do not exist or are inadequate in some way. It helps to know what solutions are being pursued by other organizations.

Contractor Q4: What are the exit criteria?

Exit criteria are requirements that must be met to successfully complete the project.

The customer may ask the following types of questions. Here are some typical questions and suggested answers.

Customer Q1: Isn't deep learning going to solve the ATR problem?

Deep learning gains attention because it is vastly overrated, overhyped, and sensationalized in magazines, books, news stories, and by startup companies and certain large companies' press releases. If billions of dollars and thousands of engineers are applied to a particular narrow problem, like driverless cars, then there is bound to be progress in solving that narrow problem. The ATR problem differs from commercial deep-learning problems in many ways. There may be insufficient target/clutter data, or sufficient data under insufficient conditions, to train a deep neural network for use in an ATR. Enemy target data is much more expensive to collect than data for cars, trucks and pedestrians. The target set is mission dependent. The ATR assumption is that the enemy is actively trying to defeat recognition. The driverless car may detect pedestrians but will not determine if they are carrying weapons. Enemy targets aren't restricted to road networks. When on roads, they don't obey traffic lights and stop signs. The deployed ATR receives a steady stream of active or passive sensor data and various forms of precise metadata for targets and confusors at long range. These data differ substantially in nature from commercial data. Some military problems, like detecting incoming missiles or munitions, differ from any commercial problem. The ATR must meet contractually specified size, weight, power, temperature, vibration, latency, and mean-time-between-failure requirements. These are often stricter than those for commercial systems.

Customer Q2: Then, isn't continuous learning going to solve the ATR problem?

An ATR that continuously learns from and adapts to its environment will outperform an ATR whose design is frozen in time. However, the introduction of continuous learning will not be cost-free. Continuous learning complicates safety, security, and configuration control issues.

Customer Q3: Why won't your ATR be able to separate targets from all types of decoys, identify targets with few pixels on them, and be able to distinguish between friendly and enemy T-72 tanks?

The ATR cannot perform miracles. If miracles are expected, then the ATR will never be deemed good enough. An ATR cannot recognize targets or differentiate them from decoys when the data provided to it is insufficient for making such decisions. It is not the job of the ATR to distinguish between a friendly and enemy T-72, or commercial and military truck of the same design. Systems outside of the ATR may use certain types of intelligence information to do that. It is the customer's responsibility to specify in great detail the performance requirements of the ATR. The ATR cannot meet those requirements without training and testing samples, extensively testing under differing conditions, and correcting deficiencies as they are revealed. If all major defense contractors determine that their ATRs cannot meet customer requirements, they either won't bid on the job or will try to convince the customer to modify requirements.

Customer Q4: Why does your ATR make mistakes?

ATR involves statistics. Like any system addressing a statistical problem, the ATR will make both Type 1 errors (incorrect rejection of a true target) and Type 2 errors (labeling a non-target as a target). The ATR will make mistakes identifying a target type if it looks very similar to a different target type under certain conditions, ranges, and viewing angles. Remember that different target types often share the same chassis or top structure. Military planes, trucks, boats, and small drones often have commercial counterparts. As a "rule of thumb," ATR performance will be roughly the same as that of a trained human observer visually processing the same data (assuming that the data is suitable for human vision). However, the ATR will not tire or be distracted, as will the human.

Customer Q5: Why can't your ATR match the 98% recognition rate that your competitor is claiming?

Their claimed performance probably results from a sloppy experimental design. A careful experimental design and good engineering practice disallows such techniques as cross-validation, use of training and test data collected by the same organization at the same location, train and test on synthetic data, or repeated self-testing against the same test set while tweaking the algorithm between each test and then only reporting best results. Independent blind test results should be taken more seriously than self-test results.

Customer Q6: How will your ATR classify a target type not in the training set?

Potential adversaries have hundreds of different types of military planes, boats, and ground vehicles in their inventories. Some adversaries may rely on commercial vehicles. It is unlikely that an ATR will be designed to identify each and every potential target type. It is up to the customer to specify performance requirements in excruciating detail. This includes specifying required performance against out-of-library target types and non-targets looking very similar to in-library targets. It is the responsibility of a contractor to bid on a program for which specified performance requirements can reasonably be met within budget and timeline. Then it is up to the customer to perform the necessary test and evaluation to determine if the winning bidder's ATR actually meets key performance requirements.

Appendix 3
Acronyms and Abbreviations

1NN	single nearest neighbor (classifier)
2D	two-dimensional
3D	three-dimensional
a.k.a.	also known as
AdaBoost	adaptive boosting (machine learning meta-algorithm)
ADAS	advanced driver assistance systems
ADU	Air Defense Unit
AFRL	Air Force Research Laboratory
AI	artificial intelligence
AIS	Automatic Identification System
AiTR	aided target recognition or aided target recognizer
ALV	autonomous land vehicle
AMRDEC	Army Aviation and Missile Research Development and Engineering Center
ANN	artificial neural network
APC	armored personnel carrier
AR	activity recognizer/recognition
ASIC	application-specific integrated circuit
ATC	automatic target cuer
ATD/C	automatic target detection/classification
ATD/R	automatic target detection/recognition
ATR	automatic target recognition or automatic target recognition system
ATRWG	Automatic Target Recognition Working Group
ATT	automatic target tracker
AUC	area under curve
AUTO-Q	Automatic Target Cuer (manufactured by Westinghouse/Northrop Grumman in 1970s and 1980s)
AvA	all-versus-all
BP	backpropagation (of error)
C	controller

C4ISR	Command, Control, Communications, Computers, Intelligence, Surveillance and Reconnaissance
CAD/CAM	computer-aided design/computer-aided manufacturing
CBO	common battlespace object
CBRN	chemical, biological, radiological, and nuclear
CCD	charge-coupled device
CCR	cue correct rate
CERDEC	Communications-Electronics Research, Development and Engineering Center
CFAR	constant false alarm rate
CIR	(commercial) color infrared (typically two bands in visible and one in near infrared)
CMOS	complimentary metal-oxide semiconductor
CNN	convolutional neural network
COMINT	communications intelligence
COMPASE	Comprehensive Assessment of Sensor Exploitation
ConOps	concept of operations
ConvNet	convolutional neural network (also CNN)
COP	common operating picture
CoT	Cursor-on-Target
COTS	commercial off the shelf
CRR	confuser rejection rate
DARPA	Defense Advanced Research Projects Agency
DCT	discrete cosine transform
DL	deep learning
DMC	digital magnetic compass
DNA	deoxyribonucleic acid (that carries genetic instructions)
DoD	Department of Defense
DoG	difference of Gaussian
DST	Dempster–Shafer theory
DTED	digital terrain elevation data
DTIC®	Defense Technical Information Center
DVE	degraded visual environment
EA	electronic attack
EKF	extended Kalman filter
ELINT	electronic intelligence
EM	electromagnetic
EO	electro-optical
EO/IR	electro-optical/infrared
EP	electronic self-protection
ERDL	Engineering Research and Development Laboratories
ES	embodied and situated; in another context can mean expert system

ESM	electromagnetic support (or surveillance) measures
EW	electronic warfare
FAR	false alarm rate
FBO or FedBiz	Federal Business Opportunities
FFT	fast Fourier transform
FLIR	forward-looking infrared
FMV	full motion video
FPGA	field-programmable gate array
GAN	generative adversarial network
GEOINT	geospatial intelligence
GMTI	ground moving-target indicator
GPS	global positioning system
GPU	graphics processing unit
GUI	graphical user interface
HMD	helmet-mounted display
HOF	histogram of optical flow (spatiotemporal domain)
HOG	histogram of oriented gradients (spatial domain)
HRR	high range resolution
HSI	hyperspectral imaging
HTM	hierarchical temporal memory
HUMINT	human intelligence
Hz	hertz
ID	identification
IE	inference engine
IED	improvised explosive device
IEEE	Institute of Electrical and Electronics Engineers
IFF	Identification Friend or Foe
IFSAR	interferometric synthetic aperture radar
IMINT	image intelligence
IMM	interacting multiple model
IMU	inertial measurement unit
INN	independent nearest neighbor
INS	inertial navigation system
IPB	intelligent preparation of the battlespace
IR	infrared
IRB	Institutional Review Board
IRST	infrared search and track
ISR	intelligence, surveillance, and reconnaissance
ISR&T	intelligence, surveillance, reconnaissance, and targeting
ITAR	International Traffic in Arms Regulations
JDL	Joint Directors of Laboratories
JPDAF	joint probablisitic data association filter

KF	Kalman filter
kNN	*k*-nearest neighbors (classifier)
KPP	key performance parameter
LADAR	light amplification for detection and ranging (also written LiDAR, LIDAR, Ladar, or ladar)
LAR	lethal autonomous robot
LAW	lethal autonomous weapon
LCD	liquid crystal display
LIDAR	light detection and ranging (also written Lidar, Ladar, or LADAR)
LLDR	Lightweight Laser Designator Rangefinder
LSOC	Layered Sensing Operations Center
LSTM	long short-term memory
LVQ	learning vector quantization
LWIR	longwave infrared
M	model
M&S	modeling and simulation
M1	M1 American tank
M60	M60 American tank
MASINT	measurement and signature intelligence
MFRF	multifunction radio frequency
MFT	mean-field theory
MHT	multiple-hypothesis tracker
MISB	Motion Imagery Standards Board
MLP	multilayer perceptron
MOE	measures of effectiveness
MOP	measures of performance
ms	milliseconds
MSC	map-seeking circuit
MST	Military Sensing Technology
MSTAR	moving and stationary target acquisition and recognition
MTI	moving target indicator
MUM-T	manned–unmanned team
MWIR	midwave infrared
NASA	National Aeronautics and Space Administration
NATO	North Atlantic Treaty Organization
NBC	naïve Bayes classifier
NCTR	non-cooperative target recognition
NGA	National Geospatial-Intelligence Agency
NIEM	National Information Exchange Model
NIIRS	National Imagery Interpretability Rating Scale
NV	night vision
NVESD	Army Night Vision and Electronic Sensor Directorate

NVL	Army Night Vision Lab
OC	operating condition
OpenCV	Open Source Computer Vision
OvA	one-versus-all
OvO	one-versus-one
P_d	probability of detection
PDAF	probability data association filter
pdf	probability distribution function
PED	processing, exploitation, and dissemination
PF	particle filter
P_{FA}	probability of false alarm
pixel	picture element
Pl	plastic
QR	quadrupole resonance
R&D	research and development
RBF	radial basis function
RDECOM	Research, Development and Engineering Command
RF	radio frequency
RL	reinforcement learning
RNN	recurrent neural network
RNNAI	recurrent neural network artificial inteligence
Rx	receiver
ROC	receiver operating characteristic
ROI	region-of-interest (image)
ROV	remotely operated (underwater) vehicle
RPA	remotely piloted aircraft
RPG	rocket-propelled grenade (launcher)
RPV	remotely piloted vehicle
RSS	root sum of squares
SAASM	Selective Availability Anti-spoofing Module
S&T	science and technology
SAIP	Semi-Automated Image Intelligence Processing
SAM	surface-to-air missile
SAR	synthetic aperture radar
SCR	signal-to-clutter ratio
Scud	NATO reporting name for Soviet tactical ballistic missiles
SDE	sensor data exploitation
SDMS	Sensor Data Management System
SEI	Software Engineering Institute
SENSIAC	Sensing Information Analysis Center
SIFT	Scale-Invariant Feature Transform
SIGINT	signal intelligence
SLOC	source lines of code

SME	subject matter expert
SPIE	Society of Photo-Optical Instrumentation Engineers
STA	surveillance and target acquisition
STI	stationary target indication
STIP	spatiotemporal interest point
SURF	Speeded-Up Robust Feature
SVM	support vector machine
SWaP	size, weight, and power
SWAP-C	size, weight and power, and cost
SWIR	short-wave infrared
T-62 or T62	T-62 Russian tank
T-72 or T72	T-72 Russian tank
T&E	test and evaluation
TEL	transporter/erector/launcher (Scuds)
THz	terahertz (10^{12} Hz)
TLE	target location error
TM	template matcher
TNA	thermal neutron activation
TRL	technology readiness level
TSV	through-silicon via
Tx	transmitter
UAS	unmanned aircraft system
UAV	unmanned air vehicle
UGS	unattended ground system
UGV	unmanned ground vehicle
UKF	unscented Kalman filter
USAF	United States Air Force
UUV	unmanned underwater vehicle
V-NIIRS	Video-National Imagery Interpretability Rating Scale
VDL	Virtual Distributed Laboratory
VHSIC	very high-speed integrated circuit
WAMI	wide-area motion imagery
XML	Extensible Markup Language
ZSL	zero-shot learning

Index

A

a priori probability, 28
activity, 150, 220, 230, 234–235
adaptive, 225, 228, 233
aided target recognition, 1
all-versus-all, 229
anomaly detection, 36
asynchronous clock, 225
automatic target tracker, 134
autonomous land vehicle, 201–202
autonomy, 249
axons, 217, 219

B

bagging, 118
Bayer pattern, 177
Bayes average, 191
Bayes belief integration, 192
binding, 167, 170, 174
biomimicry, 57
bistatic radar, 171–172
boosting, 118
Borda count, 190

C

CFAR detector, 63
change detection, 8
classification, 103, 230, 236
clock frequency, 220
clutter level, 20
clutter object, 11
cognitive radar, 199–200
committee machine, 118
common operating picture, 203–204
compressive imaging, 58
computational explanatory gap, 254
computational imaging, 58
concept of operations, 81
conceptual knowledge, 248
confidence interval, 29
confusion matrix, 25
constant false alarm rate per image, 36
continuous learning, 221, 227–228
convolutional neural network, 121, 223, 231
correlation, 36

D

decision tree, 22, 229, 230
decision tree classifier, 118
deep learning, 118
Dempster–Shafer theory, 193
dendrites, 217
detection criterion, 7
don't care object, 13

E

embodied, 222, 228, 232
ensemble classifiers, 118
events, 150, 231
experimental design, 29
expert system, 207

F
false alarm rate, 12
false alarm, 11
feature extraction, 92
feature selection, 97
feature-aided tracking, 148
feature-level fusion, 187
fingerprinting, 8
force structure, 56, 251
forensics, 152
frequency, 172

G
generative adversarial network, 235
graphics processing unit, 119, 223, 226
ground truth, 5

H
hierarchical temporal memory, 120, 229–230, 235, 238
histogram of optical flow, 100
histogram of oriented gradients, 100
human subjects, 32, 237
hyperspectral imagery, 64

I
image truth, 6, 8
Institutional Review Board, 32
intelligent preparation of the battlespace, 202–203
Integrated Product Team, 80

J
Joint Directors of Laboratories, 176

K
Kalman filter, 141

L
latency, 216, 220, 225
learning vector quantization, 112
learning-on-the-fly, 57
Lightweight Laser Designator Rangefinder, 178
linear classifier, 104
long short-term memory, 120, 228

M
majority voting, 189
manned–unmanned teaming, 205
map-seeking circuit, 116
mean-shift tracker, 149
metacognition, 254
model-based classifier, 116
moving target indication, 37
MSTAR, 116
multiclassifier fusion, 176, 187, 194
multifunction radio frequency, 199–200
multifunction radio, 172
multilayer perceptron, 114
multimodality, 227
multisensor fusion, 167, 172, 176, 195, 227, 253
mutual information, 185

N
naïve Bayes classifier, 110
narrative, 204, 206
neocortex, 216, 219
neural network, 53, 214
neuromorphic chip, 223, 225
neuron, 214, 216–217, 219–220, 222–223
neuroplasticity, 228, 233
No Free Lunch theorem, 76

O
Occam's razor, 78
one-versus-all, 112, 229
one-versus-one, 229
ontology, 21
operating conditions, 5
order of battle, 251

P
parallelism, 223, 227
pattern of life, 152
perceptron, 111
performance parameters, 33
persistent surveillance, 152
photon, 59
plastic, 228, 233
polarization, 181–183
pre-mission briefing, 247
probability of (correct) classification, 25
probability of detection, 14
probability of false alarm, 14

Q
quaternions, 159

R
Random Forest™, 118
receiver operating characteristic curve, 15
recurrent neural network, 119–120, 150, 214, 223, 226, 228, 234
Reed–Xialo algorithm, 65
region-of-interest, 12, 231
reinforcement learning, 228, 233, 236
rules of engagement, 250

S
sapient ATR, 255
Scale-Invariant Feature Transform, 61
scene gist, 245
sentient ATR, 124, 255
single-nearest-neighbor classifier, 108
situated, 228, 232
sleep, 225
spatial scale, 38
spikes, 217, 219, 222, 225
spoke filter, 52
stacking methods, 118
stationary target indication, 37
stereo camera, 180
strawman, 214, 228–230, 233
strong artificial intelligence, 255
super-intelligent ATR, 255
support vector machine, 105
surprise, 59
synapse, 216–217, 223, 225
system design, 80

T
target, 5
target classifier, 76
target detection, 7, 10
target polarity, 38
taxonomy, 21, 229
template matcher, 55
test plan, 31
track fusion, 196–197
transfer learning, 235, 249
triple window filter, 41
Turing test, 238, 242

U
Ugly Duckling theorem, 76
unattended ground sensor, 186–187

X
XPATCH®, 116

Z
zero-shot learning, 204–205, 209

Bruce J. Schachter is an engineer whose work has focused on automatic target recognition (ATR) for more than forty years. He was on the team that developed the first Automatic Target Recognizer, at the University of Maryland then later at Northrop Grumman. He has been program manager or principal investigator of a dozen ATR programs. His previous books are *Pattern Models* and the award-winning *Computer Image Generation*. The author can be contacted at Bruce.Jay.Schachter@gmail.com.